Ecological Studies, Vol. 162

Analysis and Synthesis

Edited by

I.T. Baldwin, Jena, Germany
M.M. Caldwell, Logan, USA
G. Heldmaier, Marburg, Germany
O.L. Lange, Würzburg, Germany
H.A. Mooney, Stanford, USA
E.-D. Schulze, Jena, Germany
U. Sommer, Kiel, Germany

Springer

Berlin
Heidelberg
New York
Hong Kong
London
Milan
Paris
Tokyo

The background color image is of a L7 satellite image composite of the Klamath-Siskiyou Mountains in Southern Oregon/Northern California (USA). The three black and white images 'burned' over the color image is a forest/nonforest classified L5-L7 TM satellite sequence of areas around the city of Manaus, Amazonas (Brazil). The years of these images are 1986, 1995, and 1999 (from left to right, respectively). Photographs and composition of the image by Scott Bergen

G.A. Bradshaw and P.A. Marquet (Eds.)
with the editorial assistance of Kathryn L. Ronnenberg

How Landscapes Change

Human Disturbance and Ecosystem Fragmentation in the Americas

With 64 Figures, 5 in Color, and 19 Tables

 Springer

Dr. Gay A. Bradshaw
Oregon State University
Departments of Forest Science
and Electrical Engineering
Corvallis, Oregon 97331
USA

Dr. Pablo A. Marquet
Pontificia Universidad Católica de Chile
Centro de Estudios Avanzados en Ecología y Biodiversidad
& Departamento de Ecología
Casilla 114-D
Santiago
Chile

Cover illustration: Natural and human-induced landscape fragmentation (southern Oregon - California, USA) Photograph by G. A. Bradshaw

ISSN 0070-8356
ISBN 3-540-43697-9 Springer-Verlag Berlin Heidelberg New York

Library of Congress Cataloging-in-Publication Data

How landscapes change : human disturbance and ecosystem fragmentation in the
Americas / G.A. Bradshaw and P.A. Marquet (eds.) ; with the editorial assistance of
Kathryn L. Ronnenberg.
   p. cm. -- (Ecological studies, ; v. 162)
   Includes bibliographical references.
   ISBN 3540436979 (alk. paper)
   1. Fragmented landscapes--America. 2. Ecological disturbances--America. 3.
  Nature--Effect of human beings on--America. I. Bradshaw, G.A. (Gay A.), 1959- II.
  Marquet, P. A. (Pablo A.), 1963- III. Ronnenberg, Kathryn L. IV. Ecological studies ; v.

QH101 .H69 2002
304.2'8'097--dc21   2002030229

http://www.springer.de

© Springer-Verlag Berlin Heidelberg 2003
Printed in Germany

The use of general descriptive names, registered names, trademarks, etc. in this publication does not imply, even in the absence of a specific statement, that such names are exempt from the relevant protective laws and regulations and therefore free for general use.

Cover design: *design & production* GmbH, Heidelberg
Typesetting: Kröner, Heidelberg
SPIN 10783888   31/3150 YK – 5 4 3 2 1 0 – Printed on acid free paper

Foreword

Comparisons of climatically matched but evolutionarily distinct and geographically separated regions of the world provide a powerful tool for examining the role of climate in shaping biophysical and ecosystem processes. A number of volumes of the Springer Ecological Studies Series have utilized comparisons of geophysical and biological systems of North and South America for this purpose. These previous volumes have included a spectrum of analyses such as fire ecology, biogeography, ecosystem structure, biodiversity and system functioning, and global changes among others. This volume, *How Landscapes Change*, takes a different path. It examines, in part, these matched systems that are similar due to climatic constraints, but differ in human impacts on the landscape. These chapters discuss the fundamental impacts of landscape alteration on the numerous and diverse ecological processes that characterize systems in the Americas.

The present volume reflects the dominating trend and necessity to consider human processes as part of the ecosystems we study. While it is agreed that the obvious effects of fragmentation, or land-use change, documented worldwide so vividly by satellite monitoring, is the primary driver of biotic change and extinctions, research indicates that the cumulative and indirect impacts of human activity is as pervasive and significant. Human activities are responsible for enormous shifts in the patterning of biota on the landscape, the severing of connections in resource supply as well as the interactions among different constituents of ecosystems. Understanding the cultural and social mechanisms that underlie these landscape alterations, their biotic impact and how to develop conservation strategies in face of these changes is the substance of this book. As emphasized throughout this volume, clear understanding requires a unified framework where landscape patterns and changes are viewed as emerging from the interaction between social, economic, and ecological drivers of change. The authors draw from a wide range of disciplines from the biophysical and social sciences. These diverse perspectives, represented in the essays presented here, converge upon a common theme: interdisciplinary collaboration is needed to support the agenda for all of us attempting to conserve, or enhance, the services that natural systems provide to society.

September 2002
Stanford, USA   *Harold A. Mooney*

Contents

Part II   Ecological and Evolutionary Consequences of Fragmentation

17   Bandages for Wounded Landscapes: Faunal Corridors and Their Role in Wildlife Conservation in the Americas . .   313
S.G.W. LAURANCE, W.F. LAURANCE

18   Management of the Semi-Natural Matrix   327
J.H. BROWN, C.G. CURTIN, R.W. BRAITHWAITE

Human Disturbance and Ecosystem Fragmentation in the Americas

Contributors

AIZEN, M.A.

Departamento de Ecologia, Universidad Nacional del Comahue,
Centro Regional Universitario Bariloche, Unidad Postal Universidad,
8400 San Carlos de Bariloche, Rio Negro, Argentina

ALABACK, P.

School of Forestry, University of Montana, Missoula,
Montana 59812-1063, USA

BRADSHAW, G.A.

Oregon State University, Departments of Forest Science
and Electrical Engineering, Corvallis, Oregon 97331, USA

BRAITHWAITE, R.W.

CSIRO Division of Wildlife and Ecology, Tropical Ecosystems Research
Centre, P.M.B. 44, Winnellie, Darwin, Northern Territory 0821, Australia

BROWN, J.H.

Department of Biology, University of New Mexico, Albuquerque,
New Mexico 87131, USA

BUSTAMANTE, R.O.

Departamento de Ciencias Ecologicas, Facultad de Ciencias, Universidad
de Chile, Casilla 6425, Santiago, Chile

CASE, T.J.

Department of Biology, 0116, University of California,
San Diego, 9500 Gilman Drive, La Jolla, California 92093-0116, USA

CURTIN, C.G.

Arid Lands Project, Box 29, Animas, New Mexico 88020, USA

CUNNINGHAM, M.

Department of Zoology and Co-operative Research Centre for Tropical
Rainforest and Ecology, University of Queensland,
Queensland 4072, Australia

DEBINSKI, D.M.

Iowa State University, Department of Animal Ecology,
124 Science II, Ames, Iowa 50011, USA

FEINSINGER, P.

Department of Biological Sciences, Northern Arizona University,
Flagstaff, Arizona 86011, USA

GASCON, C.

Center for Applied Biodiversity Science, Conservation International,
2501 M Street, NW, Suite 200, Washington, DC 20037, USA

GRIMWOOD, T.

P.O. Box 38, Burns, Kansas 66840, USA

GROSJEAN, M.

University of Bern, NCCR Climate, 9 Erlachstrasse,
3012 Bern, Switzerland

HOLT, R.D.

Department of Zoology, 223 Bartram Hall, University of Florida,
Gainesville, Florida 32611-8525, USA

JAKSIC, F.M.

Centro de Estudios Avanzados en Ecología y Biodiversidad &
Departamento de Ecología, Pontifica Universidad Católica de Chile,
Casilla 114-D, Santiago, Chile

JOSEPH, L.

Laboratorio de Evolución, Facultad de Ciencias, Tristán Narvaja 1674,
Montevideo 11200, Uruguay. Present address: Department of Ornithology,
Academy of Natural Sciences of Philadelphia, 1900 Benjamin Franklin
Parkway, Philadelphia, Pennsylvania 19103-1195, USA

KATTAN, G.H.

Wildlife Conservation Society, Colombia Program, Instituto de Investigationes Alexander von Humboldt, Apartado Aereo 25527, Cali, Columbia

KEITT, T.H.

Dept. of Ecology and Evolution, State University of New York
at Stony Brook, Stony Brook, New York 11794-5245, USA

KEYMER, J.E.

Departamento de Ecología, Pontificia Universidad Católica de Chile,
Casilla 114-D, Santiago, Chile. Present address: Department of Ecology
and Evolutionary Biology, Princeton University, Princeton, New York
08544-1003, USA

KILLEEN, T.J.

Center for Applied Biodiversity Science (CABS),
Conservation International and Museo de Historia Natural
Noel Kempff Mercado, Casilla 2489, Santa Cruz, Bolivia

KITZBERGER, T.

Departamento de Ecología, Universidas Nacional del Comahue,
CC 1336, 8400 Bariloche, Argentina

LARA, A.

Instituto de Silvicultura, Universidad Austral de Chile,
Casilla 567, Valdivia, Chile

LAURANCE, S.G.W.

Biological Dynamics of Forest Fragments Project,
INPA Ecologia, C.P. 478, Manaus, AM 69011-970, Brazil

LAURANCE, W.F.

Biological Dynamics of Forest Fragments Project, INPA Ecologia, C.P. 478,
Manaus, AM 69011-970, Brazil. Present address: Smithsonian Tropical
Research Institute, Apartado 2072, Balboa, Republic of Panamá

LOVEJOY, T.E.

NHB 180, Smithsonian Institution, Washington, DC 20560, USA

MARQUET, P.A.

Centro de Estudios Avanzados en Ecología y Biodiversidad &
Departamento de Ecología, Pontificia Universidad Católica de Chile,
Casilla 114-D, Santiago, Chile

MEGGERS, B.J.

Smithsonian Institution, Anth 112 (NHB-112), Washington, DC 20560, USA

MURCIA, C.

Wildlife Conservation Society, Colombia Program, Instituto de Investigationes Alexander von Humboldt, Apartado Aereo 25527, Cali, Columbia

NÚÑEZ, L.

Instituto de Investigaciones Arqueologicas y Museo, Universidad Catolica del Norte, San Pedro de Atacama, Chile

PANFIL, S.

Department of Botany, University of Georgia, Athens, Georgia 30602-7271, USA

PICKETT, S.T.A.

Institute of Ecosystem Studies, Box AB, Route 44A, Millbrook, New York 12545-0129, USA

SARRE, S.

Molecular Evolution and Systematics, Research School of Biological Sciences, Australian National University, P.O. Box 475, Canberra, ACT 2001, Australia. Present address: School of Biological Sciences, University of Auckland, PB 92019, Auckland, New Zealand

SEREY, I.A.

Departamento de Ciencias Ecológicas, Facultad de Ciencias, Universidad de Chile, Casilla 6425, Santiago, Chile

SIEBERT, S.F.

School of Forestry, University of Montana, Missoula, Montana 59812-1063, USA

SILES, T.M.

Museo de Historia Natural Noel Kempff Mercado, Casilla 2489, Santa Cruz, Bolivia

STEININGER, M.K.

NASA Goddard Space Flight Center, Code 923, Greenbelt, Maryland 20771, USA

SUAREZ, A.V.

Department of Environmental Science, Policy and Management,
Division of Insect Biology, University of California, Berkeley,
Berkeley CA 94720, USA

TIESZEN, L.L.

International Programs, USGS EROS Data Center,
Sioux Falls, South Dakota 57198, USA

TUCKER, C.J.

NASA Goddard Space Flight Center, Code 923, Greenbelt,
Maryland 20771, USA

VEBLEN, T.T.

School of Forestry, University of Montana, Missoula,
Montana 59812-1063, USA

VELASCO-HERNÁNDEZ, J.X.

Departamento de Matemáticas, UAM-Iztapalapa & Programa de
Matemáticas Aplicadas y Computación Instituto Mexicana del Petróleo,
Eje Central Lázaro Cárdenas 152 San Bartolo, Atepehuacan 07730
Mexico

VILLALBA, R.

Laboratorio de Dendrocronología,
 CRICYT – CONICET, CC 330, 5500 Menoza, Argentina

WHITLOCK, C.

Department of Geography, University of Oregon,
Eugene, Oregon 97403, USA

Introduction

G.A. BRADSHAW and P.A. MARQUET

1 Background

The rapidity of environmental change and the global nature of socio-economic forces, which affect ecosystem functions and their transformation, are shared among a spectrum of ecosystems which once appeared mutually distinct. Despite differences in historical and ecological patterns, all landscapes have sustained human-mediated change to some degree. The introduction of exotic species, rapid land conversion, habitat degradation, and fragmentation are phenomena which increasingly concern scientists and resource managers in both southern and northern hemispheres because of their direct impact upon native species diversity and ecosystem function.

North and South America share similar human and ecological histories and, increasingly, economic and social linkages. Issues of ecosystem functions and disruptions form a common thread among these societies. Knowledge about the function of biodiversity within these ecosystems, and the effect of human-induced perturbations on ecosystem functioning is critical to the sustainability of our human enterprise on earth. Ecosystem and complexity theories reinforce the importance of understanding the linkages between ecosystem processes and patterns as well as the need for achieving cross-disciplinary thinking. In this context, the objectives of this book are: (1) to synthesize the perspectives of several disciplines, in order to understand how human and ecological processes interact to affect ecosystem functions and species persistence in the landscapes of the Americas; (2) to assess the effects of diverse disruptive agents, in particular landscape fragmentation, on different ecosystems within the Americas; and 3) to provide an overview of current theory, methods, and approaches used in the analysis of ecosystem disruptions and fragmentation.

Ecological Studies, Vol. 162
G.A. Bradshaw and P.A. Marquet (Eds.)
How Landscapes Change
© Springer-Verlag Berlin Heidelberg 2003

2 Why the Americas?

North and South American landscapes mirror parallel trajectories of social and ecological histories, and are shaped by similar large-scale climatic drivers such as ENSO (El Niño Southern Oscillation). This is particularly true for the western part of North and South America where a similar latitudinal progression of major ecosystem types exists (from deserts to mediterranean-type ecosystems, and temperate evergreen forests) amid similar geographic and geological settings. With increased globalization, the ecologies and economies of these two continents are becoming more dependent upon each other. Consequently, it is not surprising to observe similarities in trends and impacts on both North and South American landscapes. Understanding how human and ecological processes interact across scales and systems in the two continents is critical to inform cogent environmental policy and land-use practices throughout these regions. Interhemispheric, comparative analyses (e.g., Mooney 1977; Fuentes et al. 1991; Mooney et al. 1993) provide an essential construct to characterize, model, and forecast the effects of human impacts on ecosystems in these two continents.

Conscious of the above similarities and the great opportunity offered by this large scale "natural experiment" in 1990, a group of North and South American scientists established a research organization called AMIGO (America's Interhemisphere Geo-biosphere Organization) aimed at understanding ecosystem responses to global change by comparing the patterns and processes in the responses of analogous ecosystems in North and South America, and examining the interactions of human and natural disruptions across the two continents. The insights provided by the scientists involved in this initiative form the core of the present book, and derive from an interdisciplinary workshop held in Maitencillo, Chile. Participants for this book were selected to represent a spectrum of disciplines reflecting the requisite perspectives and integration for problem analysis and emphasizing integration of crosscutting issues related to ecosystem functioning and disruption, landscape alteration, biodiversity declines, conservation, and management.

Here, we present a preliminary attempt to lay the conceptual foundation for a comparative approach to studying ecological disruptions by examining natural and human-induced disruptions and linkages in western temperate ecosystems in North and South America. While there are diverse sources of ecosystem disruptors, the combination of habitat loss and fragmentation has been characterized as the "single most important of the many interacting components of global change affecting ecological systems" (Vitousek 1994). For this reason, we focus more specifically on habitat loss and fragmentation by examining the drivers, consequences, and management of landscape alteration in the Americas.

3 Why Ecosystem Fragmentation?

Our history of technological advances, social changes, and current demographic pressures upon natural resources has led us to a point where humans are effecting a global change in the environment whose result, so far, is that the persistence of many species and the sustainability of several ecosystems around the world are at risk (not to mention the large number of species already extinct due to human actions). The issue is no longer limited to conservation as changes in the land, air, and water quality threaten human survival as well.

Ecosystem sustainability is threatened because many ecosystem processes have been disrupted. Further, the impact of these disruptions extends across biomes and propagates across scales, thereby increasing their effect on a wide variety of landscapes. Many of these disruptions have global implications. For instance, changes in natural fire regimes (the spatial and temporal distribution of fire occurrence and intensity), the introduction of exotic species, and habitat loss and fragmentation not only have local impacts, but can also disrupt key ecological linkages and processes so as to threaten entire ecosystems. Focusing on the source and impact of disruptors, and particularly on habitat fragmentation, provides the necessary insight into how to ameliorate their deleterious effects.

The book is organized around three main themes: (1) causes and processes of landscape fragmentation; (2) ecological and evolutionary consequences of fragmentation; and (3) the theory, methods, and implications for conservation of ecosystem fragmentation.

This collection of papers is intended to help integrate concepts from several disciplines including anthropology, ecology, theoretical ecology, and conservation ecology, all of which are necessary in order to create a common and comprehensive framework for future research and conservation policy formulation. The challenge posed to ecology today is to effectively translate basic science into applied science that can help to resolve the vexing problems posed by global environmental change. After successful decades of teasing the world apart, science is now faced with the challenge of putting it back together. This effort begins with the thoughtful integration of social sciences and ecological sciences in a common format.

Acknowledgements. The authors would like to thank the National Science Foundation, Inter-American Institute for Global Change, FONDAP-FONDECYT, and National Centre for Ecological Analysis and Synthesis for supporting the work conducted in the workshops and book preparation. We would also like to thank the workshop participants and chapter authors for their diligent patience and creativity in this project. In addition, we would like to express our gratitude to many of our colleagues who have reviewed, provided discussion and contributed to various aspects of the book in this process:

Allison Alberts, Jill Anthony, Scott Bergen, Richard Bierregaard, Robin Bjork, Jeffrey Borchers, Michael Bowers,Marcela Brugnach, R. Terry Chesser, Hernán Cofre, Cintia Cornelius, Virginia Dale, John Hayes, Cyndy Hines, James Lenihan, Mark Lomolino, Bruce Milne, Paula Minear, Carolina Murcia, Reed Noss, James L. Patton, David L. Peterson, Sir Ghillean Prance, David Pyke, Melissa Richmond, Dar Roberts, Horacio Samaniego, Celeste Silva, Paola Soublette, John Vandermeer, Nickolas Waser, John A. Wiens. Finally, we would like to thank the editorial staff at Springer-Verlag for their enthusiasm and expertise.

References

Fuentes ER, Kronberg B, Mooney HA (1991) The west coasts of the Americas as indicators of global change. Trends Ecol Evol 6:203–204
Mooney HA (ed) (1977) Convergent evolution in Chile and California. Mediterranean climate ecosystems. US/IBP Synthesis Series 5. Dowden, Hutchinson and Ross, Stroudsburg, Pennsylvania
Mooney HA, Fuentes ER, Kronberg BI (eds) (1993) Earth system responses to global change. Contrasts between North and South America. Academic Press, San Diego
Vitousek PM (1994) Beyond global warming: ecology and global change. Ecology 75:1861–1876

Part I
Causes and Processes of Landscape Fragmentation

1 Biodiversity and Human Impact During the Last 11,000 Years in North-Central Chile

L. NÚÑEZ and M. GROSJEAN

1.1 Introduction

During the pre-Columbian period (11,000 ^{14}C years B.P.–1540 A.D.), the interaction between humans and their environment can be characterized as relatively harmonious with regard to the effects of resource exploitation patterns and subsistence practices in arid and semi-arid ecosystems of north-central Chile. The strong dependency on local resources stimulated a close coupling between environmental change and human population. However, since the 16th century, ecosystems in this area have been exposed to three trends that have dramatically disrupted a long-term relationship between human populations and natural resources. First, the appearance of European settlements focused on mining activities and created urban markets for agricultural products. Second, the establishment of European land-use patterns that, by focusing on economically important and highly productive plant and animal species, out-competed and displaced local native and less productive species to marginal areas and led to the widespread degradation of soils. Third, local indigenous populations retreated into "refuge areas" where they were able to maintain their traditional pattern of resource use which is based on self-subsistence and exchange of the surplus production.

From the 16th to the 20th centuries, European human occupation substantially modified the natural composition, structure, and function of ecosystems in the area. This disruptive role of humans reached a climax with the large-scale expansion of mining after 1950. In addition to fostering the emergence of rural poverty, mining and associated activities intensified environmental degradation by exploiting the scarce and very slowly recharging water resources of the Atacama Desert. Today, available freshwater is almost exclusively used in the new cities and mines in the Atacama. We will characterize the environmental history of the Atacama through a description of the various stages of human-environment interactions through the centuries.

Ecological Studies, Vol. 162
G.A. Bradshaw and P.A. Marquet (Eds.)
How Landscapes Change
© Springer-Verlag Berlin Heidelberg 2003

1.2 Principal Phases of Human-Environment Interaction in North-Central Chile

1.2.1 Biodiversity Changes at the Pleistocene-Holocene Transition

Chronostratigraphical excavations in the Tagua-Tagua basin (34°30'S, 71°10'W, see Fig. 1.1) identified two sites with a total of 12 mastodon remains (*Stegomastodon humboldti*) associated with Fell-type projectile points dated between 10,120 and 9900 ^{14}C years B.P. The identification of cultural and natural events at the Pleistocene–Holocene transition shows that the Paleo-Indian occupation was strongly related to specialized hunting of Pleistocene mega-mammals. The habitats of these animals, and therefore hunting sites, were concentrated around lacustrine paleoenvironments as a result of increasing drought stress (Nuñez et al. 1994a).

Palynological evidence confirms the existence of a resource crisis that significantly affected the mega-fauna around 10,000 ^{14}C years B.P. In fact, in less than 1000 years, there was a marked decrease in open park vegetation formations of conifers and southern beech (*Nothofagus*) trees as a consequence of a decrease in humidity and an increase in temperature compared to present values (Heusser 1983). These vegetation patterns gave way to a herbaceous steppe dominated by Amaranthaceae under dry environmental conditions (Heusser 1983). Around lacustrine basins and wetlands, this change took place around 9300 ^{14}C years B.P. when taxa associated with marshes and aquatic habitats (e.g., *Anagallis*, Cyperaceae, *Typha*) decreased and Compositae and Ubellifera dominated in an arid landscape. This process of increasing desertification was associated with an abrupt decrease in available water, plant and animal resources. In response to this climatic shift, most faunal elements including Pleistocene mega-mammals and their main predators (i.e., humans), concentrated their activities around the few remaining water bodies (Villagrán and Varela 1990; Núñez et al. 1994a).

The extinction of mastodons and the presence of Paleo-Indian butchery sites around Lake Tagua-Tagua appear to be correlated with environmental changes, that is, increasing aridity forced mega-mammals and the humans who hunted them to occupy less stressful near-lake environments during the Pleistocene–Holocene transition. This crisis was apparently widespread and significantly affected the "ecorefuges" of North and South American proboscideans (Bryan et al. 1978; Correal 1981; Haynes 1991). Rather than targeting specialized hunting as the only cause of extinction (Martin and Klein 1984), our findings at Tagua-Tagua suggest that environmental changes were also important in enforcing the concentration of mastodon herds around lake "ecorefuges" where they were opportunistically hunted by humans. These environmental events in central Chile coincide with the disappearance of proboscideans all over the world at the end of the Pleistocene (12,000–10,000 [14]C

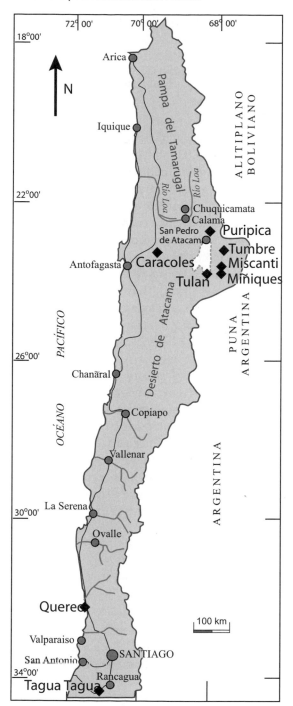

Fig. 1.1. Study area and sites discussed in the text

years B.P.), and with similar responses by Clovis groups found in North America (Haynes 1991). During this short transition period, Pleistocene hunter populations and mobility increased across North America and down to the southernmost areas of South America, opportunistically exploiting habitats with mega-mammals in a critical period of abrupt environmental change.

Evidence of similar environmental change was also found at Quereo (700 km north of Tagua-Tagua; Villagrán and Varela 1990; Veit 1994; see Fig. 1.1), suggesting that abrupt changes took place at a larger spatial scale than previously thought (Nuñez et al. 1994b). The El Niño Southern Oscillation (ENSO) may have played a major role in controlling the late Pleistocene/early Holocene precipitation pattern in this part of the world. However, little is known about the presence, function and causes of such large-scale and long-term moisture changes, and the variability of the Humboldt Current in this area (Grosjean et al. 1995; Kull and Grosjean 1998).

In contrast to increasing aridity in central Chile at the Pleistocene/Holocene transition, favorable and more humid conditions emerged and prevailed further north in the Atacama Desert during that time. The Atacama High Puna in northern Chile is presently located in an extremely arid area where the tropical summer rainfall belt and the extratropical winter precipitation belt converge and sometimes overlap. The high altitude plateau or Altiplano, between 19° and 27°S, is so arid that even today no glaciers can survive. Its sensitivity to changes in effective moisture makes this a particularly important area for studying shifts in the intensities of tropical and westerly circulation belts, and examining the phenomenon of moisture changes in the tropical Andes during the Holocene. Current research supports the hypothesis that the humid environmental conditions of the Atacama Desert during the Late Glacial and early Holocene were related to the intensification of the summer monsoon (Invierno Boliviano) which expanded its reach and caused increased precipitation as far as 24°S and resulted in favorable water, faunistic, floristic, and edaphic resources for human occupation of the area between ca. 12,000 and 8000 ^{14}C years B.P. (Messerli et al. 1993, 2000; Grosjean and Nuñez 1994; Betancourt et al. 2000; Kull et al. 2002).

Although humans may have inhabited Peru as early as 12,560 ^{14}C years B.P., the oldest dated evidence of Early Archaic hunters in the Puna de Atacama, in northern Chile, is relatively recent (10,820 ^{14}C years B.P.). Small and highly mobile groups occupied circum-lacustrine sites in the high Puna, the intermediate valleys, the depression of the Salar de Atacama, and the high-elevation sites of the Precordillera. Radiocarbon dates from seven early Archaic sites (of which Tuina, San Lorenzo, and Tambillo are the most important) suggest initial human occupation at 10,800 ^{14}C years B.P., and a rapid termination of this occupational phase at about 8000 ^{14}C years B.P. Hunters exploited modern fauna, such as camelids, deer, birds, and rodents. At the time of human arrival, the paleolakes in northern Chile were close to or already at their maximum extent (between 10,800 and <9200 ^{14}C years B.P.; Geyh et al. 1999), and the humid climate provided abundant water resources for vegetation, animals, and ultimately, humans. Recent excavations show an evident link between initial human occupation and favorable environmental conditions as recorded in paleolake sediments, fossil groundwater bodies, and paleosols (Núñez et al. 2001).

The environmental conditions at the beginning of human occupation in the Atacama Desert remain somewhat debatable. Although mega-fauna dated to late glacial times existed in the south-central Andes, a site unequivocally linking hunters and Pleistocene animals in the Atacama Desert has been missing until recently. At Barro Negro (northwestern Argentina; Fernández et al. 1991), for instance, evidence of Archaic camelid hunters is found precisely at the time when *Equus* sp. became extinct. However, recent excavations in Tuina document for the first time hunting of extinct fauna at 10,060 ^{14}C years B.P. (L. Núñez et al. 2002), suggesting that at least one Pleistocene species survived longer and overlapped for some time with Archaic human occupation. The taxonomy of this bone fragment is currently under investigation. For the Chilean part of the Altiplano, we argue that the onset of favorable environmental conditions and initial human occupation was, notwithstanding the statistically poor dating control of the first human arrival, broadly synchronous. With a few exceptions, the Pleistocene mega-fauna was possibly already largely extinct, although the modern camelids survived and served as the resource basis for the Early Archaic hunters.

Lake sediment records from the Altiplano suggest that the paleolakes disappeared very rapidly at about 8000 ^{14}C years B.P. (Grosjean et al. 1995; Geyh et al. 1999), and extremely arid conditions were established from then onwards. The severe decrease in resources in the Atacama Desert at that time is reflected in a synchronous drastic depopulation of the area ("Silencio Arqueológico"), suggesting that the favorable humid early Holocene environment was the precondition for early hunting and gathering societies in this area.

1.2.2 Camelid Domestication During the Mid-Holocene: the Rise of a New Human-Environment Interaction

Multi-proxy data for the extremely arid mid-Holocene period (8000 to around 3000 ^{14}C years B.P.) are scarce. The lake levels were extremely low, most basins were completely dry, and the lake sediments were eroded by wind and destroyed. While the paleoenvironmental conditions (i.e., active lacustrine basins) were good for the Early Archaic occupations before 8000 ^{14}C years B.P., resources were very poor in the Atacama Desert during the mid-Holocene and only available in special places along small rivers and wetlands such as the Quebradas de Puripica and Tulán (Fig. 1.1) which served as ecological refuges for hunters and gatherers in the Atacama Desert.

Twenty Archaic campsites intercalated between more than 30 alluvial deposits caused by heavy rainfall events between 6200 and 3100 ^{14}C years B.P. were discovered in Quebrada Puripica. For the first time, we found evidence of discrete human occupation in this area, filling the regional hiatus in the Atacama basin ("Silencio Arqueológico") between 8000 and 4800 ^{14}C years B.P. The unique detailed stratigraphy of this site allows a stepwise transformation of the early- and mid-Archaic tradition into the late Archaic complex culture with domesticated animals to be documented. The early hunting tradition (prior to 6200 ^{14}C years B.P.) changed by 5900 ^{14}C years B.P. into a system with large campsites, intensive exploitation of wild camelids, and an innovative lithic industry. The culmination of this process of cultural transformation occurred in the late Archaic (5100–4800 ^{14}C years B.P.) which was characterized by hunting and animal domestication (*Lama glama*), major use of local lithic materials, the development of a consolidated architecture, and rock art showing camelids in a naturalistic style (Santoro and Núñez 1987; Núñez 1992).

The interaction between the geographical setting, mid-Holocene climates of general aridity, and the existence of storm-related sediment deposition contributed to a concentration of resources at Quebrada Puripica which functioned as an "ecorefuge" in a generally hostile environment. Here, humans established more stable and permanent settlements, domesticated wild camelids, and developed pastoralism to compensate for the harsh and uncertain environment. It appears that human occupation was restricted to areas which produced a source of stable resources. All other early Archaic sites in the Atacama basin at large had apparently been abandoned by the onset of mid-Holocene aridity at 8000 ^{14}C years B.P. The cultural transitions observed within the Puripica stratigraphy do not necessarily suggest environmental forcing. However, it is interesting to note that the end of the "ecorefuge" in Puripica (i.e., the beginning of down-cutting and the cessation of alluvial deposits) coincides with a period of increasing precipitation and the onset of modern climatic conditions at about 3000 ^{14}C years B.P. (Valero-Garcés et al. 1996; Grosjean et al. 2002).

1.2.3 The Transition from Mid-Holocene to Modern Climate: Pastoralism and Agricultural Changes

The generally harsh environmental conditions peaked during the mid-Holocene. Subsequently, (i.e., after ca. 3600 ^{14}C years B.P.) precipitation rates increased, lake levels recovered, and the modern climate regime was established, reaching about 200 mm precipitation per year in the highlands (Grosjean 1994). The transition from the extremely dry mid-Holocene conditions to the slightly more humid modern climate is clearly seen in the sediments of Laguna Miscanti (Valero-Garcés et al. 1996; Grosjean et al. 2002). Lake sediments show the variable nature of this transition, suggesting that changes were probably forced by a series of alternating changes in moisture regime, and not due to a slow, progressive change.

The restoration of resources to modern conditions was synchronous with the expansion of camelid breeding in the circum-Puna area as found, for instance, in the Quebrada de Tulán (Tulán-54) some 30 km west of Lagunas de Meniques and Miscanti (Fig. 1.1). This was also the time when the altitudinal range between 3000 and 4500 m became inhabited by flocks of domestic camelids, and irrigated agriculture began to dominate the subsistence economy between 3000 and 2400 m (Núñez 1992).

During the Formative Period (4000–2000 ^{14}C years B.P.), exploitation of domesticated mammals (*Lama glama* and *Cavia* spp.), specialized hunting (*Vicugna vicugna, Lama guanicoe*, rodents and birds) and irrigated agriculture with corn (*Zea mays*), quinoa (*Chenopodium quinua*), potatoes (*Tuber*), squash (*Cucurbitas spp.*), chili pepper (*Capsicum* spp.), and gathering of tree fruits (*Prosopis* sp.) gave rise to a diverse use of Andean and Subandean animal and plant resources in the Atacama Desert. Such a situation allowed a subtle and equilibrated management of natural resources in, around and from isolated permanent settlements, which initially emerged in oases with spring water and river flow.

1.2.4 Changes During the Historic Period (16th–20th Centuries)

Pre-Hispanic hunting, gathering, agriculture, and pastoralism created a varied signature upon the landscape, while Hispanic colonization imposed a radical change on the landscape by intensifying the water and land resource use and the exploitation of local flora and fauna within only 300 years. In fact, the pre-Hispanic tropical-Andean complex of crops (major components: maize, squash, beans, quinoa, chili peppers, potatoes, etc.) was replaced by the exotic European mix of crops (grapes, olives, wheat, alfalfa, barley, vegetables, and citrus fruits such as oranges, lemons, and limes). In addition, the local domesticated fauna (camelids) had to compete with non-Andean animals such as horses, cows, sheep, pigs, donkeys, goats, rabbits, poultry and other European domesticates. At the same time, new (European) forage patterns were established which resulted in the over-exploitation and degradation of natural grasslands. A significant increase in alfalfa cultivation was also needed for the increasing exotic animal population. These animals were brought by the Europeans mainly for working in mineral ore exploitation, transportation of construction materials and agricultural products, and for meat. During this readjustment, alfalfa became the dominant crop, marginalizing the "chacras de pan llevar" (local food crops), and played a major role in the new economy, which was based on large-scale mining, traffic and transportation of goods in a trans-Andean space.

At around the middle of the 19th century, the industrial revolution first reached the Atacama Desert. New British mining technologies made large-scale smelting possible. This contrasted very much with the pre-Columbian manufactories and technologies for small-scale processing of gold, silver, and copper (Bermudez 1963; Núñez 1992). Smelters for silver and copper in the Pampa del Tamarugal, the Río Loa, in Caracoles, Chuquicamata and in the oasis of San Pedro de Atacama (Fig. 1.1) rapidly overexploited the locally available fuel wood resources (*Prosopis* forests, Algarrobo and Tamarugo; Bermudez 1963) in the area of Calama even to total deforestation (Phillipi 1860). In turn, the scarcity of local fuel wood stimulated overexploitation of many plants in remote areas of the high Andes, such as the cushion plant Yareta (*Azorella compacta*), which was gathered for fuel in the copper mine of Chuquicamata.

Increased urbanization in the Atacama Desert fostered the drastic overexploitation of local wood from the 18th to the beginning of the 20th century. Suitable wood for construction was taken from cactus and *Prosopis* trees. The limited supply was supplemented with imported pine "pino de oregón" from North America, even though many large trees were also used for fuel wood ("rodelas" de Tamarugo) and made into charcoal. Even small shrubs such as "Pingo Pingo" and others were carried with animal caravans from the oasis of San Pedro de Atacama to the town of Caracoles, the most important silver mine in former Bolivia (Phillippi 1860). A system of roads built in the early 20th century increased access to and exploitation of high altitude fuel wood sources of the Andean steppe as well as facilitating the spread of invasive species. Finally, fuel wood demands diminished with the use of electricity and petroleum. However, new pressures on these resources are likely to emerge in the future. First, because local economies (crop production, labor, and capital) in valleys and oases have historically been forced to become part of the urban markets, the termination of mining in an area will again produce serious readjustments in the relationship between the local population, their economy and their natural resources (e.g., a sudden return to self-subsistence). Second, demands for water will become even more pressing. Twentieth century mining technology is based mainly on lixiviation, which requires enormous amounts of water. Given the economic importance of mining, government policy is likely to continue to favor mining companies and urban populations over local people in rural areas in water rights conflicts. The extreme arid conditions of the area already make water a scarce and precious resource for rural people and ecosystems. The most striking case is the Loa River basin, where the river discharge has been reduced to a fraction of its natural flow during the pre-Hispanic and Hispanic agricultural periods. Surface water diversions and groundwater extractions have significantly affected wetlands, which are crucial for the survival of a highly diverse and largely endemic flora and fauna. Some wetlands have been greatly reduced in extent or have disappeared completely, severely affecting both the ecosystem and traditional patterns of resource use (Messerli et al. 1997). In short, the entire landscape of the Loa River changed dramatically due to the overexploitation of water resources.

All of these changes have significantly affected the structure and form of present-day ecosystems, habitats and landscapes of the Atacama, which are highly fragile and host a large number of endemic species adapted to this extreme environment (Villagrán et al. 1983; Marquet 1994; Marquet et al. 1998). Humans continue to play a paramount role in the modification of these landscapes. The introduction of exotic species, in particular, endangers indigenous flora and fauna and many of the bio-physico-chemical processes. The Atacama Desert has never experienced such rapid expansion of mining and urbanization as it is undergoing today, and demands on water resources will only increase as this process continues (Romero 2002).

Current patterns of land and resource use are not sustainable. Without consideration of an "ethic of survival" and requirements of local biodiversity in water resource planning and assessment, sustainability cannot be achieved (Messerli et al. 1997). Unless the amount, origin, and recharge rates are scientifically established, and water distribution is evaluated relative to local cultural and environmental conditions, we can expect the environmental crisis to intensify. This process may be accelerated by continued migration of people from the Andes into urban centers, increased mining activities, escalating tourism, and progressive desertification of arable lands due to their low productivity and high prices for water. In this sense, achieving equitable distribution of water resources among various users (mining, urban consumption, agriculture, and natural ecosystems) and a transparent water policy are the two most urgent ethical issues for the future of the Atacama Desert.

Today, it is not clear how this problem will be solved. However, the people in rural areas, their culture, their traditional methods of cultivating the land, the flora, fauna and their natural habitats deserve consideration when planning for the future. Cultural heritage and biodiversity do not have a direct economic value. Thus, cultural values and natural habitats must not compete with economic values in a global market system.

Acknowledgement. This work was made possible through FONDECYT grant 59600011. Several paleoenvironmental data were obtained within grant FONDECYT 1930022 in collaboration with the co-author M.G. from the University of Bern (NF 21.57073.99) and grants from the Dirección de Investigaciones de la Universidad Católica del Norte and the National Geographic Society.

References

Bermudez O (1963) Historia del salitre. Editorial Universitaria, Antofagasta

Betancourt JL, Latorre C, Rech J, Quade J, Rylander KA (2000) A 22,000-year record of monsoonal precipitation from northern Chile's Atacama Desert. Science 289:1542–1546

Bryan AL, Casamiquela R, Cruxent JM, Gruhn R, Ochsenius C (1978) An El Jobo mastodon kill at Taima taima, northern Venezuela. Science 20:1275–1277

Correal G (1981) Evidencias culturales y megafauna pleistocénica en Colombia. Banco de la República de Bogotá, Bogotá

Fernández J, Markgraf V, Panarello HO et al. (1991) Late Pleistocene/early Holocene environments and climates, fauna, and human occupation in the Argentine Altiplano. Geoarcheology 6:251–272

Geyh M, Grosjean M, Núñez LA, Schotterer U (1999) Radiocarbon reservoir effect and the timing of the late-glacial/early Holocene humid phase in the Atacama Desert (Northern Chile). Quat Res 52:143–153

Grosjean M (1994) Paleohydrology of Laguna Leijia (north Chilean Altiplano) and climatic implications for late-glacial times. Palaeogeogr Palaeoclimatol Palaeoecol 109:89–100

Grosjean M, Nuñez L (1994) Early and middle Holocene environments, human occupation and resource use in the Atacama (northern Chile). Geoarchaeology 9:271–286

Grosjean M, Messerli B, Ammann C, Geyh MA, Graf K, Jenny B, Kammer K, Núñez L, Schreier H, Schotterer U, Schwalb A, Valero B, Vuille M (1995) Holocene environmental changes in the Atacama altiplano and paleoclimatic implications. Bol Inst Fr Etudes Andines 24:585–594

Grosjean M, van Leeuwen JFN, van der Knaap WO, Geyh MA, Ammann B, Tanner W, Messerli B, Veit HA (2002) 22,000 14 C yr B.P. sediment and pollen record of climate change from Laguna Miscanti 23°S, northern Chile. Global Planet Change 28/1-4:35–51

Haynes G (1991) Mammoths, mastodonts, and elephants. Biology, behavior and fossil record. Cambridge University Press, Cambridge

Heusser CJ (1983) Quaternary pollen record from Laguna de Taguatagua, Chile. Science 219:1429–1432

Kull C, Grosjean M (1998) Albedo changes, Milankovitch forcing and late quaternary climate changes in the central Andes. Climate Dynam 14:871–881

Kull C, Grosjean M, Veit H (2002)Modeling modern and late Pleistocene glacio-climatological conditions in the North Chilean Andes (29°S–30°S). Climate Change 53(3):359–381

Marquet PA (1994) Diversity of small mammals in the Pacific coastal desert of Peru and Chile and in the adjacent area: biogeography and community structure. Aust J Zool 42:527–542

Marquet PA, Bozinovic F, Bradshaw GA, Cornelius CC, González H, Gutierrez JR, Hajek ER, Lagos JA, López-Cortés F, Núñez L, Rosello EF, Santoro C, Samaniego H, Standen VG, Torres-Mura JC, Jaksic FM (1998) Los ecosistemas del Desierto de Atacama y area Andina adjacente. Rev Chil Hist Nat 71:593–617

Martin PS, Klein RG (1984) Quaternary extinction: a prehistoric revolution. University of Arizona Press, Tucson, Arizona

Messerli B, Grosjean M, Bonani G, Bürgi A, Geyh MA, Graf K, Ramseier K, Romero H, Schotterer U, Schreier H, Vuille M (1993) Climate change and dynamics of natural resources in the Altiplano of northern Chile during late Glacial and Holocene time. First synthesis. Mountain Res Dev 13:117–127

Messerli B, Grosjean M, Vuille M (1997) Water availability, protected areas, and natural resources in the Andean Desert Altiplano. Mountain Res Dev 17:229–238

Messerli B, Grosjean M, Hofer T, Núñez L, Pfister C (2000) From nature-dominated to human-dominated environmental changes. Quat Sci Rev 19:459–479

Nuñez L (1992) Ocupación arcaica en la Puna de Atacama: secuencia, movilidad y cambio. {Prehistoria Sudamericana, Nuevas perspectivas} In: Meggers B (ed) Prehistoric South America, new perspectives. Taraxacum, Washington, DC, pp 283–317

Nuñez L, Varela J, Casamiquela R, Schiappacasse V, Niemeyer H, Villagrán C (1994a) Cuenca de Taguatagua en Chile: el ambiente del pleistoceno superior y ocupaciones humanas. Rev Chil Hist Nat 67:503–519

Nuñez L, Varela J, Casamiquela R, Villagrán C (1994b) Reconstrucción multidisciplinaria de la ocupación prehistórica de Quereo, Centro de Chile. Latin Am Antiquity 5:99–118

Núñez L, Grosjean M, Cartajena I (2001) Human dimensions of late Pleistocene/Holocene arid events in southern South America. In: Markgraf V (ed) Interhemispheric climate linkages. Academic Press, San Diego, pp 454

Núñez L, Grosjean M, Cartajena I (2002) Human occupations and climate change in the Puna de Atacama, Chile. Science 298:821–824

Phillippi RA (1860) Viaje al desierto de Atacama. Halle en Sajonia, Librería de Eduardo Anton

Romero HI (2002) Clash between economic and sustainable development. IHDP Newsletter 1/2002, Bonn, pp 7–9

Santoro C, Núñez L (1987) Hunters of the dry Puna and the salt Puna in northern Chile. Andean Past 1:54–109

Valero-Garcés B, Grosjean M, Schwalb A, Geyh MA, Messerli B, Kelts K (1996) Late Holocene environmental change in the Atacama Altiplano: limnogeology of Laguna Miscanti, Chile. J Paleolimnol 16:1–21

Veit H (1994) Jungquartäre Landschafts- und Klimaentwicklung am Südrand der Atacama (Norte Chico, Chile). Habilitation Thesis. University of Bayreuth, Bayreuth

Villagrán C, Varela J (1990) Palynological evidence for increased aridity on the Central Chilean coast during the Holocene. Quat Res 34:198–207

Villagrán C, Arroyo MTK, Marticorena C (1983) Efectos de la desertización en la distribución de la flora andina de Chile. Rev Chil Hist Nat 56:137–157

2 Beyond Malthus and Perverse Incentives: Economic Globalization, Forest Conversion and Habitat Fragmentation

S.F. SIEBERT

2.1 Introduction

Forest conversion and habitat fragmentation are widely believed to result from overpopulation and perverse government policies. Since the time of Malthus, demographic pressures have been identified as a cause of natural resource degradation. More recently, the role of perverse government policies (i.e., programs that undervalue and/or subsidize natural resource extraction and degradation) and market failures has been recognized as another crucial explanatory variable. While the sheer number of human beings and perverse government policies/market failures unquestionably contribute to land degradation and resource exploitation, this paper argues that these factors are inadequate to explain the rate, pattern and extent of contemporary forest degradation and habitat fragmentation. Furthermore, continued focus on Malthusian population pressures and an uncritical belief in the curative power of government policy and market reforms, obfuscates understanding other causal factors. I argue that the economic world order that has arisen in the past two decades and its associated policies and programs are an important, but largely overlooked cause of forest conversion and habitat fragmentation. Furthermore, these pressures are likely to increase in the future given contemporary economic growth rates and global institutionalization of "free" (unregulated) market principles.

In this paper, I review the role and importance of demographic pressures and perverse government policies in forest conversion and habitat fragmentation. I then briefly describe important characteristics of the new economic world order and the role that they play in forest conversion and habitat fragmentation based on an analysis of case studies from southern Mexico and Chile, two nations that have recently restructured their economies and embraced free market reforms. Finally, I identify likely future patterns and processes of forest conversion and habitat fragmentation, given the globalization of free market policies.

Ecological Studies, Vol. 162
G.A. Bradshaw and P.A. Marquet (Eds.)
How Landscapes Change
© Springer-Verlag Berlin Heidelberg 2003

2.2 Demographic Pressures

For over two centuries, demographers, politicians, and resource managers have warned that burgeoning human populations would inevitably lead to widespread famine, depletion of natural resources, and land degradation. The most recent apostles of this belief have been ecologists (e.g., Erhlich 1968; Hardin 1974; Erhlich and Ehrlich 1990; Gehrt 1996) and Malthusian perspectives continue to dominate the international conservation community as is evident from perusing recent issues of the journal *Conservation Biology*. Interestingly, however, statistical analyses have failed to establish unequivocal cause and effect relationships between population and forest conversion. For example, while numerous authors cite positive statistical relationships between human population growth and loss of biodiversity (Meffe and Carroll 1994; Forester and Machlis 1996) and between human population growth and deforestation (Southgate et al. 1991; Palo 1994; Rudel 1994), other studies find no significant relationships or even negative relationships between population and deforestation or loss of biodiversity (e.g., Kahn and McDonald 1994; Kummer and Turner 1994; Pearce and Brown 1994).

Part of the problem appears to be the uncertain reliability of population data (Forester and Machlis 1996), the fact that generalizations at the global and even regional level mask significant regional variations and factors (Kummer and Turner 1994), and variable interpretations and means of modeling deforestation and population. For example, in a rigorous econometric analysis of the causes of tropical deforestation (Pearce and Brown 1994), contributing authors model population in terms of total population density, total population growth, rural population density, and rural population growth, with one author even incorporating unequal land tenure and poverty into the population equation. Not surprisingly, reported cause and effect relationships vary.

While the sheer number of human beings putting pressure on natural resources cannot be ignored, Malthusians appear blind to political and economic factors that may be less intractable and whose amelioration could provide significant and immediate conservation benefits. When one considers the fact that human populations expanded by 15–36 % in tropical countries during the 1980s while deforestation expanded by almost 70 %, it would appear that other factors are at work (Myers 1994). Furthermore, some countries, like Belize, are currently experiencing extremely rapid forest conversion, but exhibit little net population growth (i.e., internal population growth and immigration are off-set by emigration), while others, such as Brazil, have exhibited widely fluctuating forest conversion rates, patterns and processes (Browder 1996) during which time their population growth rates have not changed (Mahar and Schneider 1994).

Some analysts have suggested that increasing human population densities not only encourage technological innovation (Boserup 1965), but lead to improved land management practices and enhanced sustainability as increased land values motivate farmers to make long-term investments in conservation farming practices (FAO 1989; Critchley et al. 1994). Other analysts (e.g., Mamdani 1972; Humphrey and Buttel 1986; Westoby 1989) argue that focusing on overpopulation blames the victims of contemporary and historical resource extraction and forest conversion, and absolves the rich and powerful beneficiaries of such policies (i.e., North Americans, northern Europeans, and Japanese) from responsibility. Furthermore, if overpopulation is the problem, then birth control in poor developing countries is the solution and there is little reason to consider other potential variables. Perhaps the most precise statement that can be made with regard to population and forest conversion is that:

"there does appear to be a positive relationship across countries between population growth (especially rural growth) and the increase in agricultural land (land extensification), land intensification in the form of increases in use of (chemical) fertilizers, and deforestation, however, in each case the relationship is weak, and dependent on the inclusion or exclusion of particular 'outlier' countries." (Bilsborrow and Geores 1994)

Even if we accept that overpopulation is a significant cause of forest conversion, a conservation policy predicated on waiting for the demographic lag effect of third world birth control to take effect seems doomed to fail given the immediacy of global forest conversion and species extinction pressures.

2.3 Perverse Incentives and Market Failures

Over the past two decades government policies and market failures have been recognized as a significant cause of forest conversion and habitat fragmentation. Some of the most pervasive examples of perverse incentives include the failure to collect realistic economic rents through royalties and taxes in resource extraction activities, unreasonably short concessions and the use of selective (i.e., high-grading) harvesting practices in the timber sector; national agricultural subsidy programs; national and international investments in mining, dams and roads; domestic tax, credit and pricing policies (e.g., generous tax holidays, subsidized credit, and direct government subsidies); national and international investment priorities biased against the small agricultural and rural sectors; and domestic land tenure policies that encourage deforestation (Gillis and Repetto 1988; Southgate 1998). Government policies of this type are widespread in both developed and developing economies.

Market failures, specifically undervaluing or failing to value natural resource assets, are another major source of environmental degradation

(Gillis and Repetto 1988; Daly 1996). In this perspective, natural resource assets will inevitably be misused or exploited until realistic long-term social and environmental costs are internalized and reflected in market prices (i.e., it is economically rational do so; Panayotou 1993). Collectively, perverse incentives and market failures stimulate forest exploitation that would otherwise likely be unprofitable and perhaps not pursued. On the other hand, the buying and selling of pollution credits through the Clean Air Act and carbon-sequestration credits are examples of how markets can be used to enhance environmental well-being.

Recognition of the importance of perverse government policies arose in large part through analyses of forest exploitation in Latin America. For example, Browder (1988) noted that almost half of the forest conversion that occurred in the Brazilian Amazon during the 1970s and 1980s could be attributed to four government subsidy programs. Similarly, Harrison (1991) concluded that forest conversion in Costa Rica was most strongly correlated with the number of cattle, and that the livestock industry was supported by economically irrational government policies and programs. Many perverse public policies persist, such as those in Belize that require land be "improved" (i.e., cleared of forest) before title or tenure is granted.

While the current emphasis on perverse incentives and market failures is important, some researchers suggest that it fails to explain the logic or persistence of forest degradation. Hecht (1993), for example, argues that the Amazonian livestock sector possesses extractive, productive and fiscal possibilities that forestry, agricultural and agroforestry alternatives lack, and that livestock ranching is actually a rational and attractive land use for small, as well as large producers. Similarly, Mahar and Schneider (1994) attributed the decline in Amazonian deforestation rates in the early 1990s to an economic recession and a decline in Brazil's rural population (and thus a reduced number of potential migrants to the Amazon). They noted, based on survey results, that 70 % of the recent migrants to Rondonia moved to the Brazilian Amazon because of agricultural mechanization (i.e., technological innovation) and land consolidation in soybean and coffee plantations in their home state of Parana in southern Brazil. They argue that governments interested in reducing deforestation should focus on policies that maximize employment opportunities in regions far from the forest and implement land reform; increase investments in health, education and family planning; and eliminate tax and credit systems that favor capital and land accumulation. In a more general review, Mendes de Carvalho and Brown (1996) conclude that the growing polarization of economic wealth within the world's 12 megadiversity countries (which are primarily tropical) and between those countries and the industrialized north is a serious and growing threat to the conservation of biological diversity in the tropics.

These analyses suggest that increased concentration of land ownership, labor displacing technological innovations, increased mobility of capital, greater capital accumulation by domestic and international elites and transnational corporations, and reduced government support of social welfare and environmental programs encourage forest conversion and habitat fragmentation. However, these policies are neither perverse, nor misguided, but rather a reflection of the values and objectives of the current global economic order. Furthermore, these and numerous related policies are avidly supported by nation states, international treaties and transnational corporations. Modest market reforms and attempts to "get the prices right" as advocated by reformers will not alter the basic operating premise of the global economy, namely: maximization of economic growth and short-term profit through unleashing "free" market forces and increasing global trade and consumption.

2.4 Economic Globalization

During the past two decades, nation states and the world's economy have undergone a radical transformation, from one in which growth and development was nationally organized to one that is globally organized (McMichael 1996). Beginning with the ascension of Reagan/Thatcher economic policies and accelerating with the collapse of East Block centralized economies and the growth of global trade agreements (i.e., WTO), the world has avidly pursued a single model of economic growth. This new economic order is predicated on several principles, including: (1) maximization of economic growth via increased consumption and trade; (2) liberalization of barriers to world trade; (3) increased participation in world markets; and (4) strict adherence to inflation-control and debt repayment schedules. At a macro-scale, these policies have led to: (1) widespread privatization of state-owned enterprises; (2) drastic reductions in state funding of social welfare and environmental programs; (3) increased influence by multinational agencies, banks, and transnational corporations in domestic economic and development priorities; (4) rapid expansion of export cash cropping and natural resource extraction; and (5) the net transfer of over $400 billion from poor southern states to rich northern nations for debt servicing during the 1980s alone (Korten 1995; Mander 1996; McMichael 1996).

The power and reach of the new global order are currently being codified through internationally binding trade treaties, in particular the WTO (World Trade Organization), and regional agreements such as NAFTA (North American Free Trade Agreement) and APEC (Asia Pacific Economic Cooperation). Significantly, these treaties facilitate the rapid flow of capital and products throughout the world, but *not* labor. That is, transnational corporations, banks and other investors enjoy greater flexibility to determine what should be produced, where and for whom, and the freedom to allocate investment and production activities as comparative advantages wax and wane. However, unregulated trade does not free labor to follow global production opportunities (i.e., to immigrate). Instead, low-income peasant farmers and workers may confront increased employment uncertainties, lower wages, minimal health benefits, and reduced subsistence agricultural opportunities as they are displaced from their lands and become wage laborers.

The policies and programs of the new global economy are strictly enforced. International Monetary Fund (IMF) structural adjustment loans, for example, stipulate specific institutional, trade and fiscal policies that nations must observe as a condition for receipt of loans and maintenance of good international credit rating. Recent developments in Indonesia graphically illustrate the effects such policies can have on forest conversion. When Indonesia faced fiscal and economic collapse in 1997 and was forced to seek emergency loans, the IMF imposed a number of policy reforms as a precondition to providing loans, including: (1) removing restrictions on foreign investments in oil palm plantations, (2) eliminating the ban on palm oil exports, and (3) reducing export taxes on logs and rattan. Oil palm plantation development has been a major cause of forest conversion in Indonesia and the primary source of regional air pollution (through widespread land-clearing by burning); and the sector in now well-positioned to grow even more rapidly as the region's economic conditions rebound (Sunderlin 1999).

Countries that fail to adhere to IMF conditions and rules, or to even play the game (e.g., Cuba) face severe economic, trade and political sanctions. Furthermore, with the exception of Cuba and perhaps the Indian state of Kerala (Franke and Chasin 1994), few alternative development models exist at the nation-state level. The social and environmental ramifications of this new global economy are apparent throughout the Americas.

2.5 The Case of Chiapas

The southern Mexican state of Chiapas is known for its natural resources, rich biological diversity, and the Zapatista rebellion. These images are not unrelated. Chiapas and other southern Mexican states are inhabited by large numbers of low-income, indigenous Maya who engage in small-scale market and subsistence-oriented agriculture and forestry, much of it on a communal basis through the legally sanctioned ejido system that was established after the Mexican revolution in the early 1900s (Collier and Quaratiello 1994). Following the 1982 debt crisis, Mexico, in adherence to structural adjustment conditions, dismantled market controls, privatized state enterprises, phased out subsidies and credits in the agrarian and forestry sectors, and backed away from historic commitments made with the peasantry after the revolution (Collier et al. 1994). Trade liberalization, deregulation, and decentralization accelerated in the early 1990s during the Salinas administration and growing international (i.e., American) pressure to secure international trade arrangements, in particular NAFTA. This included rewriting the agrarian reform section of the Mexican Constitution (Article 27), effectively ending the nation's land reform policies, in an attempt to modernize the economy through terminating government support of inefficient sectors that fail to produce for the export market (i.e., traditional peasant agriculture and ejidos; Collier and Quaratiello 1994). This economic liberalization also permitted the privatization, mortgaging and sale of ejidal land and encouraged outside investors to enter into joint ventures with peasants for the purpose of commercial production of timber and export cash crops (Collier and Quaratiello 1994; Bray and Wexler 1996).

The social and environmental implications of these reforms are potentially profound and wide-ranging. Mexico ranks fourth in the world in terms of biodiversity and much of that is found in the country's forests (Bray and Wexler 1996), particularly its tropical forests. Almost 80 % of Mexico's forest lands are held by ejidos or indigenous communities, and approximately 17 million people, many of whom are poor and politically marginalized, depend upon forest resources and land for their very survival (Bray and Wexler 1996). At a 1995 conference "Community Conservation in the Mayan Forests of Belize, Guatemala and Mexico", peasant and ejido leaders and conservationists identified economic restructuring as a threat to the region's inhabitants and forests. Specifically, they recognized that while the forests of southern Mexico currently contain several designated core protected areas, the vast majority of the region is matrix lands owned and controlled by ejidos or small farmers engaged in small-scale selective timber harvesting, subsistence agriculture and production for local markets and that these lands are important to the conservation of indigenous flora and fauna (Primack et al. 1998). With deregulation and the right to sell land, many of these matrix lands have attracted the interest of transnational corporations and wealthy elite for large-scale logging and plantation forestry and/or conversion to export cash crops (e.g., citrus) and cattle ranching. While current ejido owners are not required to sell their lands, reforms of the Mexican legal code and the huge financial windfall potentials make land sales a real possibility. As Bray (1995) observed:

"By strict free market standards, the community forest sector is just a high-cost, low-quality, uncompetitive producer of sawn timber and roundwood. Its role in forest conservation and global ecological services, in rooting people and cultures to the land they have inhabited for millennia, and in generating local employment is not valued in economic terms."

Transfer of ejido lands would likely cause massive habitat fragmentation and isolation of core protected areas as matrix forests are converted to commercial monocrop agriculture or pasture lands of little conservation value. Many indigenous inhabitants would become landless and thus required to work as plantation laborers. As Vandermeer and Perfecto (1995) have documented in Costa Rica, this is a recipe for increased forest conversion during times of low labor demand or declining international export commodity prices, as landless laborers have few alternatives, but to pursue subsistence slash and burn farming, hunting and forest extraction on the remaining forest matrix and core protected lands.

2.6 Chile – The Model of Economic Liberalization

Chile is frequently presented as an economic liberalization success story. In Chile, free market initiatives under the direction of University of Chicago economists significantly increased GNP, private consumption, foreign investment, and export earnings following the overthrow of the democratically elected Allende government and rise to power of the Pinochet dictatorship (McMichael 1996). The history of the Chilean forestry sector during this period provides revealing insights into the type and scale of forest conversion possible under unregulated markets.

During the Pinochet era (1973–1990), the export of forest products (primarily unprocessed wood and wood chips) grew from US$39 million to over $760 million (Collins and Lear 1995). Expansion of the forestry export sector was accomplished by widespread privatization (i.e., sale) of government forest lands and industries; lifting of bans on the size and concentration of land holdings; provision of generous tax credits for export-oriented industries (e.g., between 1974 and 1986, 2.5 million acres of *Pinus radiata* plantations were established with 73 % of these subsidized by the government); and effective exclusion of ordinary small farmers through the prohibition of land sales to individuals with debt (most farmers were in debt due to the previous government's small-holder loan policy and poor economic conditions; Collins and Lear 1995).

The effects of large-scale forest plantation development in Chile include the conversion of over 1 million acres of native forest to *P. radiata* monocultures, a wood chip development boom (particularly since 1987), and extensive forest land purchases by transnational corporations (particularly Japanese). The rapid expansion of forest plantations has resulted in negative social and environmental impacts, including: (1) the expulsion of thousands of rural residents from their land, many of whom remain impoverished and without alternative livelihoods; (2) serious soil erosion, particularly following planting and harvesting operations; (3) decreased in-stream flows and lowered water tables; and (4) a significant reduction in endangered native tree species and endemic wildlife populations (Lara and Veblen 1993; Meller et al. 1996; Clapp 1998). In fact, the species-rich, highly endemic temperate rainforests of Chile are now a critically threatened ecosystem (Arroyo et al. 1995). Finally, the social and environmental implications associated with the plantation economy raise questions about its long-term ecological sustainability and socioeconomic viability (Arroyo et al. 1995; Collins and Lear 1995; Clapp 1998).

2.7 Economic Globalization Effects on Forest Conversion and Habitat Fragmentation

Global economic policies affect forest land use by small and large landowners, national governments and transnational corporations. Given this fact, it may be possible to anticipate the location, type, pattern and extent of future forest conversion and habitat fragmentation based on an analysis of contemporary and historical investment patterns and policies. For example, forest conversion and habitat fragmentation in the Brazilian Amazon between 1978 and 1988 were spatially concentrated in a crescent along the southern and eastern fringe of the basin and along major transportation corridors (Skole and Tucker 1993). Forest conversion also exhibited distinctive patterns indicative of specific land uses, owners, and inherent site productivity (Browder 1996). Maps produced via remote sensing which incorporate private, state, and transnational development practices and policies, in conjunction with agricultural production potentials, topography and market access could potentially be used to predict the location, extent and type of forest conversion that may occur in a given region in the future.

To illustrate, consider the case of southern Mexico again. Market and trade deregulation, ejido privatization, and increased transnational investment in export-oriented agriculture contribute to: (1) large-scale conversion of forest matrix to monoculture citrus plantations and cattle pastures (McMichael 1996), particularly where edaphic, topographic and market (i.e., road) access conditions are favorable; (2) conversion of traditional, small-holder, species-rich, structurally diverse agroecosystems with a significant biodiversity conservation value (e.g., rustic, shade-grown coffee) to estate-grown, petrochemically dependent, species-poor, structurally uniform monocultures that are incompatible with native flora and fauna (i.e., full sun, technified coffee; Collier and Quaratiello 1994; Collier et al. 1994; Perfecto et al. 1996; Moguel and Toledo 1999); and (3) reduced connectivity between core-protected areas and increased habitat fragmentation. This last effect is of significant biodiversity conservation concern because there is extensive overlap between coffee-growing areas and high species diversity and endemism levels in southern Mexico (Moguel and Toledo 1999).

In southern Mexico, forest conversion to export agriculture is most likely to occur on lands with high agricultural production potential and these areas can be identified and mapped based upon soil nutrient and structural characteristics, the availability of water, and topography. Agricultural production is limited throughout much of southern Mexico by low soil nitrogen and phosphorus levels, limited soil moisture availability, a pronounced dry season, and steep slopes that are ill-suited to mechanized agriculture. These areas are unlikely to be converted to plantations. Existing road networks and their proximity to export processing facilities (e.g., citrus juicing plants) and urban markets may also indicate areas of potential development. In neighboring Belize, for example, Chomitz and Gray (1996) found that intensification of existing road networks was preferable to expanding roads into more distant areas for both economic and habitat fragmentation reasons, and that this could influence future World Bank lending policies. Unfortunately, species diversity is highest in fertile, lowland areas in southern and central Mexico (Moguel and Toledo 1999), which are the areas most suited to export cash crop agriculture.

As forest matrix areas are converted and bisected, edge effects are likely to become more pronounced and deleterious (Laurance and Bierregaard 1997; this volume). This may result in increased invasion by exotic plant and animal species, greater hunting pressure, and increased forest extraction throughout residual matrix lands. The expansion of plantation agriculture is also likely to increase landlessness among former peasant farmers and ejido members who sell their lands and become plantation wage laborers. As noted previously, this may increase forest conversion and hunting pressures in residual matrix and core areas when employment opportunities wane or when international market prices for cash crops decline.

Conservation biologists have been slow to recognize and support the biodiversity conservation services provided by traditional agriculturists and forest farmers. Noorgard (1988) clearly articulated some of the linkages between the loss of biological diversity and the rise of the global exchange economy. Similarly, Gliessman et al. (1981) and Altieri and Merrick (1987) empirically established that traditional farming systems can simultaneously maintain high crop and related species diversity levels *and* long-term ecological productivity. Nevertheless, only recently has the biodiversity conservation value of traditional agroecosystems become widely appreciated. For example, traditional shade-grown coffee farms in southern Mexico are now known to support a high diversity of both migratory and resident bird species (Moguel and Toledo 1999), particularly during the dry season when small farms serve as critical refuges (Greenberg 1996). Similarly, Power (1996) found that traditional coffee and cacao farms function "reasonably well as surrogates of forest" in terms of arthropod diversity, as long as shade trees and understory herbs are retained. Traditional agroforestry systems have proven particularly valuable as critical refuges for forest bird species during extreme events such as the 1998 fires that consumed thousands of hectares of Central America forest (Griffith 2000). Holloway and Stork (cited in Power 1996) go so far as to suggest that traditional shifting cultivation and agroforestry systems may be the managed ecosystems most compatible with biodiversity conservation.

The conservation potential of managed agroecosystems is reviewed by Kattan and Alvarez-Lopez (1996) who suggest that it may be possible to increase the number of species found in matrix areas close to levels expected in core protected zones by manipulating connectivity and increasing habitat heterogeneity (at least in the Colombian Andes). On a small scale, the management of traditional agroecosystems in Belize, specifically the retention of large *Ficus* spp. trees and canopy riparian corridors in small farms and pastures, has proved adequate to retain habitat and food resources for a population of endangered howler monkeys (and other forest flora and fauna; Lyon and Horwich 1996).

My point is that traditional agroecosystems and community forests provide productive habitat for many native plant and animals species, and that these lands are typically managed for a wide variety of uses by many different individuals, creating a diverse mosaic of vegetation types at the landscape level. Furthermore, species-rich, structurally diverse agroecosystems are not incompatible with high human population densities or intensive land uses. In fact, some of the most productive and diverse agroforestry systems are found in densely populated areas such as Monteverde, Costa Rica (this volume); Java, Indonesia (Soemarwoto 1987); and Nigeria (Schelhas 1996). While traditional agroecosystems and managed forests are not suitable for all species (e.g., obligate primary forest insectivorous forest-floor dwelling bird species), they are certainly preferable to the biological deserts created by large, petrochemical-based, transnational monocultures; provide critical connectivity between core protected areas; and provide rural people with productive and secure livelihoods. This latter function is particularly important when one recognizes that the activities of rural people will, in large part, determine the success or failure of biodiversity conservation efforts in matrix and core areas in the coming decades.

2.8 Conclusion

Biodiversity conservation efforts should focus more attention on matrix issues, specifically conservation *and* development activities in traditional agroecosystems, community forests and state production forest lands. In this context, issues of equity, access to resources and enhancing rural livelihoods are critical both to the conservation of specific protected areas and to the future of biodiversity conservation in general. While we must avoid simplistic notions of economic determinism, the policies and programs inherent in economic globalization appear to foster forest conversion and habitat fragmentation throughout the Western Hemisphere (e.g., Brazil, Bolivia, Peru, Canada, and the United States) just as they do in Mexico and Chile. The relationships between biodiversity conservation and local socioeconomic well-being are clear and profound, and both appear to be undermined by the new economic world order.

Acknowledgement. I am grateful for the valuable comments and suggestions provided by Jill Belsky, the editors and an anonymous reviewer on earlier drafts of this manuscript.

References

Altieri MA, Merrick LC (1987) In situ conservation of crop genetic resources through maintenance of traditional farming systems. Econ Bot 41:86–96

Arroyo MTK, Donoso C, Murua R et al. (1995) Toward an ecologically sustainable forestry project. Report made by the independent scientific commission of the Rio Condor project to Bayside Ltd, USA. Santiago, Chile

Bilsborrow R, Geores M (1994) Population, land-use and the environment in developing countries: what can we learn from cross-national data? In: Brown K, Pearce D (eds) The causes of tropical deforestation. Univ of British Columbia Press, Vancouver, BC, pp 106–133

Boserup E (1965) The conditions of agricultural growth. Aldine, Chicago

Bray DB (1995) Peasant organizations and "The Permanent Reconstruction of Nature:" grassroots sustainable development in rural Mexico. J Environ Dev 4(2):185–204

Bray DB, Wexler MB (1996) Forest policies in Mexico. In: Randall L (ed) Changing structures of Mexico: political, social and economic prospects. ME Sharpe Press, Armonk, NY, pp 217–228

Browder JO (1988) Public policy and deforestation in the Brazilian Amazon. In: Repetto R, Gillis M (eds) Public policies and the misuse of forest resources. Cambridge Univ Press, New York, pp 247–297

Browder JO (1996) Reading colonist landscapes: social interpretations of tropical forest patches in an Amazonian agricultural frontier. In: Schelhas J, Greenberg R (eds) Forest patches in tropical landscapes. Island Press, Washington, DC, pp 285–299

Chomitz KM, Gray DA (1996) Roads, land use and deforestation: a spatial model applied to Belize. Poverty, environment and growth working paper no. 3. The World Bank, Washington, DC

Clapp R (1998) Regions of refuge and the agrarian question: peasant agriculture and plantation forestry in Chilean Arucania. World Dev 26:571–589

Collier GA, Quaratiello EL (1994) Basta! Institute for Food and Development Policy, Oakland, CA

Collier GA, Mountjoy D, Nigh RB (1994) Peasant agriculture and global change. BioScience 44:398–407

Collins J, Lear J (1995) Chile's free market miracle: a second look. Institute for Food and Development Policy, Oakland, CA

Critchley WRS, Reji C, Willcocks TJ (1994) Indigenous soil and water conservation: a review of the state of knowledge and prospects for building on traditions. Land Degradation Rehabil 5:293–314

Daly H (1996) Sustainable growth? No thank you. In: Mander J, Goldsmith E (eds) The case against the global economy. Sierra Club Books, San Francisco, pp 192–196

Ehrlich PR (1968) The population bomb. Ballatine Books, New York

Ehrlich PR, Ehrlich AH (1990) The population explosion. Simon and Schuster, New York

FAO (1989) Household food security: an analysis of socio-economic issues. UN-FAO, Rome

Forester DJ, Machlis GE (1996) Modeling human factors that affect the loss of biodiversity. Conserv Biol 10:1253–1263

Franke RW, Chasin BH (1994) Kerala. Institute for Food and Development Policy, Oakland, CA

Gehrt SD (1996) The human population problem: educating and changing behavior. Conserv Biol 10:900–903

Gillis M, Repetto R (1988) Conclusion: findings and policy implications. In: Repetto R, Gillis M (eds) Public policies and the misuse of forest resources. Cambridge Univ Press, New York, pp 385–410

Gliessman SR, Garcia ER, Amador AM (1981) The ecological basis for the application of traditional agricultural technology in the management of tropical agro-ecosystems. Agro-Ecosyst 7:173–185

Greenberg R (1996) Managed forest patches and the diversity of birds in southern Mexico. In: Schelhas J, Greenberg R (eds) Forest patches in tropical landscapes. Island Press, Washington, DC, pp 59–90

Griffith D (2000) Agroforestry: a refuge for tropical biodiversity after fire. Conserv Biol 14:325–326

Hardin G (1974) Lifeboat ethics: the case against helping the poor. Psychol Today 8:38

Harrison S (1991) Population growth, land use and deforestation in Costa Rica, 1950–1984. Interciencia 16:83–93

Hecht SF (1993) The logic of livestock and deforestation in Amazonia. BioScience 43:687–695

Humphrey CR, Buttel FH (1986) Environment, energy and society. Krieger Publishing Co, Malabar, FL

Kahn J, McDonald J (1994) International debt and deforestation. In: Brown K, Pearce D (eds) The causes of tropical deforestation. Univ of British Columbia Press, Vancouver, BC, pp 57–67

Kattan GH, Alvarez-Lopez H (1996) Preservation and management of biodiversity in fragmented landscapes in the Colombian Andes. In: Schelhas J, Greenberg R (eds) Forest patches in tropical landscapes. Island Press, Washington, DC, pp 3–18

Korten D (1995) When corporations rule the world. Kumarian Press, San Francisco

Kummer DM, Turner BL (1994) The human causes of deforestation in Southeast Asia. BioScience 44:323–328

Lara A, Veblen T (1993) Forest plantations in Chile: a successful model? In: Mather A (ed) Afforestation policies, planning and progress. Belhaven Press, London, pp 118–139

Laurance W, Bierregarrd R Jr (eds) (1997) Tropical forest remnants. University of Chicago Press, Chicago

Lyon J, Horwich RH (1996) Modification of tropical forest patches for wildlife protection and community conservation in Belize. In: Schelhas J, Greenberg R (eds) Forest patches in tropical landscapes. Island Press. Washington, DC, pp 205–230

Mahar D, Schneider R (1994) Incentives for tropical deforestation: some examples from Latin America. In: Brown K, Pearce D (eds) The causes of tropical deforestation. Univ of British Columbia Press, Vancouver, BC, pp 159–171

Mamdani M (1972) The myth of population control. Monthly Review Press, New York

Mander J (1996) Facing the rising tide. In: Mander J, Goldsmith E (eds) The case against the global economy. Sierra Club Books, San Francisco, pp 3–19

McMichael M (1996) Development and social change. Pine Forge Press, Thousand Oaks, CA

Meffe GK, Carroll CR (eds) (1994) Principles of conservation biology. Sinauer Associates, Sunderland, MA

Meller P, O'Ryan R, Solimano A (1996) Growth, equity, and the environment in Chile: issues and evidence. World Dev 24:255–272

Mendes de Carvalho F, Brown I (1996) Polarization of biotic and economic wealth: the world, the tropics and Brazil. Int J Environ Pollut 6:160–171

Moguel P, Toledo V (1999) Biodiversity conservation in traditional coffee systems of Mexico. Conserv Biol 13:11–21

Myers N (1994) Tropical deforestation: rates and patterns. In: Brown K, Pearce D (eds) The causes of tropical deforestation. Univ of British Columbia Press, Vancouver, BC, pp 27–40

Noorgard RB (1988) The rise of the global exchange economy and the loss of biological diversity. In: Wilson EO (ed) Biodiversity. National Academy Press, Washington, DC, pp 206–211

Palo M (1994) Population and deforestation. In: Brown K, Pearce D (eds) The causes of tropical deforestation. Univ of British Columbia Press, Vancouver, BC, pp 42–56

Panayotou T (1993) Green markets. Institute for Contemporary Studies, San Francisco, CA

Pearce D, Brown K (1994) Saving the world's tropical forest. In: Brown K, Pearce D (eds) The causes of tropical deforestation. Univ of British Columbia Press, Vancouver, BC, pp 2–26

Perfecto I, Rice R, Greenberg R, Van der Voort M (1996) Shade coffee: a disappearing refuge for biodiversity. BioScience 46:598–608

Power AG (1996) Arthropod diversity in forest patches and agroecosystems of tropical landscapes. In: Schelhas J, Greenberg R (eds) Forest patches in tropical landscapes. Island Press, Washington, DC, pp 91–110

Primack R, Bray D, Galletti H, Ponciano I (1998) Timber, tourists, and temples: conservation and development in the Maya forests of Belize, Guatemala, and Mexico. Island Press, Washington, DC

Rudel T (1994) Population, development and tropical deforestation: a cross-national study. In: Brown K, Pearce D (eds) The causes of tropical deforestation. Univ of British Columbia Press, Vancouver, BC, pp 96–105

Schelhas J (1996) Land-use choice and forest patches in Costa Rica. In: Schelhas J, Greenberg R (eds) Forest patches in tropical landscapes. Island Press, Washington, DC, pp 258–284.

Skole D, Tucker C (1993) Tropical deforestation and habitat fragmentation in the Amazon: satellite data from 1978 to 1988. Science 260:1905–1910

Soemarwoto O (1987) Homegardens: a traditional agroforestry system with a promising future. In: Steppler HA, Nair PKR (eds) Agroforestry: a decade of development. ICRAF, Nairobi, pp 157–170

Southgate D (1998) Tropical forest conversion: an economic assessment of alternatives in Latin America. Oxford University Press, New York

Southgate D, Sierra R, Brown L (1991) The causes of tropical deforestation in Ecuador: a statistical analysis. World Dev 19:1145–1151

Sunderlin W (1999) Between danger and opportunity: Indonesia's forests in an era of economic crisis and political change. Soc Nat Resour 12:559–570

Vandermeer J, Perfecto I (1995) Breakfast of biodiversity. Institute for Food and Development Policy, Oakland, CA

Westoby J (1989) Introduction to world forestry. Basil Blackwell, Oxford

3 Forest Fragmentation and Biodiversity in Central Amazonia

C. Gascon, W.F. Laurance, T.E. Lovejoy

3.1 Introduction

Around the world, growing human populations and economic pressures are leading to widespread conversion of tropical rainforests into a mosaic of human-altered habitats and isolated remnants. The Amazon rainforest is the largest tract of undisturbed tropical forest in the world, comprising more than 30 % of the world's rainforest area. In the Amazon, forest clearing increased exponentially during the 1970s and 1980s (Fearnside 1987) and continues at alarming rates (Diário Official da União 1996). Land-use changes in Amazonia through deforestation have been shown to affect regional hydrology, the global carbon cycle, evapotranspiration rates, biodiversity loss, probability of fire, and a possible regional reduction in rainfall (Uhl et al. 1988; Uhl and Kauffman 1990; Nobre et al. 1991; Bierregaard et al. 1992; Wright et al. 1992; Nepstad et al. 1994; Vitousek 1994). Deforestation in the Tropics results in large areas of primary forest being transformed into a mosaic of pastures and forest fragments with serious consequences for biodiversity (Bierregaard et al. 1992). By understanding how and to what extent tropical rainforest ecosystems respond to deforestation, we can provide policy decision makers with information and conservation plans and thus decrease the rate of forest loss. Decreasing the rate of forest loss will, in turn, have a significant effect in reducing the effects of deforestation on global climate patterns and biodiversity loss.

3.2 Forest Fragmentation and Theory

Inevitably, tropical deforestation results in fragmentation of the forest into isolated plots of tropical rainforest surrounded by a sea of nonforest habitat. In an attempt to model and predict the consequences of forest fragmentation,

Ecological Studies, Vol. 162
G.A. Bradshaw and P.A. Marquet (Eds.)
How Landscapes Change
© Springer-Verlag Berlin Heidelberg 2003

conservation biology theory has often relied on the framework provided by MacArthur and Wilson's (1967) *The theory of Island Biogeography*. Simply stated, the theory was elaborated to predict the number of species that a given-sized island would support based on the balance between extinction of the species on the island and immigration onto the island from source populations. Because forest patches resulting from deforestation were likened to forest islands, the theory was adopted as a practical framework that would allow conservation biologists to predict the number of species a particular patch of forest could maintain. However, soon after the theory was proposed as a conservation tool (Wilson and Willis 1978), an intense debate took place regarding its relevance and significance in predicting species richness on habitat "islands" and the mechanisms responsible for the observed pattern. This controversy included questions relative to the minimum size a habitat "island" should be to conserve its integrity, and whether one large "island" would conserve more species than several small ones of the same total area [the single-large-or-several-small ("SLOSS") debate, Simberloff and Abele 1982 vs. Wilson and Willis 1978]. Although the applicability of the theory to conservation plans has been tested empirically (e.g., Zimmerman and Bierregaard 1986), few data are available to test the fundamental premise of the theory: the number of species on a habitat "island" will be determined by the size of the island and its proximity to a source of colonists. The main problem has been the absence of baseline data before habitat "islands" were created.

   The Biological Dynamics of Forest Fragments Project (BDFFP) was initiated in 1979 in the midst of the existing debate over the applicability of the theory of island biogeography. The main goal of the BDFFP is to model and predict the effects of land-use change (i.e., forest fragmentation) on the integrity of the tropical rainforest ecosystem. Forest integrity is defined to include everything from species diversity, species composition, forest structure, microclimate, forest functioning, to the ecological processes responsible for the maintenance of a dynamic tropical forest ecosystem. In this paper, the main results of over 17 years of research from the BDFFP will be briefly summarized in the context of other relevant literature to provide a better understanding of the complexity of the effects of forest fragmentation on the Amazon rainforest.

3.3 Biological Dynamics of Forest Fragments Project History and Study Sites

The BDFFP was initiated in the late 1970s and early 1980s in an area that is part of the Distrito Agropecuária under the jurisdiction of the Federal Government. The Distrito Agropecuária comprises a large area (>500,000 ha) north of Manaus, AM, and the entire area was slated for development in the

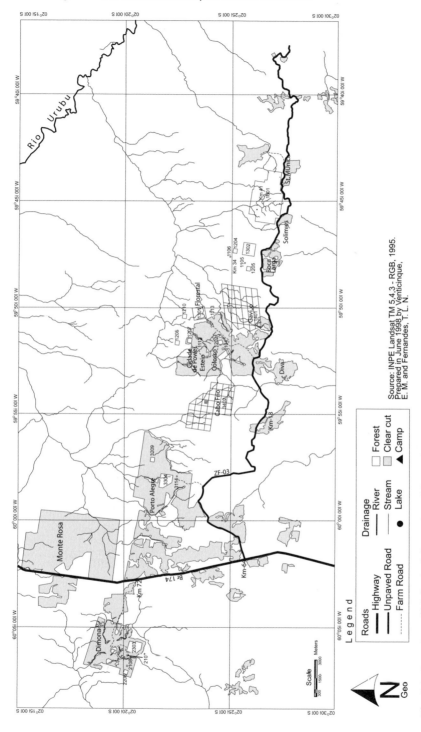

Fig. 3.1. Location of biological dynamics of forest fragment project reserves

1980s as part of a program to establish cattle ranches in the region as an economically viable solution for human settlement. With government incentives, large ranches (10,000–50,000 ha) were carved out of primary rainforest. Because of a Brazilian law that required that 50 % of all forested properties be maintained as primary forest, the project's instigators were able to gain the collaboration of the ranchers in deciding how and where to maintain some of the area protected under the 50 % provision. This allowed the BDFFP to design an experimental study consisting of a series of patches of forest (to be isolated) of different sizes, which could be surveyed before isolation and monitored after being severed from the surrounding continuous forest. In this way, a series of replicated forest patches of three different sizes (five of 1-ha, four of 10-ha, and two of 100-ha areas) were created for investigating the effects of isolation and fragment size on different components of the forest ecosystem. Several other equivalent-sized and larger reserves have been marked, but have remained part of continuous forest and are used as control areas.

The ranches and the forest patches are located 60 km north of Manaus, AM, along the secondary road ZF-3 that joins onto the BR-174 at km 65 (Fig. 3.1). Over 13,000 ha of forest in different reserves spread over a 40-km-wide area is included within the scope of the project. The forest in the area is typical terra firma forest (not subject to periodic flooding) has a canopy height of <40 m, and high tree diversity (Rankin de Merona et al. 1992). Palms dominate the understory (Scariot 1996). Rolling hills and numerous plateaus make up the topography; many small streams dissect the forest creating valleys and gullies with small flooded areas in associated lowlands. In the recent past, many of the ranches where the project's fragments are located have been abandoned. This has created a fortuitous situation, in which many areas with known land-use history are available for the study of forest regeneration and the impact of different disturbance levels on biodiversity.

3.4 Ecological Consequences of Forest Fragmentation

3.4.1 Area and Insularization

Habitat fragmentation, by definition, involves a reduction in original area and isolation of remaining patches of forest. In central Amazonia, the initial and most obvious consequence of loss of habitat is a decrease in species richness (Powell and Powell 1987; Rylands and Keuroghlian 1988; Klein 1989; Schwarzkopf and Rylands 1989; da Fonseca and Robinson 1990; Bierregaard et al. 1992; Morato 1993; Dale et al. 1994; Souza and Brown 1994; Stouffer and Bierregaard 1995). Many causes have been suggested to explain this relationship between species richness and area. The simplest explanation is that many species are lost in the resulting landscape due to a decrease in habitat heterogeneity (Williams 1943; Lack 1976; Simberloff and Abele 1982; Haila et al. 1987; Norton et al. 1995). Because habitats are naturally heterogeneous (Wiens 1976; Forman and Godron 1986), fragmentation will result in a nonrandom loss of habitats, and consequently, many specialized species may be excluded from the forest patches because of their strong association with a particular habitat type (Zimmerman and Bierregaard 1986). This problem is compounded in tropical regions where many species have small ranges, or small areas have a high percentage of endemic species (Terborgh and Winter 1983; Gentry 1986). Species with large home ranges can be excluded from patches that do not provide the minimum area for survival (Lovejoy et al. 1986; Spironello 1987). Even in cases where patch size exceeds the area requirements of a species, a valuable resource may be lacking and therefore impede a species' life cycle (Robbins 1980; Karr 1982; Zimmerman and Bierregaard 1986).

Isolation does not necessarily result in direct and immediate local extinctions. An initial increase in capture rates of birds in fragments was detected immediately following forest isolation, probably as a result of birds taking refuge in the fragments (Bierregaard et al. 1992). Capture rates and species richness in fragments eventually fell to below pre-isolation levels (Stouffer and Bierregaard 1995). In many cases, populations can persist in patches at low population density. However, small populations are much more vulnerable to a series of threats that inevitably lead to local extinction (Gilpin and Soulé 1986; Menges 1992). In addition, independent life history stages of a given species may be affected differentially as a result of isolation. Scariot (1996) showed that seedlings of most of the 36 palm species in the Manaus area are more sensitive to forest isolation than adult individuals.

Not all cases of fragmentation have led to a decline in species richness after isolation. Overall species richness has been shown to increase in isolated forest patches of central Amazonia for small mammals (Malcolm 1997), amphibians (Tocher et al. 1997), and butterflies (Hutchings 1991; Brown and Hutchings 1997), although the nature of the post-isolation community varied among these groups. Small mammals and amphibians, for example, lose few of the pre-isolation species, and most of the increase in species richness is due to the appearance of open-area associated species. Conversely, for butterflies, over 40 % of the original complement of species in a patch of forest is lost after isolation, presumably displaced by light-loving specialists (Hutchings 1991).

Regardless of whether a particular taxonomic group showed an increase or decrease in species richness after isolation, larger isolates maintained more species than smaller ones. Although these results demonstrate a clear species-area relationship, a post-isolation increase in species richness is somewhat unexpected. For example, under the theory of island biogeography, fragments should contain fewer species after isolation than before, because more extinctions and less immigration should occur in isolated patches. More importantly, however, the theory has not gained widespread acceptance because of its limited scope in predicting the entire range of biological and ecological changes associated with habitat fragmentation (Bierregaard et al. 1992). While the theory of island biogeography relies on one response variable – the number of species – to measure ecosystem change due to fragmentation, available data suggest a much more complex suite of modifications occurring as a result of habitat fragmentation.

Ecological processes can become altered as a function of forest isolation. The very diverse insect group can respond in varying ways to fragmentation (Didham et al. 1996). In the Manaus fragments, Klein (1989) showed slower decomposition of organic matter by dung and carrion beetles in fragments compared to continuous forest. This drop in decomposition rate was presumably caused by lower species richness and abundance of these beetles in the fragments. Similarly, in the dry Chaco forest of Argentina, Aizen and Feinsinger (1994) showed a decrease in reproductive output of plants in isolates compared to continuous forest. They hypothesized that this was the result of lowered rates of pollination. Malcolm (1991) showed a vertical shift in insect resources from the canopy-dominated insect biomass in continuous forest, to understory-dominated insect biomass in the fragments. This shift will certainly result in altered patterns of insect-plant interactions (i.e., pollination, herbivory, etc.). In the Manaus area, Powell and Powell (1987) observed a decrease in visitation rates of euglossine bees on flowers in isolated patches of forest, which can result in decreased reproductive success of plants as shown by Aizen and Feinsinger (1994).

Close species associations, especially in tropical systems, may lead to cascading extinctions. Harper (1987, 1989) demonstrated that the absence of army ants in small fragments also led to the disappearance of obligate ant-following birds. However, facultative ant-following birds were less affected. Zimmerman and Bierregaard (1986) hypothesized that the absence of peccaries in fragments would affect the survival of populations of certain frog species that depend on the wallows created by the peccaries for reproduction.

3.4.2 Edge Creation

The process of fragmentation entails the creation of forest edge where there was no edge previously. This edge has to be distinguished from zones of natural ecotone, the natural gradient boundary between two habitat types. The basic difference is one of degree of contrast between the two habitats. In a natural landscape, much less contrast exists between patches of adjacent habitat types, whereas in a fragmented landscape, an abrupt contrast is present (i.e., forest vs. field). Many biological consequences have been reported as a result of the creation of an edge. In continuous tropical forests, sunlight usually is restricted to vertical penetration. In a modified landscape, where forest edge exists, sunlight can penetrate laterally. This single change seriously affects the microclimatic conditions of the forest up to a certain distance from the edge (Kapos 1989; Murcia 1995). In Manaus, slight increases in temperature and decreases in relative humidity have been detected up to 40 meters from the edge inside the forest (Kapos 1989). However, these changes are not permanent and evolve with time as the edge closes up due to vegetation growth (Camargo and Kapos 1995). Microclimatic changes associated with edge creation probably are the causative factors that explain observed changes in forest structure and tree mortality and turnover of the plant community (Wandelli 1991; Ferreira and Laurance 1997).In addition, because of increased sunlight along forest edge, leaf fall and tree mortality increase and the recruitment pattern of seedlings is modified (Sizer 1992). Exposure to wind may also cause severe damage in the fragments, especially for fragment corners that are exposed on two sides (Malcolm 1994).

Faunal responses to edge may be complex as shown by the bimodal abundance response of litter invertebrates to edge in Central Amazonia (Didham 1997). Some taxa, however, use a suitable breeding habitat independently of its proximity to edge, as is the case for some frog species in central Amazonia (Gascon 1993).

3.4.3 Matrix Habitat and Landscape Configuration

The loss of primary forest results in the creation of a new matrix habitat. Until recently, however, the role of the matrix habitat in our understanding of the effects of habitat fragmentation has been seriously underestimated (nevertheless, see Harris 1984; Fahrig and Merriam 1994; Gascon et al. 1999). Matrix habitat will be important in the evolution of ecosystem dynamics in forest patches because: (1) it will act as a filter (not a barrier) for movement between landscape features; (2) disturbed area-associated species will be present and may invade forest patches and edge habitat; (3) depending on land-use, the matrix habitat will take on a different form, such as pasture, degraded pasture, or second growth forest (Williamson et al. 1997), and the nature of the matrix habitat will influence the severity of the edge effects in patches. Recent studies have begun to corroborate some of the above-mentioned characteristics. Laurance (1991) found that the best ecological correlate of small mammal extinction-proneness in forest patches of Australia was a species' abundance in the matrix habitat. Malcolm (1991), in Amazonia, found a significant positive correlation between the abundance of small mammal species in fragments and their abundance in the matrix habitat. Aberg et al. (1995) also demonstrated the importance of the matrix habitat on the occurrence of hazel grouse in isolated fragments. Angermeier (1995) showed that impeded movement (as measured by degree of isolation for aquatic habitats) was one of the main factors predicting extinction-proneness in fish. Similarly, the degree of isolation was important in predicting the abundance of euros *(Macropus robustus)* in a fragmented landscape in Australia (Arnold et al. 1995).

Obviously, the loss of primary habitat will lead to the disappearance of many forest-associated species (Borges 1995; Tocher 1997). Conversely, species associated with the new matrix habitat will colonize the area and invade the fragments (Hutchings 1991; Tocher et al. 1997), and can represent potential threats as competitors or predators for the species in fragments (Janzen 1986). Overall, it is probable that community and population dynamics of the matrix will strongly influence communities in primary fragments (Malcolm 1991).

The appearance of barriers in the modified landscape can significantly alter the metapopulation dynamics of the surviving species. The presence of a new matrix habitat (e.g., pasture) can limit dispersal, movement, and colonization as shown for small mammals in Canada (Oxley et al. 1974) and some species in Amazonia (Malcolm 1991). For species that are important in some ecological process (e.g., pollination or decomposition), these barriers may have long-term effects for the maintenance of these ecological functions. In a series of Chaco dry forest fragments in Argentina, for example, most of the 16 plant species investigated by Aizen and Feinsinger (1994) showed a decrease in the number of pollen tubes, fruit set and seed set. The authors suggested that these effects are due to the loss of important plant pollinators and their inability to recolonize the fragments. Species such as primates that depend on keystone resources during parts of the year (Terborgh 1986) may be limited in their foraging ability if they cannot move to areas where that resource is present.

The configuration of fragments within the matrix habitat is also a key factor determining long-term population survival and large-scale abundance patterns of many taxonomic groups (Askins et al. 1987; Burkey 1989; Rolstad 1991; Fahrig and Merriam 1994; Knick and Rotenberry 1995; Robinson et al. 1995; Villard et al. 1995; Flather and Sauer 1996). Lawton and Woodroffe (1991) showed that breeding water voles are less likely to be present in patches of forest surrounded by large areas of matrix habitat within a landscape. Potter (1990) also demonstrated the importance of patch configuration within the matrix habitat for population viability of brown kiwis in New Zealand. If large and small patches exist, the small ones may serve as stepping-stones for movement into larger patches. As fragmentation of the landscape increased, the distance moved by three small mammal species increased, but fewer individuals moved altogether (Diffendorfer et al. 1995).

The importance of corridors has also, until recently, remained a theoretical debate because of the scarcity of empirical data (Harris 1984; Noss 1987; Simberloff and Cox 1987; Simberloff et al. 1992; Demers et al. 1995). The few studies that do exist suggest that corridors greatly enhance movement among patches in a landscape (Bennett 1990; Merriam and Lanoue 1990; Saunders and de Rebeira 1991; Dunning et al. 1995; Haas 1995; Laurance and Laurance 2002), which in turn may decrease the probability of extinction of local populations.

3.5 Forest Fragmentation and Land Management

Results to date indicate that forest patches are highly dynamic ecological entities and that a decrease in species richness associated with isolation and fragmentation is but one component of the overall effects of changes in land use (Bierregaard et al. 1992). As seen from these studies, the biological consequences of habitat fragmentation are not limited to species extinctions, although these are usually the first and most obvious manifestations associated with fragmentation. Each of the different components of the process of habitat fragmentation entails a series of effects, which in turn can interact to determine how the biotic community of a patch of forest will evolve after fragmentation. This explains in part why the theory of Island Biogeography has not been very successful at predicting the effects of fragmentation. It is easy to imagine how a simple response variable such as species richness can lead to erroneous conclusions regarding fragmentation effects. In some cases, fragments can have approximately the same number of species of a particular taxon after isolation as they possessed before, but the turnover of species within that group may be very high (Hutchings 1991). Also, without knowledge of the ecological function of the species (i.e., pollination, decomposition, etc.), we cannot predict the long-term second-order effects associated with the loss of a given species or set of species (Klein 1989; Powell and Powell 1987; Aizen and Feinsinger 1994).

The usefulness of the Equilibrium Theory of Island Biogeography in predicting species richness as a function of patch size has been empirically questioned (Zimmerman and Bierregaard 1986). The appearance of a new matrix habitat (in this case pasture and young and old second growth) represents a barrier for many taxa (Powell and Powell 1987; Klein 1989; Malcolm 1991), which in turn may have long-term effects on ecosystem processes (Klein 1989). The presence of a barrier for certain taxa can alter behavioral patterns such as dispersal and home range size for primates (Lovejoy et al. 1986).

It is becoming increasingly clear that the effects of habitat fragmentation in the Manaus landscape are being driven by two main processes: the within-patch effects linked to the creation of forest edge, and the outside-patch influence of matrix habitat. The latter process includes a broader landscape-level interaction of habitat configuration (patches, matrix, connectivity; Fahrig and Merriam 1994). The management of landscapes should, therefore, take into account these considerations.

We believe that what we now know of ecosystem changes from different land-use patterns allows for constructive and empirically based guidelines for the management of landscapes, both fragmented and intact (Laurance and Gascon 1997). First, in modified landscapes, care should be taken to reduce the within-patch processes associated with the creation of edge. This can be achieved simply by allowing a buffer zone of forest regeneration around already existing forest patches, in turn diminishing many of the negative changes in microclimate and forest structure initially caused by isolation (Kapos 1989; Williams-Linera 1990; MacDougall and Kellman 1992; Brown 1993).

The importance of connectivity between landscape components requires the presence of corridors or some sort of passageway that is structurally similar to primary habitat to allow for movement and dispersal among patches (Hansson 1991; Harrison 1992). Both theoretical models and empirical investigations suggest that this will greatly decrease the extinction probability of local populations (Lamberson et al. 1992; Beier 1993; Haas 1995; Hill 1995). For example, in the Manaus landscape, the number of species of frogs found in different types of human-disturbed areas varies positively with the level of habitat regeneration; the older, taller second growth forests harbor more frog species than younger stands (Tocher 1997). In a fragmented landscape, it is not always possible to relink patches of isolated habitat. However, existing features of the landscape can be used to increase connectivity and opportunities for movement and dispersal within the landscape. Watercourses and steep hills are protected by law in many countries (Código Florestal 1965), and can be reforested to create linear patches of habitat that can connect larger fragments. Although the conservation value of large patches cannot be questioned, in already degraded areas, even small existing fragments are of considerable value for species conservation and sources of propagules for reforestation of other landscape components (Gascon 1993; Cortlett and Turner 1997).

Proactive land management policies based on ecological information may be much more effective for conservation of biodiversity (Zimmerman and Bierregaard 1986; Laurance and Gascon 1997). Historical land management strategies in the Amazon have not relied on ecological considerations. This results in forest fragments that often persist on steep terrain or on poor soils that would not necessarily represent the best choices for conservation purposes. In planning development of currently intact landscapes, it is also possible to take advantage of existing conservation laws and landscape features to increase patch connectivity. For example, in Amazonia, very good laws prohibit the clearing of forest along waterways and on steep slopes (Código Florestal 1965), despite little enforcement. Slight modifications of existing legislation to account for known edge effects and the importance of connectivity, and stricter enforcement, can greatly enhance the conservation value of remaining forests.

Because of the importance of edge effects and habitat connectivity, it is imperative that existing fragments or large tracts of intact land not be subdivided by the construction of roads, (Riswan and Hartanti 1995). Even narrow roads (<10 m wide) traversing a large tract of land can act as a major barrier to movement and create edge habitat with the negative associated effects (Rich et al. 1994). Human activities that can cause synergistic effects and compound species loss in patches, such as hunting of large vertebrates, must also be avoided (Mittermeier 1987).

The theory of Island Biogeography has not lived up to its promise of serving practical conservation biology. Existing knowledge about how ecosystems are affected by landscape fragmentation suggests that most of the ecological degradation can be accounted for by edge-associated effects and landscape configuration, including the lack of connectivity and inhospitable matrix habitat. A management strategy based on ecological information and proactive by nature would greatly enhance biodiversity conservation (Laurance and Gascon 1997). Although many forest species are not able to use nonforested habitats, there are some exceptions (Malcolm 1997; Tocher et al. 1997; Gascon et al. 1999). More research is needed to determine how forest species react to the presence and type of matrix habitat. This will allow us to understand which habitats act as sources and which are sinks, and can even help predict which species will be vulnerable to extinction in isolated patches (Laurance 1991; Malcolm 1991).

References

Aberg J, Jansson G, Swenson JE, Angelstam P (1995) The effect of matrix on the occurrence of hazel grouse *(Bonasa bonasia)* in isolated habitat fragments. Oecologia 103:265–269

Aizen MA, Feinsinger P (1994) Forest fragmentation, pollination and plant reproduction in a Chaco dry forest, Argentina. Ecology 75:330–351

Angermeier PL (1995) Ecological attributes of extinction-prone species: loss of freshwater fishes of Virginia. Conserv Biol 9:143–158

Arnold GW, Weeldenburg JR, Ng VM (1995) Factors affecting the distribution of Western grey kangaroos *(Macropus fuliginosus)* and euros *(M. robustus)* in a fragmented landscape. Landscape Ecol 10:65–74

Askins RA, Philbrack MJ, Sugeno DS (1987) Relationship between the regional abundance of forest and the composition of forest bird communities. Biol Conserv 39:129–152

Beier P (1993) Determining minimum habitat areas and habitat corridors for cougars. Conserv Biol 7:94–108

Bennett AF (1990) Habitat corridors and the conservation of small mammals in a fragmented forest environment. Landscape Ecol 4:109–122

Bierregaard RO Jr, Lovejoy TE, Kapos V, dos Santos AA, Hutchings RW (1992) The biological dynamics of tropical rainforest fragments. BioScience 42:859–866

Borges S (1995) Comunidade de aves em dois tipos de vegetação secundária da Amazônia central. Tese de mestrado, Instituto Nacional de Pesquisas da Amazônia,Manaus, AM, Brazil

Brown K, Hutchings RW (1997) Disturbance, fragmentation, and the dynamics of diversity in Amazonian forest butterflies. In: Laurance WF, Bierregaard RO (eds) Tropical forest remnants: ecology, management, and conservation of fragmented communities. University of Chicago Press, Chicago, pp 91–110

Brown N (1993) The implications of climate and gap microclimate for seedling growth conditions in a Bornean lowland rainforest. J Trop Ecol 9:153–168

Burkey TV (1989) Extinction in nature reserves: the effect of fragmentation and the importance of migration between reserve fragments. Oikos 55:75–81

Camargo JLC, Kapos V (1995) Complex edge effects on soil moisture and microclimate in central Amazonian forest. J Trop Ecol 11:205–221

Código Florestal (1965) Lei # 4771. Federal Government, Brasilia, Brazil

Cortlett RT, Turner IM (1997) Longterm survival in tropical forest remnants in Singapore and Hong Kong. In: Laurance WF, Bierregaard RO (eds) Tropical forest remnants: ecology, management, and conservation of fragmented communities. University of Chicago Press, Chicago, pp 333–345

Da Fonseca GAB, Robinson JG (1990) Forest size and structure: competitive and predatory effects on small mammal communities. Biol Conserv 53:265–294

Dale VH, Pearson SM, Offerman HL, O'Neill RJ (1994) Relating patterns of land-use change to faunal biodiversity in the central Amazon. Conserv Biol 8:1027–1036

Demers MN, Simpson JW, Boerner RJ et al. (1995) Fencerows, edges, and implications of changing connectivity illustrated by two contiguous Ohio landscapes. Conserv Biol 9:1159–1168

Diário Oficial da União (1996) Desmatamento na Amazônia, 1993–1994. Publicado no DOU 26-07-96, Federal Government, Brasilia, Brazil

Didham R (1997) The influence of edge effects and forest fragmentation on leaf litter invertebrates in central Amazonia. In: Laurance WF, Bierregaard RO (eds) Tropical forest remnants: ecology, management, and conservation of fragmented communities. University of Chicago Press, Chicago, pp 55–70

Didham RK, Ghazoul J, Stork NE, Davis A (1996) Insects in fragmented forests: a functional approach. Trends Ecol Evol 11:255–260

Diffendorfer JE, Gaines MS, Holt RD (1995) Habitat fragmentation and movements of three small mammals *(Sigmodon, Microtus,* and *Peromyscus)*. Ecology 76:827–839

Dunning JB Jr, Borella R Jr, Clements K, Meffe GF (1995) Patch isolation, corridor effects, and colonization by a resident sparrow in managed pine woodland. Conserv Biol 9:542–550

Fahrig L, Merriam G (1994) Conservation of fragmented populations. Conserv Biol 8:50–59

Fearnside PM (1987) Deforestation and international economic development projects in Brazilian Amazon. Conserv Biol 1:214–221

Ferreira LV, Laurance WF (1997) Effects of forest fragmentation on mortality and damage of selected trees in central Amazonia. Conserv Biol 11:797–801

Flather CH, Sauer JR (1996) Using landscape ecology to test hypotheses about large-scale abundance patterns in migratory birds. Ecology 77:28–35

Forman RTT, Godron M (1986) Landscape ecology. John Wiley, New York

Gascon C (1993) Breeding-habitat use by Amazonian primary-forest frogs species at forest edge. Biodivers Conserv 2:438–444

Gascon C, Lovejoy TE, Bierregaard RO Jr et al. (1999) Matrix habitat and species persistence in tropical forest remnants. Biol Conserv 91:223–229

Gentry AH (1986) Endemism in tropical versus temperate plant communities. In: Soulé ME (ed) Conservation biology: the science of scarcity and diversity. Sinauer Assoc, Sunderland, MA, pp 153–181

Gilpin ME, Soulé ME (1986) Minimum viable population: processes of species extinction. In: Soulé ME (ed) Conservation biology: the science of scarcity and diversity. Sinauer Assoc, Sunderland, MA, pp 19–34

Haas CA (1995) Dispersal and use of corridors by birds in wooded patches on an agricultural landscape. Conserv Biol 9:845–854

Haila Y, Hanski IK, Raivio S (1987) Breeding bird distribution in fragmented coniferous taiga in southern Finland. Ornis Fenn 64:90–106

Hansson L (1991) Dispersal and connectivity in metapopulations. Biol J Linn Soc 42:89–103

Harper LH (1987) The conservation of ant-following birds in small Amazonian forest fragments. Tese de doutorado, Universidade Estadual de Nova Iorque, New York

Harper LH (1989) Birds and army ants, observations on their ecology in undisturbed forest and isolated reserves. Acta Amazônica 19:249–263

Harris LD (1984) The fragmented forest: island biogeographic theory and the preservation of biotic diversity. University of Chicago Press, Chicago

Harrison RL (1992) Toward a theory of inter-refuge corridor design. Conserv Biol 6:293–295

Hill CJ (1995) Linear strips of rainforest vegetation as potential dispersal corridors for rainforest insects. Conserv Biol 9:1559–1566

Hutchings R (1991) The dynamics of three communities of Papilionoidae (Lepidoptera: Insecta) in forest fragments in central Amazonia. Tese de Mestrado, Instituto Nacional de Pesquisas da Amazônia, Manaus, AM, Brazil

Janzen D (1986) The eternal external threat. In: Soulé ME (ed) Conservation biology: the science of scarcity and diversity. Sinauer Assoc, Sunderland, MA, pp 286–303

Kapos V (1989) Effects of isolation on the water status of forest patches in the Brazilian Amazon. J Trop Ecol 5:173–185

Karr JR (1982) Avian extinction on Barro Colorado Island, Panama: a reassessment. Am Nat 119:220–239

Klein BC (1989) Effects of forest fragmentation on dung and carrion beetle communities in central Amazonia. Ecology 70:1715–1725

Knick ST, Rotenberry JT (1995) Landscape characteristics of fragmented shrub steppe habitats and breeding passerine birds. Conserv Biol 9:1059–1071

Lack D (1976) Island biology: illustrated by the land birds of Jamaica. Blackwell, Oxford

Lamberson RH, McKelvey R, Noon B, Voss C (1992) A dynamic analysis of northern spotted owl viability in a fragmented landscape. Conserv Biol 6:505–512

Laurance SGW, Laurance WF (2002) Utilization of linear rain forest remnants by arboreal mammals in tropical Australia. Conserv Biol (in press)

Laurance WF (1991) Ecological correlated of extinction proneness in Australian tropical rainforest mammals. Conserv Biol 5:79–89

Laurance WF, Gascon C (1997) How to creatively fragment a landscape. Conserv Biol 11:577–579

Lawton JH, Woodroffe GL (1991) Habitat and the distribution of water voles; why are there gaps in a species range? J Anim Ecol 60:79–91

Lovejoy TE, Bierregaard RO Jr, Rylands AB et al. (1986) Edge and other effects of isolation on Amazon forest fragments. In: Soulé ME (ed) Conservation biology: the science of scarcity and diversity. Sinauer Assoc, Sunderland, MA, pp 257–285

MacArthur RH, Wilson EO (1967) The theory of island biogeography. Princeton University Press, Princeton, NJ

MacDougall A, Kellman M (1992) The understory light regime and patterns of tree seedling in tropical riparian forest patches. J Biogeogr 19:667–675

Malcolm JR (1991) The small mammals of Amazonian forest fragments: pattern and process. PhD Thesis, University of Florida, Gainesville

Malcolm JR (1994) Edge effects in central Amazonian forest fragments. Ecology 75:2438–2445

Malcolm JR (1997) Biomass and diversity of small mammals in Amazonian forest fragments. In: Laurance WF, Bierregaard RO (eds) Tropical forest remnants: ecology, management, and conservation of fragmented communities. University of Chicago Press, Chicago, pp 207–221

Menges ES (1992) Stochastic modeling of extinction in plant populations. In: Fiedler PL, Jain SK (eds) Conservation biology: the theory and practice of nature conservation, preservation and management. Chapman and Hall, New York, pp 253–275

Merriam G, Lanoue A (1990) Corridor use by small mammals: field measurement for three experimental types of *Peromyscus leucopus*. Landscape Ecol 4:123–131

Mittermeier RA (1987) The effects of hunting on rainforest primates. In: Marsh C, Mittermeier RA (eds) Primate conservation in the tropical rainforest. Alan R Liss, New York, pp 109–146

Morato E (1993) Efeitos da fragmentação sobre abelhas solitárias e vespas na Amazônia central. Tese de Mestrado, Universidade Federal de Viçosa, Viçosa, Brazil

Murcia C (1995) Edge effects in fragmented forests: implications for conservation. Trends Ecol Evol 10:58–62

Nepstad DC, Carvalho CR, Davidson EA et al. (1994) The role of deep roots in the hydrologic and carbon cycles of Amazonian forest and pastures. Nature 372:666–669

Nobre C, Sellers P, Shukla J (1991) Amazonian deforestation and regional climate change. J Climate 4:957–988

Norton DA, Hobbs RJ, Atkins L (1995) Fragmentation, disturbance, and plant distribution: mistletoes in woodland remnants in the Western Australian wheatbelt. Conserv Biol 9:426–438

Noss RF (1987) Corridors in real landscapes: a reply to Simberloff and Connor. Conserv Biol 1:159–164

Oxley DJ, Fenton MB, Carmody GR (1974) The effects of roads on populations of small mammals. J Appl Ecol 11:51–59

Potter MA (1990) Movement of North Island brown kiwi between forest fragment. N Z J Ecol 14:17–14

Powell AH, Powell GVN (1987) Population dynamics of euglossine bees in Amazonian forest fragments. Biotropica 19:176–179

Rankin de Merona JM, Prance GT, Hutchings RW et al. (1992) Preliminary results of large-scale tree inventory of upland rain forest in the Central Amazon. Acta Amazônica 22:493–534

Rich AC, Dobkin DS, Niles LJ (1994) Defining forest fragmentation by corridor width: the influence of narrow forest-dividing corridors on forest-nesting birds in southern New Jersey. Conserv Biol 8:1109–1121

Riswan S, Hartanti L (1995) Human impacts on tropical forest dynamics. Vegetatio 121:41–52

Robbins CS (1980) Effect of forest fragmentation on breeding bird populations in the piedmont of the Mid-Atlantic region. Atl Nat 33:31–36

Robinson SK, Thompson FR, Donovan TM, Whitehead DR, Faaborg J (1995) Regional forest fragmentation and the nesting success of migratory birds. Science 267: 1987–1990

Rolstad J (1991) Consequences of forest fragmentation for the dynamics of bird populations: conceptual issues and the evidence. Biol J Linn Soc 42:149–163

Rylands AB, Keuroghlian A (1988) Primate populations in continuous forest and forest fragments in central Amazonia. Acta Amazonica 18:291–307

Saunders DA, de Rebeira CP (1991) Values of corridors to avian populations in a fragmented landscape. In: Saunders DA, Hobbs RJ (eds) Nature conservation 2: the role of corridors. Surrey Beatty and Sons, Norton, pp 221–240

Scariot AO (1996) The effects of rain forest fragmentation on the palm community in central Amazonia. PhD Thesis, University of California, Santa Barbara

Schwarzkopf L, Rylands AB (1989) Primate species richness in relation to habitat structure in Amazonian rainforest fragments. Biol Conserv 48:1–12

Simberloff D, Abele LG (1982) Refuge design and island biogeographic theory: effects of fragmentation. Am Nat 120:41–45

Simberloff DS, Cox J (1987) Consequences and costs of conservation corridors. Conserv Biol 1:63–71

Simberloff DS, Farr JA, Cox J, Mehlman DW (1992) Movement corridors: conservation bargains or poor investments? Conserv Biol 6:493–505

Sizer N (1992) The impact of edge formation on regeneration and litterfall in a tropical rain forest fragment in Amazonia. PhD Thesis, Cambridge University, Cambridge

Souza de OFF, Brown K (1994) Effects of habitat fragmentation on Amazonian termite communities. J Trop Ecol 10:197–206

Spironello WR (1987) Range size of a group of *Cebus apella* in central Amazônia. Am J Primatol 8:522

Stouffer P, Bierregaard RO Jr (1995) Use of Amazonian forest fragments by understory insectivorous birds: effects of fragment size, surrounding secondary vegetation, and time since isolation. Ecology 76:2429–2443

Terborgh J (1986) Keystone plant resources in tropical forests. In: Soule ME (ed) Conservation biology: the science of scarcity and diversity. Sinauer Assoc, Sunderland, MA, pp 330–344

Terborgh J, Winter B (1983) A method for siting parks and reserves with special reference to Columbia and Ecuador. Biol Conserv 27:45–58

Tocher M (1997) A comunidade de anfíbios da Amazônia central: diferenças na composição específica entre a mata primária e pastagens. In: Gascon C, Moutinho P (eds) Floresta Amazonica: dinâmica, regeneração e manejo. Instituto Nacional de Pesquisas da Amazônia, Manaus, AM, Brazil, pp 219–231

Tocher M, Gascon C, Zimmerman B (1997) Fragmentation effects on a central Amazonian frog community: a ten-year study. In: Laurance WF, Bierregaard RO (eds) Tropical forest remnants: ecology, management, and conservation of fragmented communities. University of Chicago Press, Chicago, pp 124–137

Uhl C, Kauffman JB (1990) Deforestation effects on fire susceptibility and the potential response of tree species to fire in the rain forests of the eastern Amazon. Ecology 71:437–449

Uhl C, Buschbacher R, Serrão EAS (1988) Abandoned pastures in eastern Amazonia. I. Patterns of plant secession. J Ecol 76:663–681

Villard MA, Merriam G, Maurer BA (1995) Dynamics in subdivided populations of Neotropical migratory birds in a fragmented temperate forest. Ecology 76:27–40

Vitousek PM (1994) Beyond global warming: ecology and global change. Ecology 75:1861–1876

Wandelli E (1991) Resposta eco-fisiológica da palmeira Astrocaryum sociale a mudanças micro-climáticas ligadas à borda da floresta. Tese de Mestrado, Instituto Nacional de Pesquisas da Amazônia, Manaus, AM, Brazil

Wiens JA (1976) Population responses to patchy environments. Annu Rev Ecol Syst 7:81–120

Williams CB (1943) Area and number of species. Nature 152:264–267

Williams-Linera G (1990) Vegetation structure and environmental conditions of forest edges in Panama. J Ecol 78:356–373

Williamson GB, Mesquita R de CG, Ickes K, Ganade G (1997) Estratégias de árvores pioneiras nos Neotrópicos. In: Gascon C, Moutinho P (eds) Floresta Amazonica: dinâmica, regeneração e manejo. Instituto Nacional de Pesquisas da Amazônia, Manaus, AM, Brazil, pp 131–144

Wilson EO, Willis EO (1978) Applied biogeography. In: Cody ML, Diamond JM (eds) Ecology and evolution of communities. Harvard University Press, Cambridge, MA, pp 522–534

Wright I, Gash J, da Rocha H et al. (1992) Dry season micrometeorology of central Amazonian ranchland. Q J R Meteorol Soc 118:1083–1099

Zimmerman BL, Bierregaard RO Jr (1986) Relevance of the equilibrium theory of island biogeography with an example from Amazonia. J Biogeogr 13:133–143

4 Climatic and Human Influences on Fire Regimes in Temperate Forest Ecosystems in North and South America

P. ALABACK, T.T. VEBLEN, C. WHITLOCK, A. LARA,
T. KITZBERGER, R. VILLALBA

4.1 Introduction

Fire plays an integral role in regulating ecosystem structure and processes including biodiversity, nutrient cycling, ecosystem structure, resiliency, stability, and carbon flow (Boerner 1982; Agee 1993). These effects of fire on ecosystems are extremely sensitive to all components of the disturbance regime (frequency, intensity, scale, predictability; Pickett and White 1985). To properly evaluate the impact of human activities on fire – both now and in the future – it is critical that we have an accurate baseline of past fire occurrences. In particular, we need to understand the natural role of fire in ecosystems across a range of scales from stands or sites to broad regions (Turner et al. 1994). In addition, we need to understand the controls that govern fire regimes across landscapes, so that we can better understand both how natural systems worked in the past and how fire regimes might be changed in the future (Franklin et al. 1991; Veblen and Lorenz 1991; Alaback and McClellan 1993).

There is a clear need for a broader view of fire on a larger scale in order to place the complex interactions between localized and regional human activities and climatic controls in proper perspective (Swetnam and Betancourt 1990; Veblen et al. 1999). In forests this becomes an even more challenging task because trees integrate the cumulative effects of both natural processes and human activities over time scales of years and decades to centuries (e.g., Swetnam 1993; Lertzman and Fall 1998). To understand climatic controls and their broad-scale effects on vegetation, even larger time scales (millennial) are required.

While much detailed work has been done on understanding fire history at many small sites, the question of how to address multiple-scale fire patterns has proven quite elusive. Work on landscape-level fire patterns has been initi-

Ecological Studies, Vol. 162
G.A. Bradshaw and P.A. Marquet (Eds.)
How Landscapes Change
© Springer-Verlag Berlin Heidelberg 2003

ated in several areas of the western United States and Canada. Particular emphasis has been placed on coastal forests, boreal forests, the US central Rockies, and the Southwest (e.g., Morrison and Swanson 1990; Swetnam and Betancourt 1990; Johnson and Larsen 1991; Arno et al. 1993; Turner 1994; Veblen et al. 2000). Recent studies of fire history and fire regimes in temperate South America provide a unique opportunity for bi-hemispheric comparison (Veblen and Lorenz 1988; Veblen et al. 1992, 1999; Markgraf and Anderson 1994; Kitzberger and Veblen 1997, 1999; Kitzberger et al. 1997). Large-scale work is of particular value for comparative studies. Clearly this is where new research or better-integrated studies may advance our abilities to understand human impacts and causal mechanisms of this key ecological process. Insights from this work should help clarify ecological implications of government fire policy both north and south.

Comparative studies can provide a more robust test for cause-effect relationships and for a clearer understanding of atmospheric controls and ecological responses to fire than is possible with studies done within a bioregion (Mooney 1977; Paruelo et al. 1988; Lawford et al. 1995). Cultural, economic and historical commonalities within a bioregion often constrain fire regimes over a large area or over a broad gradient of climate and topography, resulting in somewhat confounded relationships between climate and fire (e.g., Savage and Swetnam 1990; Arno and Brown 1991). Comparative studies allow evaluation of large-scale atmospheric controls on fire regimes such as the position of jet streams, effects of topography, and the movement of air masses, as well as the effects of human factors such as population growth, economics and culture on landscape patterns and the alteration of fire regimes.

The western temperate latitudes of North and South America have proven to be a particularly useful region to make continental comparisons due to the strong parallels in climate, oceanic influence, geology, vegetation and other biota, whereas a strong asymmetry exists in terms of culture, history, biogeography and economic development patterns (Alaback and McClellan 1993; Mooney et al. 1993; Lawford et al. 1995; Whitlock et al. 2000). Parallels in climate and vegetation between the semi-arid temperate grasslands and shrublands of North and South America have been well studied (e.g., Paruelo et al. 1988, 1995). Relatively little attention has been paid to comparisons between forested mountain climates of Argentina and western North America.

While fire history studies have been conducted in western North America over a longer time period than in South America, there has still been sufficient work in both hemispheres to provide a useful foundation for comparison of the two regions. In this chapter, we provide an overview of patterns of fire regimes and their controls in the western portions of North and South America to better understand the relative role of climate and human activities in determining fire regimes in the two regions, and to clarify research gaps and research opportunities in the fire ecology of these diverse regions.

4.1.1 Overview of Climate

The west coast of each continent includes a wide range of climate from that supporting lush rainforest to nearly barren desert proceeding either towards the equator or easterly across the cordillera. On the western slopes of the Andes, mediterranean scrub and forest dominate from 33–37°S, changing to temperate rainforest southwards under a humid and perhumid climate. In temperate latitudes, rainforest dominates both hemispheres from 35–40° poleward and west of the main cordillera. In the temperate rainforest region, annual rainfall ranges from approximately 2 to 7 m with fall and winter rains dominating the lower latitudes (Alaback 1991). In general, Chilean winters are milder and more likely to be snow-free than in North America, except at the most extreme latitudes (Fig. 4.1). This climatic equitability has resulted in the development of a rich and diverse flora in South America, despite a long history of geographic isolation (Arroyo et al. 1995).

Both North and South America have good representations of perhumid rainforests, but in North America seasonal rainforest is much more broadly distributed than in South America (Alaback 1995; Veblen and Alaback 1995). Subpolar rainforest is more extensive in South America due to the greater intensity of westerlies and the more oceanic climate at high latitudes. Ecologically, this difference in the distribution of rainforest leads to greater influence of wild fire and summer droughts in North America than in South America.

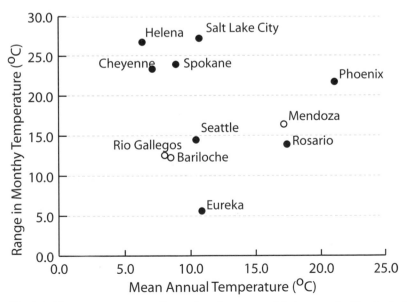

Fig. 4.1. Temperature ranges for stations in comparable temperate climates in North and South America. Rio Gallegos and Bariloche represent steppe climates; Mendoza a desert climate

Wild fires play a key role in the disturbance ecology of northern Pacific forests as far north as northern Vancouver Island, British Columbia (ca. 50°N). For a general summary of climatic influences on the disturbance ecology of temperate rainforests see Veblen and Alaback (1995).

The western slope of the Andes in central Chile (33–37°S) is characterized by a summer drought mediterranean-type climate with a strong influence of the southwest Pacific high pressure system. Further south, cooler and wetter conditions prevail and develop into west coast maritime climate with a strong influence of the westerlies. However, mediterranean-type rainfall seasonality is evident as far south as ca. 43°S. Mean annual rainfall on the windward slopes of the coastal range and the Andes ranges from 3000 to over 5000 mm as a result of orographic uplift. The coastal range produces a rain-shadow effect on the central depression of south-central Chile where at ca. 40°S mean annual rainfall is typically less than 2000 mm (Almeyda and Saez 1958).

East of the Andes, the rain shadow effect becomes extremely pronounced, developing into steppe and desert climates. At ca. 41°S, in Argentina, rainfall declines dramatically from 3000 mm to less than 800 mm over a distance of only 50 km (e.g., Veblen and Lorenz 1988). Crossing east of the coastal mountain divide the relatively equitable climate of the coast rapidly gives way to an extreme continental climate in both North and South America (although much more so in North America, Paruelo et al. 1995, Fig. 4.1). In North America continentality generally increases with latitude and distance from the Pacific coast. Along the northern coast the coast-interior contrast becomes the most extreme with permafrost and boreal forests occurring adjacent to temperate rainforests (lat. 55–60°N).

Much of the western United States is dominated by a mountain climate, leading to short growing seasons and an extreme continental climate (Weaver 1980). Forests are associated primarily with the steppe climates, but can also occur on temperate continental and oceanic climates. The occurrence of forests in grassland climates (potential evaporation exceeds annual precipitation) seems counter to the definition of climates by Köppen (1918). Forest occurs commonly in these climates in North America because of the highly localized orographic effects of mountains within dry climatic zones, the effects of localized hydrological conditions, and the exceptional genetic adaptation of many trees to grow under dry cool conditions (e.g., *Pinus ponderosa*, *Pinus flexilis* and their relatives; Potter and Green 1960; Callaway et al. 1994; Weaver 1980).

Most South American grassland types find close climatic analogs in North America, but some of the northern types, such as the mixed prairie and the Palouse prairie do not have close climatic equivalents in South America (Walter 1973; Paruelo et al. 1995). The sagebrush steppe, and semi-arid shrublands of the Great Basin, the tall grass prairies of the Midwest and the southern coastal plain have remarkably close analogs in terms of annual temperature, seasonality and summer precipitation. The forests on the eastern margin of the Andes are quite distinctive in having similar degrees and patterns of drought stress, but generally have milder winter temperatures than their North American equivalents. In the northern Rockies high late spring–early summer precipitation contrasts with more consistently dry summers in equivalent vegetation types in northern Patagonia.

4.1.2 Lightning

Lightning is a principal ignition source for wildfires. Understanding the temporal and geographic pattern of lightning now and during periods of climate change will be key to predicting changes in the incidence and role of fire in forests. Unlike temperature and precipitation, lightning is often very localized and difficult to determine accurately with standard weather stations. In general, lightning seems to require fairly high levels of energy, from both sensible heat and atmospheric mixing. Lightning is generally infrequent in cold rainy coastal climates, as in the perhumid and subpolar rainforest zones of North and South America. In southeast Alaska, for example, lightning may strike the major towns in this area only once every one to two decades. In cool temperate climates lightning seems to be generally associated with lifting and convergence of air masses over major mountain ranges, especially in dry climates. Lightning seems associated with anomalously warm dry seasonal weather in coastal regions.

In South America the temperate rainforest regions generally have a moderate occurrence of lightning in contrast to a very low incidence of lightning in the coastal Pacific Northwest. Steppe regions, by contrast, appear to have a fairly low incidence of lightning, compared to their equivalents in the central and northern Rockies. Bariloche for example has a particularly low incidence of lightning (166 reports of lightning over a 14-year period). In the four mountainous national parks around Bariloche (Nahuel Huapi, Lanín, Los Alerces and Puelo), however, lightning is a significant source of fire ignitions as they account for over 8 % of all fires and 16 % of the land area burned from 1938–1996. This is equivalent to a frequency of 3.3 lightning-ignited fires/ 100,000 ha, or about the same frequency as the boreal forests in southern Alaska (Bruno and Martin 1982). In lower latitude desert climates lightning frequency dramatically increases, as in North America. The data available at this time still suggest that lightning frequency is much lower in South American than North American sites within the same climatic zone, perhaps explaining the generally lower incidence of naturally ignited fires. This may also explain why North American vegetation has generally developed more extensive adaptations to fire than South America.

Lightning is strongly associated with mountains and dry interior climates in Western North America. In Oregon and Washington, for example, low-lying coastal regions may average as little as 1 day of lightning per year, whereas in the humid Cascades up to 5 days of lightning may occur per year, and in the foothills of the Rockies over 6 days generally occur (Agee 1993). The Montana Rockies vary in incidence of lightning from 15 to 50 days per year. The highest incidence of lightning occurs in the central Rocky Mountains from Yellowstone to Arizona where a monsoonal summer rain climate is most prominent. This may explain the extremely high incidence of fire in these dry forests (surface fires occur every 2 years or more, Swetnam and Betancourt 1990). In *Pinus ponderosa* forests in the Rockies, from 4 to 10 % of tree mortality is directly attributable to lightning (Taylor 1969). In Alaska a similar pattern is evident with the dry interior forests having the highest lightning frequency. In south-central Alaska fewer than 1.2 lightning fires per 100,000 ha were reported in 23 years, as opposed to 6.2 fires per 100,000 ha in the Yukon (interior) and 2–3/100,000 ha in western coastal Alaska (Gabriel and Tande 1983).

While lightning can be characterized by looking at general climatic parameters, more detailed predictive ability would be useful in assessing the effects of climate change. It is often assumed that storms will be more frequent and intense during periods of climate change (Price and Rind 1994). If this leads to increased incidence of lightning, especially during periods of extended drought, it could have profound implications to the ecological role of fire (Overpeck et al. 1990; Franklin et al. 1991).

4.1.3 Vegetation Patterns

Coarse-scale vegetation patterns of southern Chile and Argentina mirror the north-to-south and west-to-east climatic gradients discussed above. Central Chile (33–36°S) is characterized by evergreen sclerophyll shrublands and low forests as well as deciduous *Nothofagus* forests and scattered conifer stands dominated by the conifer *Austrocedrus chilensis*. In south-central Chile (40–45°S), Valdivian rainforests dominate with many broad-leaved evergreen myrtle, laurel, and podocarp species. East of the Andes, rainforests rapidly give way to mesic *Nothofagus-Austrocedrus* forests, *Austrocedrus* woodlands to shrub-steppe grassland (Veblen and Lorenz 1988). Subalpine forests are formed by the deciduous *Nothofagus pumilio* and *N. antarctica*. Conifers only occur in nutrient-poor sites or in droughty sites, and rarely form a dominant canopy layer (Veblen et al. 1995).

Conifers dominate forest types throughout western North America due to both historical and climatic factors (Waring and Franklin 1979; Stephenson 1990). Dry summers and cool temperatures are key factors favoring the growth of conifers relative to deciduous hardwood species. Coastal forests of the Pacific Northwest (ca. 42–57°N) are generally dominated by *Tsuga heterophylla*, *Pseudotsuga menziesii*, and *Thuja plicata*. Subalpine forests are also conifer-dominated with *Abies* spp., *Pinus contorta* and *Tsuga mertensiana*. Interior forests (45–52°N) often include a broad mixture of species including

Pinus spp., *Larix occidentalis, Pseudotsuga menzisii, Picea* spp., and *Abies* spp
(Franklin and Dyrness 1973; Habeck and Mutch 1973; Pfister et al. 1977). In
the central Rockies (38–45°N) subalpine forests (*Abies lasiocarpa* and *Picea*
spp., or seral *Pinus contorta*) dominate, but with drought-adapted *Pinus* spp.
increasing in importance southward (Peet 1981; Knight 1994). In the Canadian Rockies (50–54° northward), subalpine forests intergrade with boreal
forest northward (*Picea mariana, P. glauca*). Deciduous trees generally occur
only as early seral species or in riparian habitats, especially floodplains (e.g.,
Betula spp, *Platinus* spp., *Populus* spp., *Fraxinus* spp.).

Continuous forest generally occurs on the westward sides of the major
cordillera, with the easternmost regions having highly fragmented stringers
of forest within a matrix of grassland or tundra. Steppe-shrub or desert vegetation dominates valley bottoms and rain shadow regions throughout the
interior west and Patagonia (Weaver 1980; Veblen and Lorenz 1988). Forest
occurs primarily at higher elevations (generally 1000 m or more in North
America at 47°N to 2000 m or more in the central Rockies (Daubenmire 1956;
Peet 1981). In Patagonia, by contrast, forest is primarily distributed along precipitation gradients with less dramatic changes in elevation (Veblen et al.
1995).

In the transition between the forests of the Rockies and the Great Plains
region drought-tolerant conifers such as *Juniperus* spp. and *Pinus flexilis*
often dominate forest pockets. These latter species are rapidly invading grasslands, primarily in response to exclusion of wildfire (Potter and Green 1960;
Rebertus et al. 1991). The analogue species in Argentina would be *Austrocedrus chilensis* and *Nothofagus antarctica*. These species are also significantly
expanding their range eastward (Kitzberger and Veblen 1999).

Moisture is a key limiting factor for primary productivity east of the major
continental cordillera in both hemispheres and explains much of the variation in ecosystem structure across the region (Woodward 1987; Stephenson
1990). In western North America, low temperatures both diurnally and seasonally are key constraints to tree photosynthesis especially in interior climates (Daubenmire 1956; Emmingham and Waring 1977; Noble and Alexander 1977). In South America, however, photosynthesis of *Nothofagus* tends not
to be as restricted by low temperatures (Premoli 1994).

4.2 Relation of Climate to Fire Regimes

4.2.1 General Patterns

One of the major research challenges in ecology is to identify and quantify the
mechanisms and processes that control ecosystem responses to altered climate conditions. Such studies are required to understand how present land-

scape patterns have been influenced by climatic variation and to predict future landscape patterns that may develop in response to climatic change. Until recently, mechanisms of climate-induced vegetation changes were sought largely in the responses of individual species' performances to climatic variation. Currently, however, it is widely accepted that climatically induced variations in disturbance regimes are often the proximate causes of vegetation change (Davis and Botkin 1985; Overpeck et al. 1990).

Fire frequency in any ecosystem is a consequence of interactions among fuel accumulations, fuel moisture content, and ignition sources. Climatic variation influences fire regimes across a broad range of temporal scales (Chandler et al. 1983; Agee 1993). For example, long-term climate determines fuel types and accumulation rates; annual to seasonal variation largely determines flammability of vegetation and coarse dead fuels; weekly to daily variation controls the desiccation of fine fuels, and daily to hourly variation influences fire behavior. Analysis of fire regime responses to climate variation across a range of time scales is important for separating the effects of seasonal and annual variations on fuel desiccation and ignitions from longer-term climatic influences on fuel types and loads (Rothermel 1972; Clark 1989, 1990; Swetnam and Betancourt 1990; Johnson and Wowchuk 1993; Swetnam 1993; Millspaugh et al. 2000).

Increasingly, it has been recognized that understanding climatic influences on fire regimes is critical to predicting the effects of climatic variation on vegetation change (Malanson and Westman 1989; Baker 1990; Baker et al. 1991; Overpeck et al. 1990; Johnson and Larsen 1991; Sirois and Payette 1991; Bergeron and Archambault 1993). Climate-model simulations that examine changes in disturbance regime brought on with a doubling of atmospheric CO_2 suggest that increased disturbance will accelerate the rate of forest change compared with that which would occur from climatic change alone (Davis and Botkin 1985; Overpeck et al. 1990; Franklin et al. 1991; Alaback and McClellan 1993; Price and Rind 1994).

In studying the effects of climatic variation, it is important to consider a range of temporal scales. For example, seasonal or annual-scale droughts often increase fire frequency and extent, but multi-decade or century-long dry periods may alter vegetation patterns and fuel loadings leading to declines in land area burned (e.g., Swetnam 1993). To understand the controls of high frequency as well as low frequency variation in fire regimes, long-term records on an annual scale of resolution are required.

Similarly, the effect of climatic variation on disturbance regimes and vegetation dynamics varies across spatial scales. By specifying the spatial scale of interest, it is possible to separate the response of fire to environmental variations occurring on hemispheric, continental, regional, landscape, and local scales. On the largest scales, variations in atmospheric circulation (such as the latitudinal positions of the jet and subtropical highs, El Niño Southern Oscillation variations) determine the timing of ignition, the nature of the fire season, and the broad mosaic of vegetation. As climate varies on a regional scale, so too does the vegetation and fire frequency. Within a landscape, fire regimes are related to variation among vegetation zones and topography (e.g., dense subalpine versus open montane forests) and human settlement history. At local scales, variables such as vegetation structure and topographic position can also be important controls on fire regimes. To distinguish among controls operating at different spatial scales, observations must be made across a broad spatial scale from local to regional (Risser 1987; O'Neill 1988).

The influence of climatic variations on vegetation change is most evident at coarse spatial and temporal scales (Prentice 1992). However, at time scales of decades to one or two centuries, the effects of climate variation on vegetation dynamics have been less obvious. Within the context of global climatic change, it is precisely at this time scale of a few decades to centuries for which the need for understanding the effects of climate variation on vegetation change is most urgent, and in most ecosystems altered fire regimes will probably be a key factor.

4.2.2 Overview of Fire Disturbance Patterns in Western North America

In the past century myriad studies have been done on the fire ecology of the forests in western North America. Forests in interior Alaska, Alberta, Montana, Colorado, Wyoming, Arizona, New Mexico, and California have been studied particularly intensively (Appendix 4.1; Kilgore 1980; Heinselman 1981; Agee 1993; Veblen et al. 1999). The fire regimes of coastal mesophytic and rainforest zones within California, Oregon, and Washington have also been the subject of extensive investigations, usually on a site-specific basis (Agee 1993). Broad-scale landscape-level empirical and modeling studies are now underway in all these regions in addition to British Columbia, the Yukon, and other western regions (Morrison and Swanson 1990; Teensma et al. 1991; Wallin et al. 2002). General climatic controls of these fire regimes have been deduced by comparing intensively studied sites along climatic gradients (Agee 1993; Swetnam and Betancourt 1990; Caprio and Swetnam 1995). Long-term patterns of fire have also been examined using charcoal in sediment in the Oregon Coast Range (Long et al. 1998), northern California (Smith and Anderson 1992; Mohr et al. 2000) and in the Yellowstone area (Millspaugh et al. 2000; Millspaugh and Whitlock 2002). These long-term records suggest that fire frequency is a function of both climate and vegetation conditions.

Dry *Pinus ponderosa* and associated forests have been the most extensively studied both in the northern Rockies, California, Colorado, and Arizona (Cooper 1961; Arno 1976, 2002; Kilgore and Taylor 1979; Bonnicksen and Stone 1982; Gruell et al. 1982; Covington and Moore 1994; Mast et al. 1998; Veblen et al. 2000). Annual rainfall for these forests ranges from 350–1000 mm or more. These forests can have fire return intervals of as little as 2 years in the

Southwest or 10–15 years or more in the Northwest and Colorado, depending on elevation and moisture regime. While typically characterized as high-frequency low-intensity (mostly nonlethal) fire regimes, catastrophic fires occasionally occur, and in some moist sites, catastrophic fire may play a key role (White 1985; Arno 2002).

The more mesic *Pseudotsuga menziesii* var. *latifolia–Larix* interior sites have also been studied extensively. These forests generally occur in sites with 500 mm of precipitation or more and typically have a mixed-replacement or stand-replacement fire regime (Arno 1980). Fire return intervals generally range from 20–200 years, although 4000–8200 years B.P. charcoal evidence suggests higher frequencies (Hallett and Walker 2000). The high elevation and cooler *Pinus contorta* forests are sometimes characterized by an exclusively stand-replacement fire regime, given the adaptations of the tree (serotinous cones, limited heat tolerance of the bark). However, on dry sites, multi-aged forests are common as well, suggesting a more complex fire history (Arno 1980; Agee 1993). The *Pinus contorta* forests of Yellowstone are particularly well known as examples of extensive catastrophic wildfire, but with a complex landscape level pattern (Romme and Turner 1991; Turner et al. 1994). The intensity and frequency of fire in these forests may play a key role in regulating successional pathways and species composition. Frequent (<25 years) intense fire may lead to shrublands, whereas frequent ground fire may lead to nearly pure *Pinus contorta*. Intermediate frequencies of intense fire may lead to dominance by *Populus tremuloides* (Gabriel 1976; Kilgore 1980; Bradley et al. 1992b).

The subalpine *Picea engelmannii* and *Abies lasiocarpa* forests have also been the subject of fire history studies throughout the West. In general, the fire return intervals are relatively long (100–300 years or more), but highly variable depending on climatic gradients, microclimates, and ignition sources (Arno and Davis 1980; Romme and Knight 1981; Agee 1993). These forests also tend to be highly fragmented by physiography, disturbance (e.g., avalanches, landslides), hydrology and microclimate and, therefore, often have a complex spatial pattern of disturbance. In the drier more southerly regions (e.g., the Sierra Nevada and the Southwest), fires are generally of lesser intensity and of smaller size than in the northern Rockies because of smaller fuel loading and more clumped stand structures. Charcoal records suggest highest fire frequency in the Sierra during the early Holocene (Smith and Anderson 1992). Lightning also generally occurs in association with more rain than in the Rockies (Kilgore 1980). In Wyoming, smaller sizes of fires in the subalpine zone have also been attributed to localized snowdrifts.

The more open *Pinus-Juniperus* forests and semi-arid high elevation forests of the Great Basin and southwestern United States and Mexico also have been studied with respect to fire history. In this case, highly fragmented and sometimes low accumulations of fuels in addition to low frequencies of lightning result in low frequencies and small areas of fires as deduced from fire-scar data (Martin 1982; Agee 1993). High frequencies of light ground fires may have occurred in these sites, but were not detectable with fire-scar analysis. Some evidence suggests that fire may have been a key mechanism in response to climate change in regulating the expansion or contraction of woodland and grassland since the Pleistocene (Neilson 1987).

On the west coast, *Pseudotsuga menziesii* dominates from northern California to southern British Columbia (39–48°N). It is generally thought that at least during the past 3000 years these forests have been a product of catastrophic fire (Franklin and Dyrness 1973; Spies and Franklin 1988; Teensma et al. 1991; Agee 1993; Long et al. 1998). While the fire return intervals may be as long as 500 or 700 years for catastrophic fire, the longevity of these trees still allows them to retain dominance in these forests (Hemstrom and Franklin 1982). The only significant exception to this is on the wettest sites, where the fire-sensitive *Tsuga heterophylla* and *Thuja plicata* may dominate. In general, long fire return intervals are the rule on the coast where truly exceptional droughts or successive years of drought are needed for fire, whereas in the Cascades more frequent smaller scale mixed replacement or nonlethal fire is common (Morrison and Swanson 1990; Swanson and Franklin 1992; Long et al. 1998; Mohr et al. 2000). In the coastal redwood region in northern California ground fires can occur as often as every 2 years, resulting in an open stand structure. Major fires occur every 20–30 years or more (Kilgore 1980; Brown and Swetnam 1994; Skinner and Chang 1996; Taylor and Skinner 1998; Mohr et al. 2000). Charcoal data suggest that higher fire frequencies occurred during the medieval warm period (Mohr et al. 2000).

Frequent ground fires may play a key role in maintaining the competitive advantage of redwood in these forests, just as it does for cedar (*Thuja*) in the interior. The northern coastal pattern of infrequent fire may also characterize portions of the Chilean coast, but unraveling the complexities of anthropogenic influence and finding well preserved evidence between the long return intervals have made this a challenging undertaking (Lara, unpubl. data).

4.2.2.1 Yellowstone Region

The relation between climate, vegetational change, and fire varies depending on the time scale of interest. Fire history studies undertaken in Yellowstone National Park (YNP) offer a case study. The fires of 1988 were attributed to unusual summer drought, strong winds, and the large accumulation of fuels in the form of late-successional forest and downfallen timber. In that year eight complexes burned an area of 730,000 ha and vastly exceeded the size of any fire that had occurred since the implementation of a natural burn policy in 1972. On this time scale the 1988 fires were "unprecedented". On a decadal time scale, Balling et al. (1992) examined the fire-climate relations of the last

120 years and determined that burn area in YNP is related to low precipitation in the antecedent and current summer and high summer temperature. Any effects of fire suppression in the last 50 years did not mask the clear relation between climate and fire, and 1988 was part of a trend of decreasing winter precipitation and rising temperatures that led to large fire size. Although unprecedented, the size of the conflagration could be tied to predictable meteorological conditions.

On centennial scales, Romme (1982), Romme and Despain (1989), and Barrett (1994) examined the prehistoric fire history in different geovegetation regions of YNP using dendrochronologic methods. Data from fire-scarred tree rings and stand-age analysis suggest that large fires have a return interval of 200–400 years. The last time fires burned an area equal to that of 1988 was in the early to mid-1700s; compared with the fires of 1988, the earlier events were small and occurred over several decades (Romme and Despain 1989). Just like the 1988 fires, earlier fires were associated with debris flows that are a major source of sediment transport in parts of YNP (Meyer et al. 1995). Turner and Romme (1994) note that infrequent occurrence of high-severity crown fires imparts a coarse mosaic pattern on the landscape that has major ecological consequences. For example, seedlings of lodgepole pine, the dominant species, are most dense in areas affected by surface fires, not canopy burn. The mosaic of crown fires affects biotic interactions, community structure, and ultimately biodiversity, and sets up the landscape for further large fires.

Millspaugh and Whitlock (1995) examined the charcoal accumulation rates in lake-sediment cores to reconstruct fire history in small watershed during the last 750 years. Magnetic susceptibility of the sediments was also measured to detect periods of catchment erosion which, in conjunction with high levels of charcoal, might indicate a local fire. The records show that individual watersheds burned irregularly during the last 750 years, as a result of local variations in fuel build-up, vegetation, and weather conditions. During major fire years, for example, ca. 1988, ca. 1700, ca. 1560, and ca. 1440, more than one watershed was burned. Thus, recovery following local small fires was regularly synchronized by the large fire events.

Charcoal data also disclose variations in fire frequency as a result of climatic variations on millennial time scales. Millspaugh et al. (2000) describe the Holocene fire history from a site on the Central Plateau of Yellowstone. The fire return intervals were short between 10,000 and 7000 years ago, and gradually increased to the present interval of 200–400 years in the late Holocene. The change in fire frequency is ascribed to the regional changes that resulted from large-scale hemispheric variations in climate. Perihelion occurred in the Northern Hemisphere summer in the early Holocene, which increased summer radiation at the latitude of Yellowstone by 8%. Summers were warmer and lower effective precipitation was a direct result of the radiation changes. In addition, the East Pacific subtropical high-pressure system was stronger in the early Holocene, which also increased summer drought (Thompson et al. 1993; Whitlock 1993).

The evidence for more frequent fires in the early Holocene, in turn, suggests that burns were probably smaller than at present, that the landscape mosaic was more heterogeneous, and that more of the region was occupied by early and intermediate seral stands. In the late Holocene, the modern fire regime was established with the introduction of cool wet conditions. Pollen data show little change during these two fire regimes (Whitlock 1993). Apparently, lodgepole pine forms an edaphic climax on rhyolite plateaus and the forest composition has been fairly constant under a wide range of fire frequencies. Charcoal data also indicate no recurrent pattern or cycle in the fire return intervals of the last 10,000 years, supporting the idea of nonstationarity in fire occurrence on long time scales. Thus the lake-sediment records reveal that the fire frequency varies as a result of changes in climate on long time scales, but that within this framework, they are a response to noncyclic climate anomalies and fuel conditions.

4.2.2.2 Colorado Front Range

The synoptic-scale climate of the Colorado Front Range in the central Rockies reflects its mid-latitude, continental location interacting with atmospheric circulation patterns. During the winter, the Front Range experiences frequent intrusions of Pacific air masses, but during the summer is affected mainly by interior air masses and monsoon air from the Gulf of Mexico (Mitchell 1976). The seasonal distribution of precipitation varies with elevation and location on the western and eastern sides of the Front Range (Barry 1973). On the eastern slope of the Front Range, the precipitation-carrying storms are brought over the Front Range in winter and spring by upper westerly air flow, which results in heavy snowfall at upper elevations (Greenland 1989). Precipitation also is brought from the east, at lower elevations, by cyclonic storms centered to the east of the Continental Divide that create upslope air flows. These upslope precipitation events account for the spring maximum of precipitation. During the summer months, rainfall results from localized convectional storms. In the northern part of the Front Range at low elevations (ca. 1500–2400 m) a pronounced peak in precipitation during March and April is followed by gradually declining precipitation for May through January (Villalba et al. 1994). Thus, the fire season is mainly summer and fall.

Influences of interannual climatic variability on fire occurrence in the Ponderosa pine elevational zone (ca. 1800–2800 m) of the eastern slope of the Front Range have recently been investigated (Veblen et al. 2000). Analysis of the relatively short National Forest fire records (beginning in ca. 1920) and the instrumental climatic record indicate that the number and size of fires are highly correlated with spring (March-June) precipitation (Veblen et al., unpubl. data). To obtain long records of fire history that predate the era of modern fire suppression, tree-ring methods were used to produce a regional fire history beginning in ca. 1600 A.D. Fire history was determined from over 500 partial cross sections containing fire scars in 41 stands located over a north–south distance of ca. 40 km. Climate characteristics of years with varying extents of fires (i.e., numbers of trees and sites recording fire) were determined from tree-ring reconstructions of precipitation over the period from ca. 1550 to 1990. Both annual and spring precipitation were reconstructed.

Superposed epoch analysis (Swetnam 1993) was used to compare mean spring precipitation during years with and without fire scars in these 41 stands over the period 1549–1987. Spring precipitation is significantly below the long-term average during fire years. The severity of spring drought is substantially greater during years in which more than a single stand recorded fire and years in which two or more stands recorded fire scars on 25 % or more of the fire-scar-susceptible trees (Veblen et al. 2000). The regional scale of fire occurrence during exceptionally dry springs is indicated by the occurrence of fire in large percentages of the dispersed stands during individual dry years such as 1654, 1786, 1809 and 1859. For example, during 1654 when spring precipitation was only 46 % of the long-term average, fire was recorded in 50 % of the stands with fire-scar-susceptible trees and included the northern and southernmost stands separated by a distance of ca. 40 km.

In the ponderosa pine zone of the Front Range, above-average spring precipitation was followed 2–3 years later with extensive fires (Veblen et al. 2000). A similar pattern has been found in *Austrocedrus* woodlands in northern Patagonia (Kitzberger and Veblen 1997; Veblen et al. 1999). The abundance of fine fuels from grasses and other herbs promoted by wet spring weather is probably an important mechanism in explaining this pattern. Increased fire 2 years after above-average moisture availability also occurs in some xeric conifer forests in Arizona (Baisan and Swetnam 1990).

In general, in the forests of Arizona and New Mexico, fire activity is reduced during El Niño events because of increased winter and spring precipitation (Swetnam and Betancourt 1990). A similar pattern occurs in Ponderosa pine forests of the Colorado Front Range where El Niño events reduce fire occurrences and La Niña events favor fire occurrences. Because of the great longevity of subalpine trees in the Southwest north to Colorado, a long tree-ring record can be used to characterize the relationships of climate to fire. In more northerly regions, where the tree ring record is shorter, charcoal analysis and other techniques will be needed to determine this relationship.

4.2.3 Current State of Knowledge for Northern Patagonia

The influences of monthly-to-yearly weather variation on fire occurrence have recently been investigated along a vegetation gradient from temperate

Andean rainforests to xeric, open woodlands in northern Patagonia, Argentina (Kitzberger and Veblen 1997; Kitzberger et al. 1997; Veblen et al. 1999). Fire chronologies derived from fire scars were developed in Nahuel Huapi National Park at ca. 40°23′–41°35′S in southwestern Argentina. National Park records of fire occurrence (Bruno and Martin 1982) were analyzed for Lanín, Nahuel Huapi, Lago Puelo, and Los Alerces National Parks (39°07′–42°43′S).

Low frequency (i.e., multidecadal) variation in fire occurrence in northern Patagonia is strongly associated with human settlement and land-use changes (Veblen and Lorenz 1988; Kitzberger and Veblen 1997; Veblen et al. 1992a, 1999). However, climatic variation also differentially influences the fire regimes for different vegetation types along this steep rainfall gradient. Fire extent is highly correlated with below-average spring/summer water availability for the year of the fire season (Fig. 4.2). Additionally, the stronger climatic relationship of the fire record in wet *Nothofagus* forests, compared to the xeric vegetation zone, is apparent in both the instrumental data and the tree-ring record. Fire in *Nothofagus* rainforests is strongly favored by drought during the spring and summer of the same year in which fires occur and less strongly favored by drought during the spring of the previous year (Kitzberger and Veblen 1997). The occurrence of fire in dry vegetation types is less sensitive to climatic variation and appears to be more dependent on ignition frequency than on fuel moisture conditions.

When the southeast Pacific anticyclone is strong and located farther south, it blocks the inflow of moist westerlies into northern Patagonia and produces drier-than-normal conditions which are reflected by increased fire in the wet

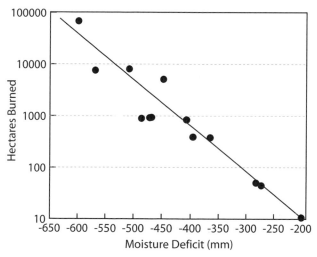

Fig. 4.2. Burn area and annual moisture deficit in national parks near Bariloche, Argentina. Moisture deficit was calculated as difference between annual precipitation and potential evapotranspiration (Thornthwaite's index). (Bruno and Martin 1982)

forests of northern Patagonia (Kitzberger and Veblen 1997). In northern Patagonia fire-promoting droughts are often related to ENSO (El Niño Southern Oscillation) events (Villalba 1994, 1995). For example, periods of frequent and severe droughts during 1910–1919, the 1950s and since 1977 are associated with more frequent El Niño events and less frequent La Niña events (Villalba 1995). Similar large-scale circulation anomalies have been shown to influence the occurrence of fire in the southern Canadian Rocky Mountains (Johnson and Wowchuk 1993) and the southwest USA (Swetnam and Betancourt 1990).

Northern Patagonia has experienced several multidecadal to centennial-scale periods of markedly cooler/wetter or warmer/drier climate over the past ca. 1000 years (Villalba 1994). For example, radiocarbon dating of glacial fluctuations for the southern Andes from ca. 41°S shows glacier expansions at ca. 1040±110 years A.D., 1330±50 years A.D., 1365±50 years A.D., and in recent centuries corresponding to the Northern Hemisphere's "Little Ice Age". Similarly, tree-ring records of climatic variation also confirm these periods of different climatic conditions (Villalba 1990; Luckman and Villalba 2002; Villalba et al. 2002). For example, tree-ring records indicate four main periods of different summer temperatures, each lasting at least 100 years. A cool period from 900 to 1070 A.D. was followed by a warm period from 1080 to 1250 A.D. (approximately coincident with the Medieval Warm Epoch of Europe). Afterwards a long, cool period followed from 1270 to 1670 A.D., peaking around 1340 and 1650 A.D. Warmer conditions then resumed in ca. 1720 A.D. Shorter periods of markedly warmer conditions are recorded in 1700–1740, 1859–1885 and 1905–1920 (Villalba 1994, 1995; Villalba et al. 1989, 2002). During such periods of distinct climatic conditions, fire occurrence is likely to have varied. Preliminary data on fire history during the past ca. 400 years in northern Patagonia indicate that over 50-year time periods the frequency of years of more widespread fire coincides with periods of increased ENSO activity (Kitzberger and Veblen 1997). Periods of increased ENSO activity are characterized in northern Patagonia by both increased summer drought and more frequent lightning.

4.3 Anthropogenic Influences on Fire Regime – Land Use and Fire Regimes in the Rocky Mountain Region and Northern Patagonia

The most dramatic effects of humans on fire regimes are, of course, changes in the frequency of ignitions and in fire suppression. In this section we briefly review the recent history of human influences on fire regimes for the forests that border steppe ecosystems in both North and South America, since it has been in these ecosystems where the most severe alteration of fire regimes has occurred. In the north, these are primarily woodlands of *Pinus ponderosa* and *Pseudotsuga menziesii,* and in the south they are *Nothofagus* spp. and *Austrocedrus chilensis.* The recent history of anthropogenic fire is generally similar for ecosystems of the Rocky Mountain region and northern Patagonia in southwestern Argentina. Fire history in these two areas can be generalized into three periods (Veblen and Lorenz 1991; Veblen et al. 1992b): (1) the Native American Period, (2) the Euro-American Settlement Period, and (3) the Modern Land Use Period.

4.3.1 Native American Period

The impact of Native American burning practices on natural fire regimes is controversial and requires critical evaluation. The question is how much did Native American burning exceed the natural fire regime. Support for widespread burning is suggested by ethnographic observations in grassland and forests across North America (Pyne 1982; Boyd 1999), and the reasons for burning are as varied as the ecosystems themselves: early accounts describe Native Americans setting fires to open the landscape for travel, hunting, seed harvest, farming, and insect and berry gathering. Some have claimed that every acre of the West was burned by Native American-set fires at some point, although the assertion is highly speculative and not supported adequately by data (Vale 1998). Modern and historic documents, however, indicate that the Selway Indians of the northern Rocky Mountains regularly burned low-elevation forest to clear understory vegetation to improve hunting and berry gathering (Leiberg 1900). Barrett and Arno (1982) cite dendroclimatologic evidence to suggest that fire frequency was highest near settlements and that more distant locations in the same vegetation type had lower fire return intervals. The Kalapuya of western Oregon enlisted burning in a highly patterned way that affected various areas at regular times (Boyd 1986). These fires were used for circle hunts of deer and gathering of tarweed. In Yosemite National Park studies suggest that indigenous people primarily burned valley bottom areas and had little impact on high elevation forests in the area (Vale 1998).

Many of the early descriptions of Native American burning date to periods when indigenous populations were being decimated by disease or removed from native territory, and the use of fire may not be representative of prehistoric practices. Annual burning by the Kalapuya Indians of western Oregon, for example, is credited for the existence of the present-day oak savanna (Johannessen et al. 1971; Boyd 1986). However, observations of burning come from journals written after 1825; surprisingly, no mention of this practice can be found during the first 12 years of Euro-American contact (Whitlock and Knox 2002). Most accounts of burning were written after 1850, when Euro-Americans were actively engaged in forest clearing and agriculture through deliberate burning. Moreover, pollen records indicate that this biome was present for the last 10,000 years in western Oregon, and it has expanded and contracted in a predictable way with climatic variations in the Holocene (Whitlock and Knox 2002). Native Americans may have acted as an ignition source in prehistoric times, but the vegetation and fire were driven by variations in the intensity of summer drought that are known from independent lines of evidence (Whitlock 1992; Thompson et al. 1993).

The occurrence of fires in the US has dropped sharply in the last few centuries with Euro-American settlement. This decrease in fire frequency can be seen in lake-sediment records that show less charcoal in the last 100 years than before (e.g., Clark and Royall 1995; Millspaugh and Whitlock 1995), dendrochronologic data, and historic documents (Arno 1980; Barrett et al. 1991; Taylor and Halpern 1991; Agee 1993; Swetnam 1993; Skinner and Chang 1996; Veblen et al. 2000). The expansion of late-successional forests in the western US is generally attributed to the elimination of Native American burning and early settlement burning, changes in fire regime with grazing, increasingly effective efforts to suppress fires since World War II, and changes in climate in the twentieth century (Fig. 4.3). In the southwest, years of high fire occurrence in the American Southwest correspond with the high phase of the Southern Oscillation when springs are anomalously dry (Swetnam and Betancourt 1990). The area burned in the last century during the high phase has been smaller than in previous centuries, as a result of fire suppression. Of concern now is the fact that ecosystems that are maintained by frequent fires naturally may be subjected to severe stand-replacement burns in the future with increasing drought and fuel accumulation (Agee 1994; Jensen and Bourgeron 1994). How far the current fire regime exceeds the variability of natural disturbance is a topic of debate, and the need for more information on fire history is clear.

During the Native American Period (i.e., previous to permanent Euro-American settlement), hunters frequently set fires to drive game and/or to improve forage to attract game animals (Barrett and Arno 1982; Boyd 1986; Covington and Moore 1994; Veblen and Lorenz 1988, 1991; Veblen et al. 1992b). There is considerable debate over the temporal and spatial scale at which this occurred, but at least in valley bottoms and other landscapes conducive to human occupancy there is growing evidence of a significant anthropogenic signal. This pattern of enhanced ignitions was most typical of habitats transitional between forests and grasslands (e.g., *Pinus ponderosa* woodlands in North America and along the ecotone between Andean forests and the Patagonian steppe in South America; Veblen and Lorenz 1988; Markgraf and Anderson 1994). Thus, for many areas, intentionally set fires in combination with lightning-ignited fires killed tree seedlings and maintained open, park-like stands of trees at sites that otherwise would support denser forests. In the Colorado Front Range, for example, during the Native American Period composite mean fire return intervals (sensu Dieterich 1980; Romme 1980) in open *Pinus ponderosa* stands near the ecotone with the Great Plains

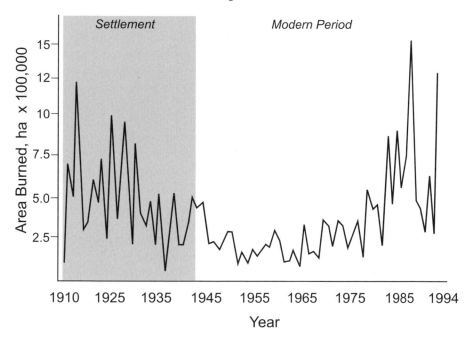

Fig. 4.3. Average area burned annually in the western United States, 1916–1994. Data compiled by Interagency Fire Center, Boise, Idaho, and the US Forest Service. (Adapted from Hardy and Arno 1996). Note that data prior to 1931 do not include National Park or Indian reservation lands; data prior to 1926 only include forested land so actual burn area is underestimated in this portion of the record

grassland were as short as 7 years at a spatial scale of ca. 2 km² (Veblen et al. 1996). In western Montana, this resulted in the establishment of extensive grasslands or steppe vegetation and open park-like *Pinus ponderosa* with little hiding cover or vegetation structure to support ungulates or canopy bird species (Gruell 1983).

Although most fires in northern Patagonia are set by humans today, lightning-ignited fires also occur. For example, in the Argentine Andes for the period 1938–1982 between ca. 39° and 43°30'S, of the fires with known origins recorded by National Park authorities, more than 15 % were started by lightning strikes (Bruno and Martin 1982). In southern Patagonia, lightning storms infrequently occur on Tierra del Fuego (A. Rebertus, pers. observ.). Although the incidence of naturally ignited fires in Patagonia is believed to be low, it appears to have been great enough to have played a selective role in the evolution of the flora (e.g., post-fire sprouting of many species, thick bark of *Araucaria*, *Fitzroya* and *N. glauca*) and also to have affected the structures of *Nothofagus* forests (Veblen et al. 1995, 1996).

The long-term impact of aboriginal burning is unclear. Charcoal particles in sedimentary samples from Neuquén (ca. 39°S) to Tierra del Fuego, document the antiquity of fire in Patagonia (Heusser 1987; Heusser et al. 1988; Markgraf and Anderson 1994). For example, charcoal has been dated at earlier than 10,000 years B.P. for several sites in southern Patagonia as far south as Puerto Williams south of Tierra del Fuego (Heusser 1987; Markgraf and Anderson 1994). Charcoal is found in association with pollen assemblages indicating xeric, open vegetation, and the highest levels are often associated with times of climatic variability (Markgraf and Anderson 1994). However, there is no doubt that with the arrival of aboriginal humans the frequency of burning in Patagonian forests increased. In northern Patagonia early explorers describe Indians using fires to hunt rheas and guanacos in the steppe adjacent to forests (Veblen and Lorenz 1988). In the Chilean archipelago south of 47°S the Alacalufe native people used to burn the conifer *Pilgerodendron* for fuelwood reserves. In Chile, on-going research is indicating that *Fitzroya* forests in the coastal range have been affected by fire during the last 500 years. The earliest fire dated from a fire scar occurred in 1397 (Lara, unpubl. data). Nevertheless, it is not clear whether these fires were natural or set by humans.

4.3.2 The Euro-American Settlement Phase

During the Euro-American settlement phase (e.g., mid-nineteenth century for the southern Andes and most of western North America), ranchers and prospectors burned vast areas of forest from temperate rain forests to relatively xeric forests. For example, over most of the Colorado Rocky Mountains prospectors burned large areas of coniferous forest during the 1860s–1900 (Veblen and Lorenz 1991). In the northwestern states and adjacent Canada, particularly extensive fires occurred in 1910. Smoke generated from these fires was carried high enough in the atmosphere to have broad climatic consequences. The combination of drought and logging practices appear to have worked synergistically to cause these massive fires (Morris 1934; Agee 1993). Interestingly, this exceptional fire year appears to have played a pivotal role changing human perceptions of fire in these ecosystems and led to the establishment of strict fire suppression policies (e.g., Agee 1993, Fig. 4.3). In the Ponderosa pine zone of the Front Range, mean fire return intervals computed for three major watersheds for the Euro-American settlement period (1859–1916) are consistently ca. 50% shorter than during the previous 57-year segment of the Native American period (Veblen et al. 2000).

Similarly, in south-central Chile as well as the southern Andes of Argentina, large areas of temperate rain forest were burned in failed attempts to increase the area of habitat suitable for cattle ranching (Veblen and Lorenz 1988). This is reflected by a major increase in the frequency of fires in the mesic forest zone of northern Patagonia beginning in the late nineteenth century (Veblen et al. 1992b; Kitzberger et al., unpubl. data). Vast areas of forest were deliberately burned to increase pasture and to permit agriculture (Veblen and

Lorenz 1988). The phase of burning from ca. 1884 to 1920 is reflected in char-coal layers up to 50-cm thick near the surface of sedimentary records and by the abundance of 70–100-year-old even-aged *Nothofagus* forests in northern Patagonia (Veblen and Lorenz 1987; Veblen and Markgraf 1988; Markgraf and Anderson 1994). In south-central Chile vast areas of forests were burned and converted to agriculture and pasture land, leading to one of the fastest rates of resource depletion reported in Latin America.

Due to extensive burning during the late nineteenth century by Euro-American settlers in the Rocky Mountain, south-central Chile (36–39°S) and northern Patagonian regions, the forests are now ca. 100–120-year-old post-fire stands that in some cases have become susceptible to insect pests and fun-gal diseases (Veblen and Lorenz 1991; Donoso et al. 1993). For example, in the Colorado Front Range, vast areas of forests burned during the mining booms of the late nineteenth century, which today is reflected by the abundance of even-aged stands of Douglas fir and lodgepole pine dating from ca. 1900 (Veblen and Lorenz 1986, 1991; Hadley and Veblen 1993). During recent decades the increased severity and regional synchroneity of western spruce budworm outbreaks affecting Douglas fir forests of the southern Rocky Mountains may reflect the existence of relatively homogeneous post-fire forests that originated during the nineteenth century settlement period (Swetnam and Lynch 1993). Similarly, stand-level dieback in the *Nothofagus* forests of the southern Andes has been attributed partially to the existence of even-aged post-fire stands that originated during the Euro-American settle-ment period there (Veblen and Lorenz 1988; Rebertus et al. 1993; Kitzberger et al. 1995).

4.3.3 Modern Land Use Period

During the Modern Land Use Period (i.e., post-1920 in North America and post-1940 in South America), intentional fires have largely been eliminated and the frequency of natural fires has dramatically declined (Fig. 4.3). In some ecosystems, fire frequency during this period has been substantially less than it was during the centuries immediately prior to Euro-American settlement (Arno 1980; Veblen et al. 1992a; Agee 1993; Swetnam 1993). For example, in the Colorado Front Range near the ecotone with the Plains grassland, in Pon-derosa pine woodlands which had a mean fire return interval of ca. 7 years from 1648 to 1916, no fire scars date from 1917 to 1994 (Veblen et al. 2000). One of the consequences of this reduction in fire frequency has been the increased density of trees in areas where previously frequent surface fires pre-vented the development of dense stands (Gruell et al. 1982; Veblen and Lorenz 1988; Covington and Moore 1994). For example, the increased tree density and cover in ponderosa pine forests in the southwestern US have been blamed for many forest management problems such as overstocking of small trees, reduced tree growth, insect outbreaks, decreased decomposition rates, and stagnation of nutrient cycles (Covington and Moore 1994).

Similarly, in northern Patagonia, *Austrocedrus* woodlands have dramatically increased in density and have expanded eastwards into the Patagonian steppe as shown by the comparison of historical (ca. 1900) and modern landscape photographs over a north-to-south area from ca. lat. 39–43°S (Veblen and Lorenz 1988). With the demise of the Indian population following the late 19th century Indian wars, frequent burning of the steppe and adjacent open woodlands for guanaco hunting ceased. With the elimination of fires on the steppe, invasion of grasslands by *Austrocedrus* and other trees continues to the present (Veblen and Lorenz 1988).

In Chile, the area affected by fire and changes in fire frequency in recent decades is probably higher compared to the Native American period, and lower than the Euro-American settlement phase. This contrasts with what has been reported for Argentina where fire suppression has dramatically reduced fire frequency. Fire records indicate that since 1966, despite significant year-to-year fluctuations due to climate variability, there has been a slow increase in area burned, and a more rapid increase in number of fires. This has led to the reduction of the area per fire as a result of fire suppression policies (Soto 1995). Even in North America, where the US government spends from 300–800 million dollars annually on fire suppression, there has been a significant increase in area burned each year (Fig. 4.3). This has been generally interpreted as evidence that increased accumulation of fuels from fire suppression has generally changed fire behavior from ground fires and mixed-replacement to stand-replacement (catastrophic) fires (Arno and Brown 1991).

There have been major changes in cover type associated with land use that have direct implications for fire behavior. For example, forest clearing and conversion to agricultural use has dramatically altered the connectivity of native forests so that the behavior and size of fires are likely to differ from those expected in a less fragmented landscape (Turner and Romme 1994). Similarly, the fragmentation of forested landscapes potentially can alter ecosystem response to fire by changing the accessibility of biota to post-fire sites. For example, the size of continuous patches of a particular vegetation type is expected to alter the animal communities, which potentially can influence the course of plant succession following fire. Furthermore, the introduction of naturalized exotic plant species has fundamentally altered the nature of post-fire vegetation changes, which in turn can influence flammability, water use efficiency and other aspects of ecosystem function. For example, in south-central Chile extensive areas of native *Nothofagus* forests and some agricultural land has been converted to *Pinus radiata* plantations. This has led to greater landscape homogeneity and susceptibility to fire. A similar pattern has been documented in the Oregon Coast Range (Teesma et al. 1991; Wimberly et al. 2000; Wallin et al. 2002).

Logging practices are another major impact of people on landscapes which affects the relation of landscapes to fire. Throughout the inland West, high-grading (selective cutting of the largest, healthiest trees in a forest patch) as well as extensive clear-cutting has exacerbated the effects of fire exclusion. Fire exclusion has led to the invasion of *Pinus* and *Pseudostuga* forests by *Abies* spp. and *Pseudotsuga* (Gruell 1983; Hardy and Arno 1996). These dense multilayered forests have much higher leaf areas and nutrient accumulation than the open fire maintained forests. These alterations of nutrient and water balance have been proposed to make these forests very susceptible to invasion by root pathogens, insects and other vectors of forest decline and mortality (e.g., Waring and Pitman 1985). High-grading of these stands has resulted in forests which are mostly composed of these short-lived shade-tolerant invasive trees, precluding the option of a restoration thinning treatment (since there are no old-growth *Pinus* dominants left).

4.4 Effects of Fire on Landscape Patterns

Although much has been learned of the general relationships of climate to fire behavior and the adaptations of ecosystems to specific fire regimes, translating this information to the landscape scale has often been quite problematic (Lertzman and Fall 1998). At the landscape scale other processes become important which are difficult to assess at the stand level. First, there is often a close interaction between ecosystem types or between landscape elements depending on slope, aspect and overall landscape context. High frequency fires in grasslands, for example, lead to higher fire frequencies and damage in upslope forests. Highly localized microclimates, such as cold air drainage, winds coming off lakes and basins with inversions can influence both fire intensity and rate of spread. The nature of ecosystem boundaries or edges also plays a key role in fire movement and behavior across landscapes. Open abrupt edges, for example, allow for much greater penetration of dry open climatic influences on forests than do well-sealed or gradual edges (Forman 1995). Flows of air masses and moisture can also be influenced by landscape structure, in particular the spatial structure of grassland or recently logged areas and forests, the existence of inherent (geomorphic or hydrologically controlled) or induced (artifacts of disturbance events) corridors. A key issue in landscape patterns of fire is understanding interactions between burns, fuels and successional development. Climatic anomalies appear to play an important role in regulating landscape pattern (Turner et al. 1994). Recent advances in remote sensing and a better defined conceptual foundation for characterizing landscapes should provide major advances in this field when better integrated and applied to a broad range of natural and anthropogenically influenced landscapes (e.g., Forman 1995; Kitzberger and Veblen 1999).

A lot of attention has been paid to landscape-scale problems by pooling stand-level data together to characterize some aspects of spatial patterns and by constructing mechanistic models of fire behavior. This research has provided many insights on the sensitivity of landscapes to subtleties in landscape structure and microclimate as well as the extreme sensitivity of fire behavior to fuel structure and moisture. It is really beyond the scope of this review to detail the state of research in spatial modeling of fire. However, it is becoming clear from this research that parameterization of these models to apply to real landscapes is limited by the quality and spatial distribution of empirical data on landscape structure and fuel distribution and moisture. This is particularly true when forests are a product of variable intensity burns. The sharp decline in fire frequency during most of the twentieth century period of fire suppression has important consequences for fuel accumulation and fire hazard that need to be considered in modeling potential landscape consequences of changes in fire regimes (Baker 1992).

Most remote imagery, for example, has limited abilities to characterize forest stand structure. Since fire is extremely sensitive to stand structure (fuel loading, ladder fuels, density, etc.), major advances in realism in portraying landscape dynamics will occur if this can be done more expeditiously through remote sensing. Some forms of multispectral radar, such as spherical aperture radar (SAR) or light detection and ranging (LIDAR) using a laser transmitter, may be able to provide some of this kind of information in the future (Dobson et al. 1995).

4.5 Summary

Extensive research on fire ecology has been ongoing for many decades in both hemispheres. While many advances have been made in understanding the role of fire in ecosystems and general patterns between ecosystem structure and dynamics and fire, there remain many challenges. Clearly, fire plays an overwhelmingly fundamental role in the ecology of most temperate forests. Precise and accurate knowledge of disturbance ecology in general, and fire ecology in particular, and its determinants will be vital in developing strategies for the restoration and management of fire-dependent ecosystems throughout the temperate zone (Hardy and Arno 1996).

Anthropogenic changes in landscape structure and dynamics have had profound implications in changing the relationship of fire to the functioning of forests both in North and South America. The efficiency and tremendous investment in fire control by the United States Forest Service, for example, has resulted in a fundamental transformation of the relationship of fire to landscapes throughout the western United States. Solving the imbalance in the relationship of stand structure, landscape structure and the role of fire to stand dynamics and nutrient cycling will likely require a commitment equal in magnitude to the original government fire control program (Arno and Brown 1991).

Detailed work on historic and contemporary fire regimes both in terms of empirical studies, experiments, and modeling studies are needed to better understand how climatic controls influence fire regimes and their ecological effects. This work has the potential of providing a solid foundation for evaluating policy options and for developing more cost-effective strategies for landscape scale restoration.

4.6 Research Needs

Under the sponsorship of Americas Interhemispheric Geobiophysical Organization (AMIGO), the temperate ecosystems of the western North and South America have been chosen for comparative studies of the interactions of fire and global change because of the comparability of the climates and ecosystems of these regions. These regions include areas that are broadly similar in climate and vegetation physiognomy, but differ in other ways that may provide insights into understanding the sources of variability in fire regimes and their ecological consequences. For example, there are important differences in the timing of human settlement and the application of different land management practices. Recent impacts of livestock and other large introduced herbivores have been substantially greater in the forests of the south than in the north. Similarly, there are important differences in the biota of the regions reflecting distinct evolutionary history. For example, the much greater abundance and dominance of conifers in the north has potentially important implications both in recording fires and in response to fire. Comparative studies can effectively exploit the important differences in settlement history, land use, and resource technologies between temperate South and North America to disentangle the influences of climatic variation and human activities on fire regimes.

To understand an ecosystem's response to wildfires requires information on the natural variability of fire under different vegetational and climatic conditions as they have developed over long-term periods. The sources of these records of fire are tree rings (covering the past ca. 500–1000 years), and sedimentary charcoal covering longer time spans. Generally, these types of data are more abundant for North than for South America, which reflects both differences in numbers of researchers working in the two regions as well as possible technical limitations. For example, most tree-ring records of fire are derived mainly from conifers, and in southern South America conifers have a relatively restricted distribution (Veblen et al. 1995). In southwestern Argentina the principal species exploited for fire history has been the conifer

Austrocedrus. The study of fire regimes and vegetation responses in *Austroce-drus* forests in Chile as well as *Fitzroya* and *Araucaria* forests in Chile and Argentina is incipient and needs to be pursued further.

Studies of long-term vegetation change based on fossil pollen are available for the temperate ecosystems of both North and South America, but studies which focus on changes in fire frequency based on charcoal data are comparatively scarce for both regions (Mehringer et al. 1977; Smith and Anderson 1992; Millspaugh and Whitlock 1995; Long et al. 1998; Mohr et al. 2000; Millspaugh et al. 2000;).

An important objective of future research on fire and climate in the western Americas should be a comparison of the significance to fire regimes of broad-scale climatic episodes over the past 1000 years. For example, a time period of particular concern is the Medieval Warm Period (ca. 900–1300 A.D.) and Little Ice Age (ca. 1650–1890 A.D.), which have been documented in both western North and South America (Luckman and Villalaba 2002). During the Little Ice Age, for example, paleoenvironmental records suggest the Pacific Northwest was colder by 1 °C and more humid than today (Graumlich and Brubaker 1986). Comparative studies of fire history across a range of biome types for North and South America would be extremely useful in evaluating the impact of these late-Holocene periods of climatic variation on vegetation patterns. Furthermore, by selecting study areas that have a common pattern of regional climatic variation, but differ in the timing and intensity of human activities, it should be possible to disentangle the influences of climate and humans on low frequency variation in fire regimes.

Some of the broad questions which should define the future research agenda on fire and global change are:

1. What is the relationship of fire to climate variation across a range of biomes and at a range of temporal scales from seasonal to millennial? How are climate-induced changes in fire regimes linked to broad-scale atmospheric circulation patterns and mechanisms?
2. How have fire regimes been altered by land-use practices by Native Americans, fire suppression, logging, landscape fragmentation, exotic invasions, and conversion to forest plantations?
3. What is the role of landscape heterogeneity in influencing how fire regimes respond to climate variation and human impacts?
4. What is the role of scale in evaluating fire-mediated processes in forest landscapes?

For example, general circulation models (GCM) characterize climate change at a very large coarse scale (thousands of km^2).

Most ecological studies have been conducted at extremely small scales (<1 ha). Scaling up empirical field-oriented studies or scaling down remote imagery or GCM simulations is critical to studying linkages over varying spatial and temporal scales over a broad array of possible forcing functions and feedback loops. Research needs to address what are the appropriate scales for looking at different processes or linkages between climate, fire, and human land use.

An important component of future research comparing fire regimes in North and South America is to explore how tree-ring methods and sedimentary methods of studying fire regimes can be applied in a variety of ecosystem types and physical settings. In particular, studies should be designed to permit the linkage of high resolution tree-ring dating of fires with long-term high resolution sedimentary records of fire. Such retrospective studies should be combined with the monitoring of future changes in ecosystem parameters and processes across a broad range of spatial scales (Stohlgren et al. 1995).

Acknowledgement. The research on which this review is based was supported by grants to T.T.V., C.W., T.K. and R.V. from the National Science Foundation, the National Geographic Society, the National Aeronautics and Space Administration, the National Biological Survey, the City of Boulder, the University of Wyoming-National Park Service Research Center, and the University of Colorado. Support was also provided by grants to AMIGO from NSF and from the USDA Forest Service Pacific Northwest Research Station (PA-PNW-94-0450, CW-PNW-98-5122-ICA), Pacific Southwest Research Station (CW-PSW-99-0010, 95-0022CA) and the University of Montana, to A.L. and R.V. from FONDECYT (Grant 93-049) and to A.L. from the National Geographic Society.

Appendix 4.1: Summary of dendrochronological fire history studies of the western United States and Canada, arranged by forest cover type

State/province	Forest type	Eleva-tion	MFI	Fire regime[a]	Fire size	Reference
Utah	*A. concolor*	M		6	Medium	Bradley et al. (1992b)
California-E	*A. concolor*	M	10–60	7	Medium	Agee (1993)
Montana-W	*A. grandis*	M		11	Medium	Fischer and Clayton (1983); Fischer and Bradley (1987)
California-E	*A. magnifica*	H	10–50	7	Medium	Taylor and Halpern (1991); Taylor (1994)
Washington-W	*Juniperus*	L	25	1	Small	
Oregon-E	*Juniperus*	L	25	1	Small	Agee (1993)
Canada-NWT	*P. banksiana*	L	100/25	4/3	Large	Rowe et al. (1974)
Utah	*P. engelmanii*	H		8–12	Medium	Bradley et al. (1992a)
Alaska, Yukon	*P. mariana–P.glauca*	L	50–200	4	Very Large	Barney (1971); Viereck (1973); Johnson and Strang (1982)
Montana-W	*P. menziesii*	M	10–150	4–6	Medium	Gabriel (1976); Davis (1980); Davis et al. (1980); Antos and Habeck (1981)
California-I	*P. menziesii*	M	10–50	4	Medium	Wills and Stuart (1994)
Utah	*P. menziesii*	M		4–5	Medium	Bradley et al. (1992a)
Washington-C	*P. menziesii*	M	10–50	6	Medium	Agee (1993)
Oregon-W, C	*P. menziesii*	M	10–50	6	Medium	Agee (1993)
Washington-W	*P. menziesii m.*	L	217	11	Small	Fahnestock and Hare (1983); Morrison and Swanson (1990); Agee (1993)
Oregon-W	*P. menziesii m.*	L	150	11	Small	Fahnestock and Agee (1983); Agee (1990)
Utah	*P. pungens*	M		6	Medium	Bradley et al. (1992a)
Washington-W	*P. sitchensis*	L	1146	11	Medium	Fahnestock and Agee (1983)
Oregon-W	*P. sitchensis*	L	200–400	11	Small	Agee (1993)
Canada-NWT	*P.glauca/mariana*	L	100	4	Very large	Rowe et al. (1974)
Canada-BC	*Picea glauca*	M	103	5	Large	Heinselman (1981)
Montana-NC	*Picea glauca*	H		5	Medium	Fischer and Clayton (1983); Fischer and Bradley (1987)
Canada-NWT	*Picea glauca*	L	120	5	Medium	Johnson and Rowe (1977)
Alaska[b]	*Picea glauca fp.*	L	200	6	Small	Barney (1971; Viereck (1973)

State/province	Forest type	Elevation	MFI	Fire regime[a]	Fire size	Reference
Canada-NWT[b]	*Picea glauca fp.*	L	200	5	Small	Rowe et al. (1974)
Alaska	*Picea glauca/lich.*	L	130	5	Large	Barney (1971); Viereck (1973)
Alberta-W	*Picea-Abies*	M	55	3-4	Medium	Tande (1979)
Alberta-W	*Picea-Abies*	H	100	4	Large	Tande (1979)
Utah	*Picea-Abies*	H		8-12	Medium	Bradley et al. (1992b)
Montana-W	*Picea-Abies*	H	70-300/30	5/3	Medium	Arno (1976, 1980); Arno and Peterson (1983)
Yellowstone	*Picea-Abies*	H	150	5	Medium	Loope and Gruell (1973); Romme and Knight (1981); Romme (1982)
Colorado	*Picea-Abies*	M	100?	4	Large	Clements (1910); Veblen (1986); Rebertus et al.(1991)
Alberta-W	*Pinus contorta*	M	25	2-3	Medium	Tande (1979)
Alberta-W	*Pinus contorta*	H	50	3-4	Large	Tande (1979)
Montana-NW	*Pinus contorta*	M	150/125	5/2	Medium	Gabriel (1976)
Montana-NW	*Pinus contorta*	H	150	5	Large	Gabriel (1976)
Montana-W	*Pinus contorta*	M	25	1-4	Large	Arno (1976)
Yellowstone	*Pinus contorta*	L	25	2/3	Medium	Houston (1973)
Yellowstone	*Pinus contorta*	M	200-500	3-4	Large	Loope and Gruell (1973); Romme (1982); Barrett (1994)
Canada-BC	*Pinus contorta*	L	50	4	Large	Smith (1979)
Utah	*Pinus contorta*	H		8	Medium	Bradley et al. (1992b)
Washington-W	*Pinus contorta*	H	110	7	Large	Agee et al. (1990)
Oregon-W	*Pinus contorta*	H	80	7	Large	Agee et al. (1990)
Montana-NC	*Pinus flexilis*	L		1	Small	Fischer and Clayton (1983); Fischer and Bradley (1987)
Utah	*Pinus flexilis*	L		9	Small	Bradley et al. (1992b)
Colorado-FR	*Pinus flexilis*	H		8?	Medium	Rebertus et al. (1991)
Montana-W	*Pinus ponderosa*	L	10-20	2	Small	Arno (1976, 1980, 2002); Barrett and Arno (1982)
Utah	*Pinus ponderosa*	M	5-25	3	Small	Madany and West (1980); Bradley et al. (1992b)
Colorado-C	*Pinus ponderosa*	M	30-65	3	Small	Laven et al. (1990)

State/province	Forest type	Elevation	MFI	Fire regime[a]	Fire size	Reference
Washington-C	Pinus ponderosa	L	11–16	3	Small	Agee (1993)
Oregon-C	Pinus ponderosa	L	4–16	3	Small	Weaver (1959); Agee (1993)
California-E	Pinus ponderosa	M	11	3	Small	Kilgore and Taylor (1979); Warner (1980)
Arizona/NM	Pinus ponderosa	M	2–6	3	Small	Pickett and White (1985); Savage and Swetnam (1990)
Utah	Pinus-Juniperus	L	13–20	1	Small	Kilgore (1980; Bradley et al. (1992b)
Colorado	Populus tremuloides	M	100	3	Large	Heinselman (1981)
Utah/Wyoming	Populus tremuloides	M	50	7	Medium	Loope and Gruell (1973); Bradley et al. (1992b)
Utah	Quercus	L		2	Small	Bradley et al. (1992b)
California-I	S. sempivirens	L	10–50	4	Small	Agee (1993)
California-W	S. sempivirens	L	5–10	3	Small	Veirs (1982); Swetnam (1993); Brown and Swetnam (1994)
California-E	S. gigantea	M	100–500 2–30	6	Medium	Kilgore and Taylor (1979); Swetnam et al. (1991); Swetnam (1993); Caprio and Swetnam (1995)
BC and Washington-W	T. mertensiana	H	800–1500	11	Small	Fahnestock and Hare (1983); Lertzman and Krebs (1991)
Montana-W	Thuja-Tsuga	M	50–1000	11	Small	Habeck (1977, 1978); Arno and Davis (1980); Fischer and Clayton (1983); Fischer and Bradley (1987)
Washington-W	Thuja-Tsuga	L	937	11	Small	Fahnestock and Agee (1983)
Oregon-W	Thuja-Tsuga	L	400	11	Small	Fahnestock and Agee (1983)

[a] Using the expanded system of Bradley et al. (1992a).
[b] River floodplain sites.

References

Agee JK (1990) The historical role of fire in Pacific Northwest forest. Natural and pre-scribed fire in Pacific Northwest forests. In: Walstad JD, Radosevich S, Sandberg D (eds) Oregon State University Press, Corvallis, OR pp 25–38

Agee JK (1993) Fire ecology of Pacific Northwest forests. Island Press, Washington, DC

Agee JK (1994) Fire and weather disturbances in terrestrial ecosystems of the eastern Cascades. General Technical Report PNW-GTR-320. USDA Forest Service, Pacific Northwest Res Stn, Portland, OR

Alaback P (1991) Comparative ecology of temperate rainforests of the Americas along analogous climatic gradients. Rev Chil Hist Nat 64:399–412

Alaback PB (1995) Biodiversity patterns in relation to climate in the temperate rain-forests of North America. In: Lawford R, Alaback P, Fuentes ER (eds) High latitude rain forests of the west coast of the Americas: climate, hydrology, ecology and conser-vation. Springer, Berlin Heidelberg New York, pp 105–133

Alaback PB, McClellan M (1993) Effects of global warming on managed coastal ecosys-tems of western North America. In: Mooney HA, Fuentes ER, Kronberg BI (eds) Earth system response to global change. Contrasts between North and South America. Aca-demic Press, New York, pp 299–327

Almeyda AE, Saez SF (1958) Recopilación de datos climáticos de Chile y mapas sinópti-cos respectivos. Ministerio de Agricultura, Santiago, Chile

Antos JA, Habeck JR (1981) Successional development in *Abies grandis* (Dougl.) Forbes forests in the Swan Valley, western Montana. Northwest Sci 55:26–39

Arno SF (1976) The historical role of fire on the Bitterroot National Forest. Research Paper INT-187). USDA Forest Service, Intermountain Res Stn, Ogden, UT

Arno SF (1980) Forest fire history in the northern Rockies. J For 78:460–465

Arno SF (2002) Structure of old-growth Ponderosa pine stands on dry and moist sites as influenced by past fires. Can J For Res (in press)

Arno SF, Brown JK (1991) Overcoming the paradox in managing wildland fire. West Wildlands 17:40–46

Arno SF, Davis DH (1980) Fire history of western redcedar/hemlock forests in northern Idaho. In: Stokes MA, Dieterich JH (eds) Proceedings of the fire history workshop, October 20–24, 1980, Tucson, AZ. Gen Tech Rep GTR-RM-81. USDA Forest Service, Rocky Mountain Forest and Range Exp Sta, Ft. Collins, CO, pp 21–26

Arno SF, Reinhardt ED, Scott JH (1993) Forest structure and landscape patterns in the subalpine lodgepole pine type: a procedure for quantifying past and present condi-tions. Gen Tech Rep No INT-294). USDA Forest Service, Intermountain Res Stn, Ogden, UT

Arroyo MTK, Riveros M, Peñaloza A, Cavieres L, Faggi AM (1995) Phytogeographic rela-tionships and regional richness patterns of the cool temperate rainforest flora of Southern South America. In: Lawford R, Alaback P, Fuentes ER (eds) High latitude rain forests of the west coast of the Americas: climate, hydrology, ecology and conser-vation. Springer, Berlin Heidelberg New York, pp 134–172

Baisan CH, Swetnam TW (1990) Fire history on a desert mountain range: Rincon Moun-tain Wilderness, USA. Can J For Res 20:1559–1569

Baker WL (1990) Climatic and hydrologic effects on the regeneration of *Populus angus-tifolia* James along the Animas River, Colorado. J Biogeogr 17:59–73

Baker WL, Egbert SL, Frazier GF (1991) A spatial model for studying the effects of cli-matic change on the structure of landscapes subject to large disturbances. Ecol Mod-eling 56:109–125

Balling RC Jr, Meyer GA, Wells SG (1992) Climate change in Yellowstone National Park: is the drought related risk of wildfires increasing? Climatic Change 22:34–35

Barney RJ (1971) Wildfires in Alaska–some historical and projected effects and aspects. In: Fire in the northern environment–a symposium. USDA Forest Service, Pacific Northwest Res Stn, Portland, OR, pp 51–59

Barrett SW (1994) Fire regimes on andesitic mountain terrain in Yellowstone National Park, Wyoming. Int J Wildland Fire 4:65–76

Barrett SW, Arno SF (1982) Indian fires as an ecological influence in the northern Rockies. J For 80:647–651

Barrett SW, Arno SF, Key CH (1991) Fire regimes of western larch-lodgepole pine forests in Glacier National Park, Montana. Can J For Res 21:1711–1720

Barry RG (1973) A climatological transect on the east slope of the Front Range, Colorado. Arctic Alpine Res 5:89–110

Bergeron Y, Archambault S (1993) Decreasing frequency of forest fires in the southern boreal zone of Quebec and its relation to global warming since the end of the 'Little Ice Age'. Holocene 3:255–259

Boerner REJ (1982) Fire and nutrient cycling in temperate ecosystems. BioScience 32:187–192

Bonnicksen TM, Stone EC (1982) Reconstruction of a presettlement giant sequoia-mixed conifer forest community using the aggregation approach. Ecology 63:1134–1148

Boyd R (1986) Strategies of Indian burning in the Willamette Valley. Can J Anthropol 5:65–86

Boyd R (ed) (1999) Indians, fire and the land. Oregon State Univ Press, Corvallis, OR

Bradley AF, Fischer WC, Noste NV (1992a) Fire ecology of the forest habitat types of eastern Idaho and western Wyoming. Gen Tech Rep No INT-290. USDA Forest Service, Intermountain Res Stn, Ogden, UT

Bradley AF, Noste N, Fischer WC (1992b) Fire ecology of forests and woodlands in Utah. Gen Tech Rep No INT-287. USDA Forest Service, Intermountain Res Stn, Odgen, UT

Brown JK, Arno SF, Barrett SW, Menakis JP (1994) Comparing the prescribed natural fire program with presettlement fires in the Selway-Bitterroot wilderness. Int J Wildland Fire 4:157–168

Brown PM, Swetnam TW (1994) A crossdated fire history in a coast redwood forest near redwood National Park, California. Can J For Res 24:21–31

Bruno J, Martin G (1982) Los incendios forestales en los Parques Nacionales. Administración de Parques Nacionales, Buenos Aires

Callaway RM, Delucia EH, Schlesinger WH (1994) Compensatory responses of CO_2 exchange and biomass allocation and their effects on the relative growth rate of ponderosa pine in different CO_2 and temperature regimes. Oecologia 98:159–166

Caprio AC, Swetnam TW (1995) Historic fire regimes along an elevational gradient on the west slope of the Sierra Nevada, California. In: Brown JK, Mutch RW, Spoon CW, Wakimoto RH (eds) Proceedings: symposium on fire in wilderness and park management, March 30-April 1, 1993. Gen Tech Rep INT-GTR-320. USDA Forest Service, Intermountain Res Stn, Missoula, MT, pp 173–179

Chandler C, Cheney P, Thomas P, Trabaud L, Williams D (1983). Forest fire behavior and effects. Fire in forestry, vol I. Wiley, New York

Clark JS (1989) Effect of climate change on fire regimes in northwestern Minnesota. Nature 334:233–235

Clark JS (1990) Patterns, causes, and theory of fire occurrence during the last 750 year in northwestern Minnesota. Ecol Monogr 60:135–169

Clark JS, Royall PD (1995) Transformation of a northern hardwood forest by aboriginal (Iroquois) fire: charcoal evidence from Crawford lake, Ontario, Canada. Holocene 5:1–9

Cooper CF (1961) Pattern in ponderosa pine forests. Ecology 42:493–499

Covington WW, Moore MM (1994) Southwestern ponderosa forest structure: changes since Euro-American settlement. J For 92:39–47

Daubenmire R (1956) Climate as a determinant of vegetation distribution in eastern Washington and northern Idaho. Ecol Monogr 26:131–154

Davis KM (1980) Fire history of a western larch/Douglas-fir forest type in northwestern Montana. In: Stokes MA, Dieterich JH (eds) Proceedings of the fire history workshop, October 20–24, 1980, Tucson, AZ. Gen Tech Rep GTR-RM-81. USDA Forest Service, Rocky Mountain Forest and Range Exp Stn, Ft Collins, CO, pp 69–74

Davis KM, Clayton BD, Fischer WC (1980) Fire ecology of Lolo National Forest habitat types Gen Tech Rep INT-79. USDA Forest Service, Intermountain Res Stn, Ogden, UT

Davis MB, Botkin DB (1985) Sensitivity of cool-temperate forest and their fossil pollen record to rapid temperature change. Quat Res 23:327–340

Dieterich JH (1980) The composite fire-interval – a tool for more accurate interpretation of fire history. In: Stokes MA, Dieterich JH (eds) Proceedings of the fire history workshop, October 20–24, 1980, Tucson, AZ. Gen Tech Rep GTR-RM-81. USDA Forest Service, Rocky Mountain Forest and Range Exp Stn, Ft Collins, CO, pp 8–14

Dobson MC, Fawwaz FT, Pierce LE et al. (1995) Estimation of forest biophysical characteristics in northern Michigan with SIR-C/X-SAR. IEEE Trans Geosci Remote Sensing 33:877–895

Donoso C, Sandoval V, Grez R, Rodriguez J (1993) Dynamics of *Fitzroya cupressoides* forests in southern Chile. J Vegetation Sci 4:303–312

Emmingham WH, Waring RH (1977) An index of photosynthesis for comparing forest sites in western Oregon. Can J For Res 7:165–174

Fischer WC, Bradley AF (1987) Fire ecology of western Montana forest habitat types. Gen Tech Rep INT-GTR-223. USDA Forest Service, Intermountain Res Stn, Ogden, UT

Fischer WC, Clayton BD (1983) Fire ecology of Montana forest habitat types east of the Continental Divide. Gen Tech Rep INT-141. USDA Forest Service Intermountain Res Stn, Ogden, UT

Forman RTT (1995) Land mosaics. Cambridge University Press, New York

Franklin JF, Dyrness CT (1973) The natural vegetation of Washington and Oregon. Gen Tech Rep PNW-8. USDA Forest Service Pacific Northwest Forest and Range Exp Stn, Portland, OR

Franklin JF, Swanson FJ, Harmon ME et al. (1991) Effects of global climatic change on forests in northwestern North America. Northwest Environ J 7:233–254

Gabriel HW, Tande GF (1983) A regional approach to fire history in Alaska (Tech Rep No. 9). USDI Bureau of Land Management, Anchorage, Alaska

Gabriel WHI (1976) Wilderness ecology: the Danaher Creek drainage, Bob Marshall Wilderness, Montana. PhD Thesis, University of Montana, Missoula, MT

Graumlich LJ, Brubaker LB (1986) Reconstruction of annual temperature (1590–1979) for Longmire, Washington, derived from tree rings. Quat Res 25: 223–234

Greenland D (1989) The climate of Niwot Ridge, Front Range, Colorado. Arctic Alpine Res 21:380–394

Gruell GE (1983) Fire and vegetative trends in the Northern Rockies: interpretations from 1871–1982 photographs. Gen Tech Rep INT-GTR-158. USDA Forest Service, Intermountain Res Stn, Ogden, UT

Gruell GE, Schmidt W, Arno S, Reich W (1982) Seventy years of vegetative change in a managed Ponderosa pine forest in western Montana. Gen Tech Rep INT-130. USDA Forest Service, Intermountain Res Stn, Odgen, UT

Habeck JR (1977) Forest succession in the Glacier Park cedar-hemlock forests. Ecology 49:872–880

Habeck JR (1978) A study of climax western redcedar (Thuja plicata) forest communities in the Selway-Bitterroot wilderness, Idaho. Northwest Sci 52:67–76

Habeck JR, Mutch RW (1973) Fire dependent forests in the Northern Rocky Mountains. Quat Res 3:408–424

Hadley KS, Veblen TT (1993) Stand response to western spruce budworm and Douglas-fir bark beetle outbreaks, Colorado Front Range. Can J For Res 23:479–491

Hallett DJ, Walker RC (2000) Paleoecology and its application to fire and vegetation management in Kootenay National Park, British Columbia. J Paleolimnol 24:401–414

Hardy CC, Arno SF (1996) The use of fire in forest restoration. A general session at the annual meeting of the Society for Ecological Restoration, Seattle, WA, Sept. 14–16, 1995. Gen Tech Rep INT-GTR-341. USDA Forest Service, Intermountain Res Stn, Ogden, UT

Heinselman ML (1981) Fire intensity and frequency as factors in the distribution and structure of northern ecosystems. In: Mooney HA, Bonnicksen TM, Christensen NL, Lotan JE, Reiners WA (eds) Fire regimes and ecosystem properties. Gen Tech Rep WO-26. USDA Forest Service, Washington, DC, pp 7–57

Hemstrom MA, Franklin JF (1982) Fire and other disturbances of the forests in Mount Rainier National Park. Quat Res 18:32–51

Heusser CJ (1987) Fire history of Fuego-Patagonia. Quat S Am Antarct Peninsula 5:93–109

Heusser CJ, Rabassa J, Brandani A, Stuckenrath R (1988) Late-Holocene vegetation of the Andean Araucaria region, Province of Neuquén, Argentina. Mountain Res Dev 8:53–63

Houston DB (1973) Wildfires in northern Yellowstone National Park. Ecology 54:1111–1117

Jensen ME, Bourgeron PS (1994) Volume II: ecosystem management: principles and applications. Gen Tech Rep PNW-GTR-318. USDA Forest Service, Pacific Northwest Res Stn, Portland, OR

Johannessen CL, Davenport WA, Millet A, McWilliams S (1971) The vegetation of the Willamette Valley. Ann Assoc Am Geogr 61:286–302

Johnson AH, Strang RM (1982) Forest fire history in the Central Yukon Canada. For Ecol Manage 4:155–159

Johnson EA, Larsen CPS (1991) Climatically induced change in fire frequency in the southern Canadian Rockies. Ecology 72:194–201

Johnson EA, Wowchuk DR (1993) Wildfires in the southern Canadian Rocky Mountains and their relationship to mid-tropospheric anomalies. Can J For Res 23:1213–1222

Kilgore BM (1980) Fire in ecosystem distribution and structure: western forests and scrublands. In: Mooney HA, Bonnicksen TM, Christensen NL, Lotan JE, Reiners WA (eds) Fire regimes and ecosystem properties. Gen Tech Rep WO-26. USDA Forest Service, Washington, DC, pp 58–59

Kilgore BM, Taylor D (1979) Fire history of a sequoia-mixed conifer forest. Ecology 60:129–142.

Kitzberger T, Veblen TT (1997) Influences of humans and ENSO on fire history of *Austrocedrus chilensis* woodlands in northern Patagonia, Argentina. Ecoscience 4:508–520

Kitzberger T, Veblen TT (1999) Fire-induced changes in northern Patagonian landscapes. Landscape Ecol 14:1–15

Kitzberger T, Veblen TT, Villalba R (1995) Tectonic influences on tree growth in northern Patagonia, Argentina: the roles of substrate stability and climatic variation. Can J For Res 25:1684–1696

Kitzberger T, Veblen TT, Villalba R (1997) Climatic influences on fire regimes along a rainforest-to-xeric woodland gradient in northern Patagonia, Argentina. J. Biogeogr 24:35–47

Knight DH (1994) Mountains and plains: the ecology of Wyoming landscapes. Yale University Press, New Haven, CT

Köppen W (1918) Klassifikation der Klimate Nach Temperatur Niederschlag, und Jahreslauf. Petermann's Mitt 64:193–203

Laven RD, Omi PN, Wyant JG, Pinkerton AS (1980) Interpretation of fire scar data from a ponderosa pine ecosystem in the central Rocky Mountains, Colorado. In: Stokes MA, Dieterich JH (eds) Proceedings of the fire history workshop, 20–24 October 1980, Tucson, AZ. Gen Tech Rep GTR-RM-81. USDA Forest Service, Rocky Mountain Forest and Range Exp Stn, Ft Collins, CO, pp 46–49

Lawford R, Alaback P, Fuentes ER (1995) High latitude rain forests of the west coast of the Americas: climate, hydrology, ecology and conservation. Ecological Studies 116. Analysis and Synthesis. Springer, Berlin Heidelberg New York

Leiberg JB (1900) Bitterroot forest reserve. 20th Annual report, part 5. US Geological Survey, pp 317–410

Lertzman KP, Fall J (1998) From forest stands to landscapes. In: Peterson DL, Parker VT (eds) Ecological scale: theory and applications. Columbia University Press, New York, pp 339–367

Long CJ, Whitlock C, Bartlein PJ, Millspaugh SH (1998) A 9000-year fire history from the Oregon Coast Range, based on a high-resolution charcoal study. Can J For Res 28:774–787

Loope LL, Gruell GE (1973) The ecological role of fire in the Jackson Hole area, northwestern Wyoming. Quat Res 3:425–443

Luckman B, Villalba R (2002) Assessing the synchronicity of glacier fluctuations in the western cordillera of the Americas during the last millennium. In: Markgraf V (ed) Interhemispheric climate linkages: present and past interhemispheric climate linkages in the Americas and their societal effects. Academic Press, New York (in press)

Madany MH, West NE (1980) Fire history of two montane forest areas of Zion National Park. In: Stokes MA, Dieterich JH (eds) Proceedings of the fire history workshop, October 20–24, 1980, Tucson, AZ. Gen Tech Rep GTR-RM-81. USDA Forest Service, Rocky Mountain Forest and Range Exp Stn, Ft Collins, CO, pp 50–62

Malanson GP, Westman WE (1989) Modeling the interactions of fire regime, air pollution, and CO_2-induced climate change on Californian coastal sage scrub. Climate Change :

Markgraf V, Anderson L (1994) Fire history of Patagonia: climate versus human cause. Rev 16 São Paulo 15:35–47

Martin RE (1982) Fire history and its role in succession. In: Means JE (ed) Forest succession and stand development in the Northwest. Oregon State University, Forest Research Laboratory, Corvallis, OR, pp 92–99

Mast JN, Veblen TT, Linhart YB (1998) Disturbance and climatic influences on age structure of ponderosa pine at the pine/grassland ecotone, Colorado Front Range. J Biogeogr 25:743–755

Mehringer PJ, Arno SF, Peterson KL (1977) Postglacial history of Lost Trail Pass Bog, Bitterroot Mountains, Montana. Arctic Alpine Res 9:345–368

Meyer GA, Wells SG, Jull AJT (1995) Fire and alluvial chronology in Yellowstone National Park: climatic and intrinsic controls on Holocene geomorphic processes. Geol Soc Am Bull 107:1211–1230

Millspaugh SH, Whitlock C (1995) A 750-year history based on lake sediment records in central Yellowstone National Park, USA. Holocene 5:283–292

Millspaugh SH, Whitlock C (2002) Postglacial fire, vegetation, and climate history of the Yellowstone-Lamar and Central Plateau provinces, Yellowstone National Park. In: Wallace L (ed) After the fires: the ecology of change in Yellowstone National Park. Yale University Press, New Haven, CT (in press)

Millspaugh SH, Whitlock C, Bartlein PJ (2000) Variations in fire frequency and climate over the last 17,000 year in Yellowstone National Park. Geology 28:211–214

Mitchell VL (1976) The regionalization of climate in the western United States. J Appl Meteorol 15:920–927

Mohr JA, Whitlock C, Skinner CN (2000) Postglacial vegetation and fire history, eastern Klamath Mountains, California, USA. Holocene 10:587–601

Mooney HA (1977) Convergent evolution in Chile and California. Mediterranean climate ecosystems. US/IBP Synthesis Series 5. Dowden, Hutchinson and Ross, New York

Mooney HA, Fuentes ER, Kronberg BI (eds) (1993) Earth system response to global change: contrasts between North and South America. Academic Press, New York

Morris WG (1934) Forest fires in western Oregon and western Washington. Oregon Hist Q 35:313–339

Morrison P, Swanson FJ (1990) Fire history and pattern in a Cascade Range landscape. Gen Tech Rep PNW-GTR-254. USDA Forest Service, Pacific Northwest Res Stn, Portland, OR

Neilson RP (1987) On the interface between current ecological studies and the paleobotany of pinyon-juniper woodlands. In: Proceedings, Pinyon-juniper conference, Reno, NV, 13–16 January 1986. Gen Tech Rep INT-215. USDA Forest Service, Intermountain Res Stn, Ogden, UT, pp 93–98

Noble DL, Alexander RR (1977) Environmental factors affecting natural regeneration of Engelmann spruce in the central Rocky Mountains. For Sci 23:420–429

O'Neill RV (1988) Hierarchy theory and global change. In: Rosswall RGW a PGRT (ed) Scales and global change. Wiley, New York, pp 29–45

Overpeck JT, Rind D, Goldberg R (1990) Climate-induced changes in forest disturbance and vegetation. Nature 343:51–53

Paruelo JM, Jobbagy EG, Sala OE; Lauenroth WK, Burke IC (1988) Functional and structural convergence of temperate grassland and shrubland ecosystems. Ecol Appl 8:194–206

Paruelo JM, Lauenroth WK, Epstein HE et al. (1995) Regional climatic similarities in the temperate zones of North and South America. J Biogeogr 22:915–925

Peet RK (1981) Forest composition of the Colorado Front Range: composition and dynamics. Vegetatio 45:3–75

Pfister RD, Kovalchik BL, Arno SF, Presby RC (1977) Forest habitat types of Montana. Gen Tech Rep INT-34. USDA Forest Service, Intermountain Forest and Range Exp Stn, Ogden, UT

Pickett STA, White PS (1985) The ecology of natural disturbance and patch dynamics. Academic Press, New York

Potter LD, Green DL (1960) Ecology of a northeastern outlying stand of *Pinus flexilis*. Ecology 45:866–868

Premoli AC (1994) Genetic, morphological, and ecophysiological variation in geographically restricted and widespread species of *Nothofagus* from southern South America. PhD Thesis, University of Colorado, Boulder, CO

Prentice C (1992) Climate and long-term vegetation dynamics. In: Glenn-Lewin DC, Peet RA, Veblen TT (eds) Plant succession: theory and prediction. Chapman and Hall, New York, pp 293–339

Price AJ, Rind D (1994) The impact of 2 X CO_2 climate on lightning-caused fires. J Climate 7:1484–1494

Pyne SJ (1982) Fire in America: a cultural history of wildland and rural fire. Princeton University Press, Princeton, NJ

Rebertus AJ, Burns BR, Veblen TT (1991) Stand dynamics of *Pinus flexilis* dominated subalpine forests in the Colorado Front Range. J Vegetation Sci 2:445–458

Rebertus AJ, Veblen TT, Kitzberger T (1993) Gap formation and dieback in Fuego-Patagonian *Nothofagus* forests. Phytocoenologia 23:581–599

Risser PG (1987) Landscape ecology: state of the art. In: Turner M (ed) Landscape heterogeneity and disturbance. Springer, Berlin Heidelberg New York, pp 3–14

Romme WH (1980) Fire history terminology: report of the Ad Hoc Committee. In: Stokes MA, Dieterich JH (eds) Proceedings of the fire history workshop, October 20–24, 1980, Tucson, AZ. Gen Tech Rep GTR-RM-81. USDA Forest Service, Rocky Mountain Forest and Range Exp Stn, Ft Collins, CO, pp 135–137

Romme WH (1982) Fire and landscape diversity in the subalpine forests of Yellowstone National Park. Ecol Monogr 52:199–221

Romme WH, Despain DG (1989) Historical perspective on the Yellowstone fires of 1988. Bioscience 39:695–698

Romme WH, Knight DH (1981) Fire frequency and subalpine forest succession along a topographic gradient in Wyoming. Ecology 62:319–26

Romme WH, Turner MG (1991) Implications of global climate change for biogeographic patterns in the Greater Yellowstone Ecosystem. Conserv Biol 5:373–386

Rothermel RC (1972) A mathematical model for predicting fire spread in wildland fuels. Res Pap INT-115) USDA Forest Service, Intermountain Res Stn, Ogden, UT

Rowe JS, Bergsteinsson JL, Padbury GA, Hermesh R (1974) Fire studies in the Mackenzie Valley. INA Publ. No. QS-1567-000-EE-AL. ALUR 73–74–61, Canada Dept Indian and Northern Affairs

Savage M, Swetnam TW (1990) Early nineteenth-century fire decline following sheep pasturing in a Navajo ponderosa pine forest. Ecology 71:2374–2378

Sirois L, Payette S (1991) Reduced postfire tree regeneration along a boreal forest-tundra transect in northern Quebec. Ecology 72:619–629

Skinner CN, Chang C (1996) Fire regimes: past and present. In: Sierra Nevada ecosystem project: final report to congress, vol. II. Assessments and scientific basis for management options. University of California, Davis, Centers for Water and Wildland Resources, Davis, CA, pp 1041–1069

Smith SJ, Anderson RS (1992) Late Wisconsin paleoecologic record from Swamp Lake, California. Quat Res 38:91–102

Soto L (1995) Estadisticas de ocurrencia y daño de incendios forestales temporadas 1964 a 1995. Informes Estadistico N0 44. Gerencia Tecnica. Departamento de Manejo del Fuego, Corporacion Nacional Forestal (CONAF), Santiago

Spies TA, Franklin JF (1988) Old growth and forest dynamics in the Douglas-fir region of western Oregon and Washington. Nat Areas J 8:190–201

Stephenson NL (1990) Climatic control of vegetation distribution: the role of the water balance. Am Nat 135:649–670

Stohlgren TJ, Binkley D, Veblen TT, Baker WL (1995) Attributes of reliable long-term landscape-scale studies: malpractice insurance for landscape ecologists. Environ Monitoring Assess 36:1–25

Swanson FJ, Franklin JF (1992) New forestry principles from ecosystem analysis of Pacific Northwest forests. Ecol Appl 2:262–274

Swetnam TW (1993) Fire history and climate change in Giant Sequoia groves. Science 262:885–889

Swetnam TW, Betancourt JL (1990) Fire-Southern Oscillation relations in the southwestern United States. Science 249:1017–1020

Swetnam TW, Lynch AM (1993) Multi-century, regional-scale patterns of western spruce budworm history. Ecol Monogr 63:399–424

Swetnam TW, Touchan R, Baisan CH, Caprio AC, Brown PM (1991) Giant sequoia fire history in the Mariposa Grove, Yosemite National Park. In: Yosemite Centennial Symposium: prospects for the future. Yosemite National Park, October 13–20th, 1990, NPS D-374. National Parks Service, Denver Service Center, Denver, CO, pp 249–255

Tande GF (1979) Fire history and vegetation pattern of coniferous forests in Jasper National Park, Alberta. Can J Bot 57:1912–1931

Taylor AH (1994) Fire history and structure of red fir (*Abies magnifica*) forests, Swain Mountain Experimental Forest, Cascade Range, northeastern California. Can J For Res 23:1672–1678

Taylor AH, Halpern CB (1991) The structure and dynamics of *Abies magnifica* forests in the southern Cascade Range, USA. J Vegetation Sci 2:189–200

Taylor AH, Skinner CN (1998) Fire history and landscape dynamics in a late-successional reserve, Klamath Mountains, California, USA. For Ecol Manage 111:285–301

Taylor AR (1969) Lightning effects on the forest complex. In: Proceedings of the eighth annual Tall Timbers Fire Ecology Conference. Tall Timbers Res Sta, Tallahassee, FL, pp 127–150

Teensma PDA, Rienstra JT, Yeiter MA (1991) Preliminary reconstruction and analysis of change in forest stand age classes of the Oregon Coast Range from 1850 to 1890. USDI Bureau of Land Management Tech Note T/N OR-9

Thompson RS, Whitlock C, Bartlein PJ, Harrison S, Spaulding WG (1993) Climatic changes in the western United States since 18,000 year B.P. In: Wright JHE, Kutzbach JE, Webb T III et al. (eds) Global climates since the last glacial maximum. University of Minnesota Press, Minneapolis, MN, pp 468–513

Turner MG, Romme WH (1994) Landscape dynamics in crown fire ecosystems. Landscape Ecol 9:59–77

Turner MG, Romme WH, Gardner RH (1994) Landscape disturbance models and the long-term dynamics of natural areas. Nat Areas J 14:3–11

Vale TR (1998) The myth of the humanized landscape: an example from Yosemite National Park. Nat Areas J 18:231–236

Veblen TT (1986) Age and size structure of subalpine forests in the Colorado Front Range. Bull Torrey Bot Club 113:225–240

Veblen TT, Alaback PB (1995) A comparative review of forest dynamics and disturbance in the temperate rainforests in North and South America. In: Lawford R, Alaback P, Fuentes ER (eds) High latitude rain forests of the west coast of the Americas: climate, hydrology, ecology and conservation. Springer, Berlin Heidelberg New York, pp 173–213

Veblen TT, Lorenz DC (1986) Anthropogenic disturbance and recovery patterns in montane forests, Colorado Front Range. Phys Geogr 7:1–24

Veblen TT, Lorenz DC (1987) Post-fire stand development of *Austrocedrus-Nothofagus* forests in northern Patagonia. Vegetatio 71:113–126

Veblen TT, Lorenz DC (1988) Recent vegetation changes along the forest/steppe ecotone in northern Patagonia. Ann Assoc Am Geogr 78:93–111

Veblen TT, Lorenz DC (1991) The Colorado Front Range: a century of ecological change. Univ Utah Press, Salt Lake City, UT

Veblen TT, Markgraf V (1988) Steppe expansion in Patagonia? Quat Res 30:331–338

Veblen TT, Kitzberger T, Lara A (1992a) Disturbance and forest dynamics along a transect from Andean rain forest to Patagonian shrubland. J Vegetation Sci 3:507–520

Veblen TT, Mermoz M, Martin C, Kitzberger T (1992b) Ecological impacts of introduced animals in Nahuel Huapi National Park, Argentina. Conserv Biol 6:71–83

Veblen TT, Burns BR, Kitzberger T, Lara A, Villalba R (1995) The ecology of the conifers of southern South America. In: Enright NJ, Hill RS (eds) Ecology of the southern conifers. Smithsonian Inst Press, Washington, DC, pp 120–155

Veblen TT, Donoso C, Kitzberger T, Rebertus AJ (1996) Ecology of southern Chilean and Argentinean *Nothofagus* forests. In: Veblen TT, Hill RS, Read J (eds) Ecology and biogeography of *Nothofagus* forests. Yale University Press, New Haven, CT, pp 293–353

Veblen TT, Kitzberger T, Villalaba R, Donnegan J (1999) Fire history in northern Patagonia: The roles of humans and climatic variation. Ecol Monogr 69:47–67

Veblen TT, Kitzberger T, Donnegan J (2000) Climatic and human influences on fire regimes in ponderosa pine forest in the Colorado Front Range. Ecol Appl 10:1178–1195.

Veirs SD (1982) Coast redwood forest: stand dynamics, successional status, and the role of fire. In: Means J (ed) Forest succession and stand development research in the Northwest. Oregon State University Press, Corvallis, pp 119–141

Viereck LA (1973) Wildfire in the taiga of Alaska. Quat Res 3:465–495

Villalba R (1990) Climatic fluctuations in northern Patagonia during the last 1000 years as inferred from tree ring records. Quat Res 34:346–360

Villalba R (1994) Tree-ring and glacial evidence for the Medieval warm epoch and the Little Ice Age in southern South America. Climatic Change 26:183–197

Villalba R (1995) Climatic influences on forest dynamics along the forest-steppe ecotone in northern Patagonia. PhD Thesis, University of Colorado, Boulder, CO

Villalba R, Boninsegna JA, Cobos DR (1989) A tree-ring reconstruction of summer temperature between A.D. 1500 and 1974 in western Argentina. In: Third international conference of southern hemisphere meteorology and oceanography, Buenos Aires, Argentina, pp 196–197

Villalba R, Veblen TT, Ogden J (1994) Climatic influences on the growth of subalpine trees in the Colorado Front Range. Ecology 75:1450–1462

Villalba R, D'Arrigo RD, Cook ER, Wiles G, Jacoby GC (2002) Decadal-scale climatic variability along the extra-tropical western coast of the Americas inferred from tree-ring records. In: Markgraf V (ed) Interhemispheric climate linkages: present and past interhemispheric climate linkages in the Americas and their societal effects. Academic Press, New York (in press)

Wallin DO, Swanson FJ, Marks B, Cissel JH, Kertis J (2002) Comparison of managed and pre-settlement landscape dynamics in forests of the Pacific Northwest, USA. For Ecol Manage (in press)

Walter H (1973) Vegetation of the Earth. Springer, Berlin Heidelberg New York

Waring RH, Franklin JF (1979) Evergreen coniferous forests of the Pacific Northwest. Science 204:1380–1386

Waring RH, Pitman GB (1985) Modifying lodgepole pine to change susceptibility to mountain pine beetle attack. Ecology 66:889–897

Warner TE (1980) Fire history in the yellow pine forest of Kings Canyon National Park. In: Stokes MA, Dieterich JH (eds) Proceedings of the fire history workshop, October 20–24, 1980, Tucson, AZ. Gen Tech Rep GTR-RM-81. USDA Forest Service, Rocky Mountain Forest and Range Exp Stn, Ft Collins, CO, pp 89–92

Weaver H (1959) Ecological changes in the ponderosa pine forest of the Warm Springs Indian Reservation in Oregon. J For 57:15–20

Weaver TA (1980) Climates and vegetation types of the Northern Rocky Mountains and adjacent plains. Am Midland Nat 103:392–399

White AS (1985) Presettlement regeneration patterns in a southwestern ponderosa pine stand. Ecology 66:589–594

Whitlock C (1992) Vegetational response to environmental changes in the Pacific Northwest during the last 20,000 years. Northwest Environ J 8:5–28

Whitlock C (1993) Postglacial vegetation and climate of Grand Teton and southern Yellowstone National Park. Ecol Monogr 63:173–198

Whitlock C, Knox MA (2002) Prehistoric burning in the Pacific Northwest. In: Vale TR (ed) Fires in western wilderness. Island Press, Washington, DC (in press)

Wills RD, Stuart JD (1994) Fire history and stand development of a Douglas-fir/hardwood forest in northern California. Northwest Sci 68:205–212

Wimberly MC, Spies TA, Long CJ, Whitlock C (2000) Simulating historical variability in the amount of old forests in the Oregon Coast Range. Conserv Biol 14:167–180

Woodward FI (1987) Climate and plant distribution. Cambridge University Press, New York

5 Natural Versus Anthropogenic Sources of Amazonian Biodiversity: the Continuing Quest for El Dorado

B.J. MEGGERS

5.1 Introduction

Growing concern that human activities are producing unanticipated and often undesirable changes in biota and climate is stimulating efforts to document their extent and reduce or reverse their impact. Achieving this goal is hampered by differences in the perception, motivation, expertise, education, goals, and other characteristics of the interested parties; by deficiencies in the comprehensiveness and representativeness of the data, and by insufficient interaction among biologists, ecologists, climatologists, geologists, and anthropologists. Furthermore, specialists who attempt to integrate diverse kinds of evidence often place greater confidence on that from disciplines other than their own.

These kinds of factors have produced conflicting interpretations of the origin and functioning of Amazonian ecosystems and their vulnerability to intensive human exploitation. Biologists disagree over the relative importance of climatic fluctuations, geographical and ecological barriers, and environmental remodeling in creating the greatest biodiversity on the planet. Anthropologists dispute the potential of the tropical rainforest for sustainable intensive exploitation, one faction denying the existence of any intrinsic obstacles and the other pointing to indigenous practices implying environmental constraints. Political, developmental, and commercial interests take advantage of these uncertainties to deny any adverse impact from their activities and to challenge the need for moderation. Although similar conflicts exist at various scales throughout the planet, the magnitude of Amazonia and the accelerating pace of devastation make resolution of differences of paramount concern. In the following pages, I will point out some of the discrepancies in interpreting the evidence that must be resolved if we are to advance our understanding of the genesis, functioning, and vulnerability of this remarkable region.

Ecological Studies, Vol. 162
G.A. Bradshaw and P.A. Marquet (Eds.)
How Landscapes Change
© Springer-Verlag Berlin Heidelberg 2003

5.2 Significant Characteristics of the Amazonian Environment

The Amazonian rainforest extends over a larger area than any similar forma-tion elsewhere on the earth. Its equatorial location minimizes seasonal differ-ences in temperature and precipitation and provides continuously warm and humid conditions for the growth of plants. These benefits are offset by the exceptional deficiency of soil nutrients, which is compensated by a remark-able recycling system that combines a diverse array of decomposers with plants having differing nutritional requirements. The dispersed distribution of trees of the same species creates a bottleneck for terrestrial vertebrates in the form of low concentration and abundance of subsistence resources, which limit their size, density, and diversity. The water-absorbing capacity of the vegetation maximizes intake of soluble nutrients to the plants at the cost of impoverishing the foundation of the aquatic food chains in the rivers that drain the leached topography of the Guayana and Brazilian shields.

Although diverse local habitats can be recognized, they can be assigned to two major ecosystems: (1) the várzea or floodplain of the Amazon and those tributaries originating in the Andes, which maintain the fertility of inundated soil by annual deposition of sediment, and (2) the terra firme, representing the remaining 90 % of the lowlands, which is dominated by impoverished soils and nutrient-poor rivers. Adaptations by the terra firme flora to cope with annual variability in the timing, duration, and intensity of rainfall are sources of subsistence uncertainty for the fauna. The potentially greater pro-ductivity of the várzea for intensive human exploitation is similarly dimin-ished by fluctuations in the flood regime, which deplete wild and cultivated food resources at unpredictable intervals and cannot be offset by storage.

These environmental constraints must be minimized or neutralized to achieve the sustainable local food supply essential for settlement permanence and population concentration, which in turn are prerequisite to the emer-gence and maintenance of sociopolitical complexity. This correlation has important implications for the future of Amazonia. If urban civilizations could be sustained on local resources prior to European contact, the prerequi-site level of population density can be restored. If the small and scattered vil-lages of surviving indigenous groups optimize sustainable exploitation, how-ever, the prospect is quite different. The degree to which pre-Columbian Amazonians succeeded in moderating the environmental constraints is thus a matter of more than academic importance.

5.3 Evidence for Dense Pre-Columbian Populations

Proponents of the existence in pre-Columbian Amazonia of population density and political complexity that "rivaled or even exceeded" that of sixteenth century Europe (Whitehead 1994), not only along the várzea, but also in the interior of the Guianas, base their interpretations on the same biological, ecological, ethnohistorical, archeological, and ethnographic evidence that receives different evaluations by specialists in each of these disciplines. In exemplifying some of the conflicting assessments, I will emphasize those by anthropologists.

5.3.1 Botanical Evidence

Hecht and Cockburn (1989) regard babassu palm forests as anthropogenic and conclude that "much of the Amazon's forests may very well reflect the intercessions of man". The existence of large uniform stands of bamboo and extensive caatinga forests leads Balée (1989) to estimate that "at least 11.8 % of the *terra firme* forest in the Brazilian Amazon alone is of archaic, cultural origin". He also attributes the liana forest of eastern Amazonia to "late-successional residue of prior cultures", although his Araweté and Asuriní informants consider it primary (Balée and Campbell 1990). The indigenous view is compatible with Gentry's observation that African liana forests "tend to cluster at what would be considered in the Neotropics to be the transition from moist to dry forest, exactly where the most liana-rich Neotropical forests occur" (Gentry 1993; Nelson 1994).

Since African elephants are reported to be able to transform woodland into savanna in 10 years at densities of $1/km^2$, Cooke and Ranere (1992) argue that humans could accomplish similar results at densities of only $1/mile^2$. However, the concentration of these presumably unnatural formations on the margins of the rainforest is compatible with disturbance by early European colonists rather than indigenous groups. Humans are not responsible for the ability of palms and grasses to maintain large disease-resistant single stands in contrast to most other tropical plants. The rapid growth of palms under exceptionally warm and rainy conditions in the Riobamba region of eastern Peru during the fall of 1997, which far exceeded that of other trees, suggests that repeated episodes of this kind could have produced the extensive palm forests of southwestern Amazonia (Abelardo Sandoval, pers. comm.).

The origin of charcoal encountered in the soils beneath normally nonflammable rainforest is another source of disagreement. Whereas an increase in particulate carbon, secondary forest taxa, and weedy species in pollen cores from highland Panama ca. 11,500 B.P. has been attributed by anthropologists to human intervention (Piperno et al. 1990; Ranere 1992), similar changes in the flora of the eastern highlands of northern Peru have been interpreted by biologists as a natural response to increased aridity (Hansen and Rodbell 1995). The absence in Paleo-Indian tool kits of artifacts suitable for felling trees, the rarity of habitation sites outside rock shelters, the failure of contemporary forest-dwelling hunter-gatherers to create clearings, and the assertions by surviving shifting agriculturalists that their gardens were significantly smaller prior to obtaining metal tools (Colchester 1984; Hill and Kaplan 1989) also favor natural causes. Indeed, the amount of effort required to fell trees using stone axes has led some observers to question the antiquity of slash-and-burn agriculture (Denevan 1992).

Biologists in general appear to regard human agency as unpersuasive, given the responses by the vegetation to natural disturbance (Bailey et al. 1991) and the demonstration that catastrophic fires occur only in rainforests subjected to severe drought (Leighton and Wirawan 1986Turcq et al 1998) and destructive windfall (Goldamner 1991; Nelson 1994). Experimental efforts in northern Amazonia indicate that fires will not spread from gaps to adjacent closed-canopy forest under normal climatic conditions, when relative humidity exceeds 65 % (Uhl et al. 1988). The fact that human efforts would have been successful only under circumstances that would have favored natural ignition makes an anthropogenic explanation "the unlikely hypothesis" (Russell and Forman 1984).

The occurrence of natural fires raises questions concerning their ecological consequences and their impacts on human communities. Trans-Amazonian discontinuities in archeological sequences throughout the lowlands correlate with mega-Niño events during the past two millennia, implying repeated dislocation of semi-sedentary horticulturists. The most immediate impetus for dispersal was depletion of local subsistence resources as a consequence of drought (Meggers 1994), which would have been aggravated by fire. The few eyewitness accounts of recent lesser events that I have encountered are instructive.

The relatively low normal rainfall in north-central Amazonia makes that region especially susceptible to catastrophic conflagrations. During the 1912 El Niño, a fire lasting several months is said to have caused the deaths of hundreds of rubber gatherers in the Rio Negro basin (Carvalho 1952; Sternberg 1987). During 1926, fires burned throughout the same region for more than a month, killing animals and birds, and even fish. Fires resulting from the 1972 drought destroyed the gardens of the Yanomami in the forest on the Brazil-Venezuela border, forcing the Indians to abandon their villages and subsist for a year on wild foods (Lizot 1974). The devastating fires during the 1998 ENSO event, which may approximate the intensity of the prehistoric mega-Niños, received international media attention. An Associated Press release dated April 1 reported 13,000 square miles had been charred in Rondônia alone. Consequences included loss of crops, livestock and homes, shortage of food and water, respiratory diseases, smoke pollution, and unbearable heat.

The impact of indigenous humans on the landscape must be assessed in the context of other human and nonhuman sources of perturbation. In contrast to recent European invaders, they experienced at least ten millennia of coevolution with the rest of the biota, during which groups that failed to develop a sustainable relationship would not have survived. They enhanced the distributions of some kinds of plants and affected the densities of others, but there is no evidence that their behavior was more detrimental ecologically than that of other organisms (Janzen 1983). Like them, humans may often have been victims rather than perpetrators of environmental alterations.

5.3.2 Ethnohistorical Evidence

Roosevelt argues that "the sources for Greater Amazonia contain indisputable evidence of large-scale, very populous regional societies comparable to complex chiefdoms and small states known in other parts of the world" (1993). Some "regional Amazonian cultures...had territories tens of thousands of square kilometers in size, larger than those of many recognized prehistoric states." On Marajó alone, she estimates "the population could have been up to one million people, and the density could have been as great as 50 people per square kilometer" (1991). Whitehead contends that "chieftaincy developed as much in the interfluvial as in the floodplain areas" and that particularly in the interior of the Guianas, "we are dealing with civilizations of considerable complexity, possibly even protostates". He envisions "a powerful polity that straddled the Amazon and Orinoco drainage basins in the area of the Sierra Acarai/Tumuc Humac, linking the Corentyn and Berbice with the Paru and Trombetas rivers", which controlled the manufacture and trade of gold objects and other trans-Guyana commerce (Whitehead 1991, 1994).

In contrast to these archeologists, historians have long denied the credibility of the early chronicles. At the beginning of the twentieth century, Rothery (1910) observed that "The fifteenth-century European was dominated by the Greek spirit. He went West, not to discover a new world, but to find a short cut to India, with its boundless wealth and all its wonders and monsters, as recorded by the ancients. Dreams of the lost Atlantis, the superb island-continent, the home of the Elysian fields, which had formed in imagination a mysterious and golden bridge between Africa and India, was a constant obsession to them.... Consequently, it is quite natural that the early explorers from all countries, but more especially those who had come into closer contact with and had received an intellectual stimulus from Arabic civilization, should have seen things with a distorted vision, the result of preconceived ideas, unfailing credulity, and an abundant superstition." This assessment has been reaffirmed by Gheerbrant (1992): "the first Europeans to set foot on Amazonia let their imaginations run away with them and claimed actually to see and hear everything they had hitherto only imagined: from the works of Pliny to

Herodotus, from the words of Arabian storytellers to Mogul writers, from the tales of knightly derring-do to medieval hagiographies.... Seldom have reality and fantasy complemented each other so well."

The predispositions provided by Greek mythology were soon conflated with the myth of El Dorado, stimulating rumors of a fabulous kingdom with its capital on the shore of the vast saline Lake Parima in the center of the Guianas. Here, the Amazons lived in stone buildings, wore woolen garments, raised "sheep," and – most importantly – controlled vast quantities of gold. The legend of El Dorado "became one of the most famous chimeras in history, a legend that lured hundreds of hard men into desperate expeditions" (Hemming 1978). In spite of repeated failure to verify its existence, Lake Parima was depicted on maps until the end of the 18th century, a situation that has been characterized as "by far the biggest and most persistent hoax ever perpetrated by geographers" (Fig. 5.1; Gheerbrant 1992; cf. Alès and Pouyllau 1992).

The credibility attached to the accounts of the early explorers of Amazonia by anthropologists also contrasts with the skepticism accorded such "eye-witness" reports on other regions. Conaty (1995) argues that the worldview of European males "has significantly biased our understanding of 17th-, 18th-, and 19th-century indigenous [Blackfoot] cultures". Schrire (1984) finds it "instructive to realize how persistently scholars still accept at face value the messages of historical drawings", a theme reiterated by Trigger (1976) for the Huron of eastern North America and by Smith (1960) for the south Pacific. Lorenz (1997) downgrades the authority of Natchez chiefs on the basis of archeological evidence and Lightfoot (1995) contends that "if every student of North American archaeology better understood the biases and limitations of different sources of written records, then many of the most flagrant abuses of direct historic analogy would probably cease, and the privileging of written records over archaeological materials might be curtailed". Finally, "archaeological naiveté in the face of historical documents" has been noted by Galloway (1992), who warns that the authors "wrote stories for self-justification and glory; it was not necessary that they portray the places they went and the people they saw accurately – just that they do it convincingly."

Why are these criteria not applied to assess the credibility of early European descriptions of Amazonia, especially in the Guianas? Archeological surveys along rivers draining from the Guayana Shield in the Brazilian states of Pará (Meggers and Evans 1957; Chmyz, pers. comm. Miller et al. 1992) and Roraima (Ribeiro et al. 1996; Miller, pers.comm. 1998), in the upper Essequibo forest and Rupununi savanna of Guyana (Evans and Meggers 1960; Williams 1979, 1985), and along the upper Orinoco, Ventuari, Manipiare, and Casiquiare in southern Venezuela (Evans et al. 1960; Cruxent, pers.comm.) have recorded only habitation sites compatible with small semi-sedentary villages, with no indication of occupational specialization or hierarchical social organization. The only evidence of metallurgy is an ornament typical of the Tairona style of northern Colombia, dredged from the Mazuruni River (Whitehead 1990;

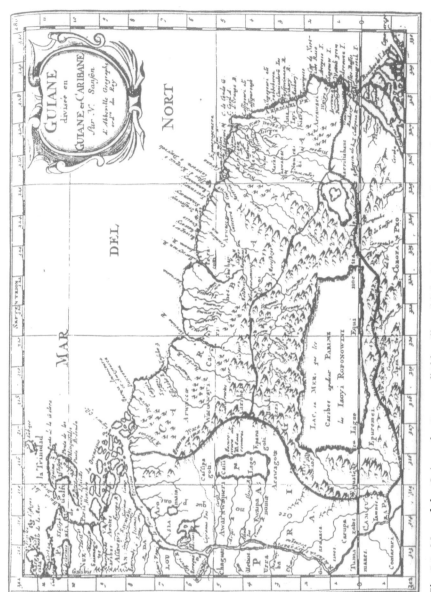

Fig. 5.1. Map of the Guianas by Sanson d'Abbeville published in 1734, one of many showing the mythical Lake Parima. (After Alès and Pouyllau 1992, Fig. 10)

Meggers 1995a). Ethnographic studies verify the existence of autonomous communities linked by exchange networks based on egalitarian relations rather than elite coercion (Colson 1985; Arvalo-Jiménez and Biord 1994). Ecological investigations in the upper Negro region of southern Venezuela indicate that fire was a more important source of forest disturbance than human activity and that carrying capacity is limited by the availability of protein (Clark and Uhl 1987).

5.3.3 Archeological Evidence

Two principal kinds of archeological features have been cited in support of the existence of large sedentary populations in Amazonia during the pre-Columbian period: (1) earthworks and (2) habitation sites extending several hundred meters along riverbanks. Both are subject to conflicting interpretations of their social and settlement implications.

5.3.3.1 Earthworks

According to Roosevelt, "the scale and extent of Amazonian earthworks and occupation sites are extraordinary," "covering many hundreds of square kilometers," implying that "much of the topography on the lowlands is man-made" (Roosevelt 1993). On Marajó, "there are so many Marajoara sites that a lifetime would not be enough to visit and sample all the known sites" (Roosevelt 1991). Tests conducted in the upper 2 m of Teso dos Bichos indicate that "occupations on the mounds were dense and continuous" (Roosevelt 1991) and that "the estimated population for modest mounds would be between 300 and 500, more than 1000 for larger mounds, and several thousand or more for the large multimound sites... If reported sites are only a fraction of those that existed, the population could have been up to one million people" (Roosevelt 1991).

These statements are not supported by existing archeological evidence. First, with the exception of eastern Ecuador, artificial mounds have been reported only from lowland Bolivia and the western llanos of the Orinoco, where conditions oscillate seasonally between inundation and desiccation and where the dominant vegetation is savanna. Second, it has not been demonstrated that all or even most of the mounds in these locations were constructed simultaneously and occupied continuously; on the contrary, limited excavations on Marajó indicate that habitation was temporally and spatially discontinuous and that some mounds were used principally or exclusively as cemeteries (Meggers and Evans 1957). Third, experiments conducted in North America more than a century ago showed that the amount of labor needed to produce large mounds "did not surpass the common industry of the savages" (McCoy 1840; also Carr 1883; Krause 1995). Observations and experiments have also shown that the construction and maintenance of mounds and drained fields are well within the capacity of small family groups (Heider 1970; Clay 1988; Dillehay 1990; Erickson 1992; Graffam 1992).

Even stone constructions of the magnitude achieved by the Maya were within the capability of relatively small populations (Erasmus 1965). At Copán, for example, calculations of the time required for quarrying, transporting, and placing stone and fill imply that "Temple 1 could have been built by 130 people, each working 100 days during seven successive dry seasons" (Webster and Kirker 1995). The maximum population at Tikal, where pyramids, platforms, and other constructions are distributed over some 120 km², has been estimated between 40,000 (Haviland 1972) and 72,000 (Willey 1989). Extensive excavations at Cahokia, which consists of more than 100 mounds including the largest one in North America, and which covers 6.5 miles², have reduced the estimates of the population to ca. 10,000 and the social organization to the level of a simple chiefdom (Williams 1991; Milner 1998).

Whereas estimates of prehistoric population size are increasing in magnitude in Amazonia, they are being reduced in other regions as a consequence of detailed documentation of living space, water sources, subsistence productivity, and other variables. Applying these criteria to Xculoc in Campeche led to a 30–50 % reduction in the estimated number of inhabitants and the suggestion that "most estimates of Maya population made during the past 20 years need to be reexamined" (Becquelin and Michelet 1995). Similar conclusions have been reached for the Maya city of Tikal (Webster 1997) and for the 16th-century population of Basin of Mexico and the Teotihuacan Valley (Sanders 1992). In Egypt, where estimates have ranged up to 20 million, a population of 6 million "must be regarded as very close to the maximum and may have been approached only at rare peak periods" (Hassan 1994). Estimates extrapolated from presumed post-contact decimation are also being revised downward, based on more detailed assessments of the impact of European diseases (Snow 1995).

The existence of natural processes of biotic or abiotic origin that produce regularly spaced mounds and ridges also indicates that caution should be exercised in assigning all such features to human initiative, especially when they occur in tropical wetlands and savannas. Micromounds produced by competition between savanna woodland and open grassland communities in the Brazilian pantanal "may often show a remarkable spatial regularity when viewed from the air" (Ponce and da Cunha 1993). On the central Argentina pampa, fossorial rodents construct mounds up to 30 m in diameter and 3 m high, which are distributed singly, in irregular clusters or in chains up to 250 m long and occur in densities up to 20/ha (Cox and Roig 1986). Regularly spaced mounds are also produced on floodplains by termites (Oliveira Filho 1992). Parallel "ripples" created by variation in grass growth (Hills 1969) and fossil ridges produced by beach action (Watters 1981) may be mistaken for degraded ridged fields (Fig. 5.2; Klausmeier 1999). The opportunistic use of natural mounds for habitation may also give the impression of artificial construction. These considerations indicate that the existence even of substantial earthworks does not necessarily imply large, sedentary, and hierarchically organized populations.

5.3.3.2 Habitation Sites

Habitation refuse extending a kilometer or more along the Amazon and its tributaries has been considered confirmation of ethnohistorical reports that the banks were "literally lined by settlements, some of which appear to have been of urban scale and complexity" (Roosevelt 1991; also Heckenberger 1992; Whitehead 1994). As has been noted for the Andean area, however, a correlation between the surface area of a habitation site and the size of the population must be demonstrated rather than assumed, and investigation often shows "a distressingly weak relationship between these two variables" (Schreiber and Kintigh 1996). Demonstration is particularly relevant in Amazonia, where small and frequently moved villages remain the typical pattern of indigenous settlement.

Extensive surveys conducted during the past two decades along the main tributaries of the Amazon and portions of the floodplain indicate that all but the smallest habitation sites are the product of multiple reoccupations over centuries or millennia by villages within the dimensions reported among surviving indigenous groups (Meggers et al. 1988; Miller et al. 1992; Meggers 1996). Discontinuities within and among stratigraphic excavations in the same site permit detection of episodes of abandonment and reoccupation of each location. Intrasite correlations identify the area corresponding to each episode and intersite correlations identify the number of contemporary settlements. The resulting reconstructions of village area, village movement, territorial boundaries, and social organization are compatible with ethnographic descriptions (Meggers 1990, 1995b, 1999).

Multiple carbon-14 dates from excavations in different parts of a site provide additional support for discontinuous occupation. In the case of RO-PV-35, a typical extensive habitation site of the Jamarí Phase on the Rio Jamarí in the Brazilian state of Rondônia, carbon-14 dates from the same depth in different excavations differ by as much as 2000 years (Fig. 5.3; Miller et al. 1992). The existence of dense populations is also challenged by the rarity or absence of pottery and other evidence of habitation in most of the large extensions of dark soil (terra preta) in the lower Tapajós region, which implies that they were not dwelling sites (Woods 1995).

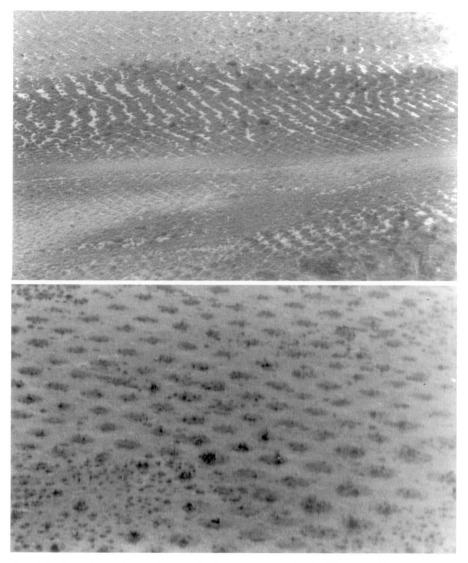

Fig. 5.2. Parallel ripples and rows of micromounds of probable natural origin on the Llanos de Moxos, northeastern lowlands of Bolivia. (Courtesy Bernardo Dougherty)

5.3.4 Ethnographic Evidence

The relevance of contemporary indigenous practices for estimating the impact of pre-Columbian humans on the rainforest has been questioned by anthropologists on the assumption that surviving groups are decimated and decultured relicts that underexploit their resources. Alvard (1994) argues that "low human population densities, lack of markets, and limited technology

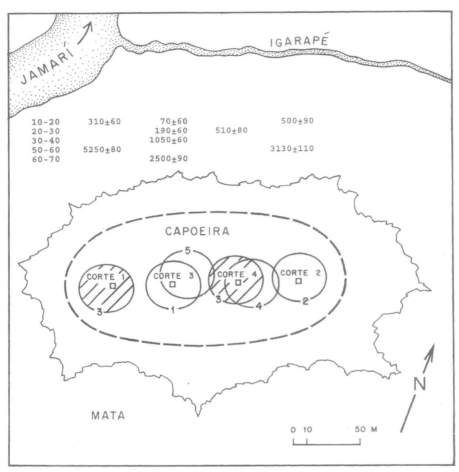

Fig. 5.3. A typical habitation site of the Jamarí Phase on the Rio Jamarí, a right-bank tributary of the Rio Madeira. Ceramic discontinuities in four stratigraphic excavations imply five noncontemporaneous episodes of occupation in the locations sampled, one with two houses. The large differences in carbon-14 dates from the same depths in different locations also support lack of correlation between site area and village size, as well as the existence of multiple reoccupations. (After Miller et al.1992)

more parsimoniously explain the equilibrium enjoyed by native groups than does a putative harmonious relationship with nature" and his assessment is widely shared. Lizot (1980) concludes that the Yanomama of southern Venezuela use only a third of the agricultural potential within their territory. Thomas (1972) asserts that "even in the eastern reaches of the Gran Sabana, where cultivation is limited to small gallery forests, there is sufficient space available for much more extensive cultivation than is current among the Pemon". Allen and Tizón (1973) believe that the territory occupied by the Campa of the Alto Pachitea in eastern Peru could support a much larger population than now exists. Descola (1994) speaks of "flagrant underutilization of garden produce" by the Achuar of eastern Ecuador and Wagley (1977) reaches the same conclusion for the Tapirapé.

By contrast, these and other indigenous Amazonians consider subsistence shortage a constant threat. The Achuar regard "the task of gardening as a chancy and hazardous undertaking" (Descola 1994). The Machiguenga become nervous when the surplus shrinks (Johnson 1983) and the Tapirapé limit a woman to three offspring to avoid exceeding carrying capacity (Wagley 1977). Indeed, most of the behavior of traditional communities constitutes classic risk-avoidance strategy (Halstead and O'Shea 1989; Cashdan 1990). Small and frequently moved villages, permanent territories, multiple varieties of primary cultigens, detailed knowledge of edible wild plants, temporary and permanent taboos on consumption of specific game, mandatory sharing of prey, population stabilization, periodic abandonment of the village by some or all of the inhabitants for days or weeks, warfare, intra-community visiting, long-distance exchange networks, and egalitarian social organization all fall into this category (Meggers 1996).

In addition to pursuing strategies that minimize overexploitation, indigenous groups maximize the availability of subsistence resources by various unintentional and intentional kinds of manipulation. Small garden clearings increase the abundance of secondary vegetation and its constituent species of desirable plants and herbivores (Unruh 1990; Balée 1994). Seed dispersal is enhanced by snacking on fruits during travel and by transporting them to the village for consumption and subsequent discard (Politis 1996).

Intentional management is also practiced. Plants used for medicine, paint or condiment are often relocated near the house (Frikel 1978). Some groups engage in selective weeding during early stages of succession to enhance the density of useful species (Irvine 1989). Frikel (1978) asserts that the peach palm (*Guilielma gasipaes*) was cultivated "from remote times" and Prance (1984) has reviewed evidence suggesting the dispersal of fruit trees by humans. The antiquity of these practices is difficult to assess, but the Yanomama attribute stands of bamboo to such behavior among their predecessors (Lizot 1980) and the modern Guyanese consider clumps of large cane to have been planted by "old-time people" (Evans and Meggers 1960). In Rondônia, several species of palms are more abundant adjacent to prehistoric habitation sites and some appear to occur exclusively in such locations (Miller et al. 1992).

None of these practices is conducive to major ecological disturbance because they work with rather than against the environmental constraints. Small clearings separated by forest retard the spread of pests and accelerate the rate of regeneration. Planting multiple varieties of cultigens maximizes uptake of nutrients and reduces vulnerability to crop loss by unseasonable weather. Relief from predation provided by frequently moved villages allows recuperation of preferred game. The detailed knowledge of plants and animals and their interactions possessed by indigenous forest dwellers greatly exceeds that of natural scientists and changes are monitored on a daily basis (Descola 1994; Kane 1995; Politis 1996). Disruptive behavior is further constrained by the view that humans are part of nature, rather than apart from it, and have an obligation to treat it with respect (Chernela 1994; Politis 1996; Reichel-Dolmatoff 1996). The contention that indigenous Amazonians are not "conservationists" fails to recognize that much of their conservational behavior is built into their social fabric rather than intentional, and that its abandonment by decultured groups reflects the substitution of commercial values for traditional ones.

5.4 Conclusion

During the first two centuries following the discovery of the Americas, hundreds of Europeans lost their fortunes, their health, even their lives searching for El Dorado, inspired by greed, encouraged by reports from the Indians, and undeterred by the failures of their predecessors (Hemming 1978). At the end of the 17th century, the quest was abandoned and Lake Parima was stricken from the maps. Now, after three centuries of dormancy, the myth is being revived by anthropologists, in the form of vivid descriptions of urban populations with powerful rulers that controlled the manufacture and trade of gold throughout northern South America. As before, no physical evidence has been found. Although humans have selectively eliminated, dispersed, modified, and otherwise affected distributions, densities, abundances, interactions, and habitats of other animals and plants for thousands of years, the extent to which their pre-Columbian behavior was more influential than that of other keystone species or more destructive than catastrophic natural events has yet to be established.

Whereas anthropologists treat humans as independent of environmental constraints and attribute most vegetational changes to human agency (e.g., Piperno and Pearsall 1998), natural scientists investigate ecosystems as if they were independent of human intervention and attribute their characteristics to short and long-term climatic fluctuations, complex interactions among the biota, and other natural processes. These extreme positions prevent understanding either the natural sources of development and maintenance of Amazonian biodiversity or the extent to which humans can intensify its sustainable exploitation. In Amazonia, the anthropocentric perspective is popular because it contributes to humanity's fascination with "lost worlds" and furthers national and international economic and political interests. Although an experienced tropical forest ecologist warned more than a decade ago that "poor advice from anthropologists...can only result in further confusion and inadequate resource planning" (Lamb 1987), their credibility remains unchallenged. Meanwhile, "developers" have increased their momentum and proposals to conserve critical environmental and ecological resources continue to be ignored. Until these conflicting assessments of the past and present impact of humans are resolved, we cannot hope to understand the extent of our ability to influence the future course of life in Amazonia or elsewhere on the earth.

References

Alès C, Pouyllau M (1992) La conquête de l'inutile: les géographies imaginaires de l'El Dorado. L'Homme 32:271-308

Allen WL, Tizón JH (1973) Land use patterns among the Campa of the Alto Pachitea, Peru. In: Lathrap DW, Douglas J (eds) Variation in anthropology. Illinois Archaeological Survey, Urbana, pp 137-153

Alvard MS (1994) Conservation by native peoples; prey choice in a depleted habitat. Hum Nat 5:127-154

Arvalo-Jiménez N, Biord H (1994) The impact of conquest on contemporary indigenous peoples of the Guiana shield. In: Roosevelt AC (ed) Amazonian Indians. University of Arizona Press, Tucson, pp 55-78

Bailey RC, Jenike M, Rechtman R (1991) Reply to Colinvaux and Bush. Am Anthropol 93:160-162

Balée W (1989) The culture of Amazonian forests. Adv Econ Bot 7:1-21

Balée W (1994) Footprints in the forest: Ka'apor ethnobotany. Columbia University Press, New York

Balée W, Campbell DG (1990) Evidence for the successional status of liana forest (Xingu River basin, Amazonian Brazil). Biotropica 22:36-47

Becquelin P, Michelet D (1995) Demografía en la zona puuc: el recurso del étodo. Latin Am Antiquity 5:289-311

Carr L (1883) The mounds of the Mississippi Valley historically considered. Memoirs of the Kentucky Geological Survey, vol. 2. Frankfurt, KYCarvalho JCM (1952) Notas de viagem ao Rio Negro. Publicações Avulsas 9. Museu Nacional, Rio de Janeiro

Cashdan E (ed) (1990) Risk and uncertainty in tribal and peasant economies. Westview Press, Boulder, Colorado

Chernela J (1994) Tukanoan know-how: the importance of the forested margin to neotropical fishing populations. Res Explor 10:440-457

Clark K, Uhl C (1987) Farming, fishing, and fire in the history of the upper Rio Negro region of Venezuela. Hum Ecol 15:1-26

Clay JW (1988) Indigenous peoples and tropical forests: models of land use and management. Cultural Survival Inc, Cambridge

Colchester M (1984) Rethinking stone age economics: some speculations concerning the pre-Columbian Yanomama economy. Hum Ecol 12:291-314

Colson AB (1985) Routes of knowledge: an aspect of regional integration in the circum-Roraima area of the Guiana Highlands. Antropológica 63-64:103-149

Conaty GT (1995) Economic models and Blackfoot ideology. Am Ethnol 22:403-412

Cooke RG, Ranere AJ (1992) Precolumbian influences on the zoogeography of Panama: an update based on archaeofaunal and documentary data. Tulane Stud Zool Bot Suppl 1:21-58

Cox GW, Roig VG (1986) Argentine Mima mounds occupied by ctenomyid rodents. J Mammal 67:428-432

Denevan WM (1992) Stone vs. metal axes: the ambiguity of shifting cultivation in pre-historic Amazonia. J Steward Anthropol Soc 20:153–165

Descola P (1994) In the society of nature: a native ecology in Amazonia. Cambridge University Press, New York

Dillehay TD (1990) Mapuche ceremonial landscape, social recruitment and resource rights. World Archaeol 22:223–241

Erasmus CJ (1965) Monument building: some field experiments. Southwest J Anthropol 21:277–301

Erickson CL (1992) Applied archaeology and rural development: archaeology's potential contribution to the future. J Steward Anthropol Soc 20:1–16

Evans C, Meggers BJ (1960) Archeological investigations in British Guiana. Bureau of American Ethnology Bull 177. Smithsonian Institution, Washington, DC

Evans C, Meggers BJ, Cruxent JM (1959) Preliminary results of archeological investigations along the Orinoco and Ventuari Rivers, Venezuela. In: Actas del 331 Congreso Internacional de Americanistas. Lehman, San José, Costa Rica, July 1958, pp 359–369

Frikel P (1978) Areas de arboricultura pré-agrícola na Amazônia, notas preliminares. Rev Antropol 21:45–52

Galloway P (1992) The unexamined habitus: direct historical analogy and the archaeology of the text. In: Gardin JD, Peebles CS (eds) Representations in archaeology. University of Indiana Press, Bloomington, pp 178–195

Gentry AH (1993) Diversity and floristic composition of lowland tropical forest in Africa and South America. In: Goldblatt P (ed) Biological relationships between Africa and South America. Yale University Press, New Haven, CT, pp 500–547

Gheerbrant A (1992) The Amazon: past, present, and future. HN Abrams Inc, New York

Goldammer JG (1991) Tropical wild-land fires and global changes: prehistoric evidence, present fire regimes, and future trends. In: Levine JS (ed) Global mass burning: atmospheric, climatic, and biospheric implications. MIT Press, Cambridge, MA, pp 83–91

Graffam G (1992) Beyond state collapse: rural history, raised fields, and pastoralism in the south Andes. Am Anthropol 94:882–904

Halstead P, O'Shea J (eds) (1989) Bad year economics: cultural responses to risk and uncertainty. Cambridge University Press, Cambridge

Hansen BCS, Rodbell DT (1995) A late-glacial/Holocene pollen record from the eastern Andes of northern Peru. Quat Res 44:216–227

Hassan FA (1994) Population ecology and civilization in ancient Egypt. In: Crumley CL (ed) Historical ecology. School of American Research, Santa Fe, NM, pp 155–181

Haviland W (1972) Family size, prehistoric population estimates and the ancient Maya. Am Antiquity 37:135–139

Hecht S, Cockburn A (1989) The fate of the forest: developers, destroyers and defenders of the Amazon. Verso, London

Heckenberger M (1992) A conquista da Amazônia. Cienc Hoje 15:62–67

Heider KG (1970) The Dugum Dani: a Papuan culture in the highlands of west New Guinea. Viking fund publications in anthropology 49. Aldine, Chicago

Hemming J (1978) The search for El Dorado. Michael Joseph Ltd, London

Hill K, Kaplan H (1989) Population and dry-season subsistence strategies of the recently contacted Yora of Peru. Natl Geogr Res 5:317–334

Hills T (1969) The savanna landscapes of the Amazon Basin. McGill University, Montreal

Irvine D (1989) Succession management and resource distribution in an Amazonian rain forest. Adv Econ Bot 7:223–237

Janzen DH (1983) Food webs: who eats what, why, how, and with what effects in a tropical forest? In: Golley FB (ed) Tropical rain forest ecosystems: structure and function. Elsevier, Amsterdam, pp 167–181

Johnson A (1983) Machiguenga gardens. In: Hames RB, Vickers WT (eds) Adaptive responses of native Amazonians. Academic Press, New York, pp 29–63

Kane J (1995) Savages. Knopf, New York

Klausmeier CA (1999) Regular and irregular patterns in semiarid vegetation. Science 284:1826–1828

Krause RA (1995) Great Plains mound building: a postprocessual view. In: Duke P, Wilson MC (eds) Beyond subsistence. University of Alabama Press, Tuscaloosa, pp 129–142

Lamb FB (1987) The role of anthropology in tropical forest ecosystem resource management and development. J Dev Areas 21:429–458

Leighton M, Wirawan N (1986) Catastrophic drought and fire in Borneo tropical rain forest associated with the 1982–1983 El Niño Southern Oscillation event. In: Prance GT (ed) Tropical rain forests and the world atmosphere. Westview Press, Boulder, CO, pp 75–102

Lightfoot KG (1995) Culture contact studies: redefining the relationship between prehistoric and historic archaeology. Am Antiquity 60:199–217

Lizot J (1974) El Río de los Periquitos: breve relato de un viaje entre los Yanomami del Alto Siapa. Antropológica 37:2–23

Lizot J (1980) La agricultura Yanomama. Antropológica 53:3–93

Lorenz KG (1997) A re-examination of Natchez sociopolitical complexity: a view from the Grand Village and beyond. Southeast Archaeol 16:97–112

McCoy I (1840) History of the Baptist Indian missions. WM Morrison, Washington; H & S Raynor, New York

Meggers BJ (1990) Reconstrução do comportamento locacional pré-histórico na Amazônia. Bol Mus Para Emílio Goeldi 6:183–203

Meggers BJ (1994) Archeological evidence for the impact of mega-Niño events on Amazonia during the past two millennia. Climatic Change 28:321–338

Meggers BJ (1995a) Amazonia on the eve of European contact: ethnohistorical, ecological, and anthropological perspectives. Rev Arqueol Am 8:91–115

Meggers BJ (1995b) Judging the future by the past: the impact of environmental instability on prehistoric Amazonian populations. In: Sponsel LE (ed) Indigenous peoples and the future of Amazonia. University of Arizona Press, Tucson, pp 15–43

Meggers BJ (1996) Amazonia: man and culture in a counterfeit paradise. Revised edition. Smithsonian Institution Press, Washington, DC

Meggers BJ (1999) La utilidad de secuencias cerámicas seriades para inferir conducta social prehistórica. El Caribe Arqueol 3:2–19

Meggers BJ, Evans C (1957) Archeological investigations at the mouth of the Amazon. Bureau of American Archeology Bulletin 167. Smithsonian Institution, Washington, DC

Meggers BJ, Dias OF, Miller ET, Perota C (1988) Implications of archeological distributions in Amazonia. In: Vanzolini PW, Heyer WR (eds) Proceedings of a workshop on neotropical distribution patterns. Academia Brasileira de Ciências, Rio de Janeiro, pp 275–294

Miller ET et al. (anon.) (1992) Archeology in the hydroelectric projects of Eletronorte, preliminary results. Centrais Elétricas do Norte do Brasil SA, Brasilia

Milner GR (1998) The Cahokia chiefdom: the archaeology of a Mississippian society. Smithsonian Institution Press, Washington, DC

Nelson BW (1994) Natural forest disturbance and change in the Brazilian Amazon. Remote Sensing Rev 10:105–125

Oliveira Filho AT de (1992) Floodplain 'murundus' of central Brazil: evidence for the termite-origin hypothesis. J Trop Ecol 8:109

Piperno DR, Pearsall DM (1998) The origins of agriculture in the lowland tropics. Academic Press, San Diego

Piperno DR, Bush MB, Colinvaux PA (1990) Paleoenvironments and human occupation in late-glacial Panama. Quat Res 33:108–116

Politis GG (1996) Nukak. Instituto Amazónico de Investigaciones Científicas Sichi, Bogotá

Ponce VM, da Cunha CN (1993) Vegetated earthmounds in tropical savannas of Central Brazil: a synthesis. J Biogeogr 20:219–225

Prance GT (1984) The pejibaye, *Guilielma gasipaes* (HBK) Bailey, and the papaya, *Carica papaya* L. In: Stone D (ed) Pre-Columbian plant migration. Papers of the Peabody Museum of archaeology and ethnology 76. Harvard University, Cambridge, MA, pp 85–104

Ranere AJ (1992) Implements of change in the Holocene environments of Panama. In: Ortiz-Troncoso OR, Van der Hammn T (eds) Archaeology and environment in Latin America. Universiteit van Amsterdam, Amsterdam, pp 25–44

Reichel-Dolmatoff G (1996) The forest within: the world-view of the Tukano Amazonian Indians. Themis Books, Devon

Ribeiro PM, Ribeiro CT, Guapindaia VLC, Machado AL (1996) Pitture rupestri nel Territorio de Roraima, Brasile. World J Prehist Primitive Art Oct:151–157

Roosevelt AC (1991) Moundbuilders of the Amazon: geophysical archaeology on Marajó Island, Brazil. Academic Press, San Diego

Roosevelt AC (1993) The rise and fall of the Amazon chiefdoms. L'Homme 33:255–283

Rothery GC (1910) The Amazons (1995 edition). Studio Editions, London

Russell EWB, Forman RTT (1984) Indian burning,'the unlikely hypothesis'. Bull Ecol Soc Am 65:281–282

Sanders WT (1992) The population of the Central Mexican symbiotic region, the Basin of Mexico, and the Teotihuacan Valley in the sixteenth century. In: Denevan W (ed) The native population of the Americas in 1492. University of Wisconsin Press, Madison, pp 85–150

Schreiber KJ, Kintigh KW (1996) A test of the relationship between site size and population. Am Antiquity 61:573–579

Schrire C (1984) Wild surmises on savage thoughts. In: Schrire C (ed) Past and present in hunter gatherer studies. Academic Press, San Diego, pp 1–25

Smith B (1960) European vision and the South Pacific, 1768–1850: a study in the history of art and ideas. Oxford University Press, Oxford

Snow DR (1995) Microchronology and demographic evidence relating to the size of pre-Columbian North American Indian populations. Science 268:1601–1604

Sternberg HOR (1987) Aggravation of floods in the Amazon River as a consequence of deforestation. Geogr Ann Ser A 69A:201–219

Thomas DJ (1972) The indigenous trade system of southeast Estado Bolivar, Venezuela. Antropológica 33:3–37

Trigger BG (1976) The children of Aataentsic: a history of the Huron People to 1660. McGill-Queen's University Press, Montreal

Turcq B, Sifeddine A, Martin L, Absy ML, Soubles F, Suguio K, Volkmer-Ribeiro C (1998) Amazonia rainforest fires: a lacustrine record of 7000 years. Ambio 27:139–142

Uhl C, Kaufman JB, Cummings DL (1988) Fire in the Venezuelan Amazon 2: environmental conditions necessary for forest fires in the evergreen rainforest of Venezuela. Oikos 53:176–184

Unruh JD (1990) Iterative increase of economic tree species in managed swidden fallows of the Amazon. Agrofor Syst 11:175–197

Wagley C (1977) Welcome of tears; the Tapirapé Indians of central Brazil. Oxford University Press, New York

Watters DR (1981) Linking oceanography to prehistoric archaeology. Oceanus 24:11–19

Webster D (1997) City-states of the Maya. In: Nichols DL, Charlton TH (eds) The archaeology of city-states. Smithsonian Institution Press, Washington, DC, pp 135–154

Webster D, Kirker J (1995) Too many Maya, too few buildings: investigating construction potential at Copán, Honduras. J Anthropol Res 51:363–389

Whitehead NL (1990) The Mazaruni pectoral: a golden artifact discovered in Guyana and the historical sources concerning native metallurgy in the Caribbean, Orinoco and northern Amazonia. Archaeol Anthropol 7:19–38

Whitehead NL (1991) Los señores de los epuremei; un examen de la transformación del comercio y la política indígenas en el Amazonas y Orinoco. In: Jorna P, Malaver L, Oostra M (eds) Etnohistoria del Amazonas. Quito, Abya-Yala, Georgetown, pp 255–263

Whitehead NL (1994) The ancient Amerindian polities of the Amazon, the Orinoco, and the Atlantic coast. In: Roosevelt AC (ed) Amazonian Indians. University of Arizona Press, Tucson, pp 33–53

Willey GR (1989) Settlement pattern studies and evidences for intensive agriculture in the Maya lowlands. In: Lamberg-Karlovsky CC (ed) Archaeological thought in America. Cambridge University Press, Cambridge, pp 167–182

Williams D (1979) A report on preceramic lithic artifacts in the south Rupununi savannas. Archaeol Anthropol 2:10–53

Williams D (1985) Petroglyphs in the prehistory of northern Amazonia and the Antilles. Adv World Archaeol 4:335–387

Williams S (1991) Fantastic archaeology: the wild side of North American prehistory. University of Pennsylvania Press, Philadelphia

Woods WT (1995) Comments on the black earths of Amazonia. In: Schoolmaster FA (ed) Papers Proc Appl Geogr Conf 18:159–165

Part II
Ecological and Evolutionary Consequences of Fragmentation

6 Bees Not to Be? Responses of Insect Pollinator Faunas and Flower Pollination to Habitat Fragmentation

M.A. Aizen and P. Feinsinger

6.1 Introduction

Many plants rely on animal pollinators to set seed. Therefore, plant-pollinator mutualisms can be critical to the functioning and maintenance of native ecosystems (Bawa 1974, 1990; Gilbert 1980; Bawa et al. 1985; Bullock 1985; Feinsinger 1987, Nabhan and Fleming 1993). Some such mutualisms involve "charismatic microvertebrates" such as hummingbirds and bats, but ca. 90 % of animal-pollinated plants are serviced by less charismatic, often unnoticed insects. Flies, butterflies, moths, beetles, and most importantly bees are responsible for a large proportion of all seeds produced by the earth's wild and cultivated plants (Barth 1991; Buchmann and Nabhan 1996; Kearns and Inouye 1997; Kearns et al. 1998).

Evidence is accumulating that anthropogenic habitat alteration can strongly affect the diversity and composition of vertebrate assemblages. Much less data exist, however, regarding the effects of habitat alteration – for example, fragmentation – on assemblages of insects and other invertebrates (Rathcke and Jules 1993; Didham et al. 1996; Kearns et al. 1998). Here, we review the evidence on responses of insect pollinator faunas to fragmentation; propose a variety of possible mechanisms behind the responses; and then evaluate the links between those responses on the part of pollinators – in terms of abundance, species diversity, and assemblage composition – and subsequent responses on the part of the plants – in terms of pollination levels and seed production. Along the way, we identify important gaps in knowledge and suggest areas for further research.

Ecological Studies, Vol. 162
G.A. Bradshaw and P.A. Marquet (Eds.)
How Landscapes Change
© Springer-Verlag Berlin Heidelberg 2003

6.2 Patterns of Change in Pollinator Faunas Due to Habitat Fragmentation

Plant-pollinator interactions display considerable spatial and temporal variability. The abundance and composition of pollinator assemblages changes not only among localities and years, but also from plant to plant and day to day (Herrera 1988, 1995; Horvitz and Schemske 1990; Eckhart 1992; Roubik 1992; Aizen 1997). One might expect that, given this high intrinsic variability, insect pollinator assemblages would not be particularly susceptible to moderate levels of habitat fragmentation, or that any effect would be difficult to detect. Likewise, as many plants are pollinated not by a single animal species, but rather by a diverse array of taxa (Feinsinger 1983; Jordano 1987; Roubik 1992; Waser et al. 1996), one might expect plant reproduction to be somewhat buffered from any fragmentation-induced changes in the nature of pollinator assemblages. Most studies conducted to date, however, indicate that at least the animal side of the plant-pollinator mutualism – specifically, the insect pollinator fauna – responds markedly to habitat fragmentation, changing in abundance, diversity, and species composition with increasing fragmentation (Rathcke and Jules 1993; Murcia 1996; Kearns et al. 1998).

Studies comparing insect pollinator faunas in fragments with those in continuous expanses of natural habitat have found consistent decreases in diversity and abundance of native insect pollinators with increasing fragmentation. Butterfly inventories in 12.0-, 2.1-, and 1.4-ha urban remnants of tropical deciduous forest in SE Brazil listed respectively 78, 47, and 46 species (Rodrigues et al. 1993), a conspicuous trend although the number of species still persisting in small fragments is surprisingly large (see also Turner and Corlett 1996). Likewise, abundance of euglossine bees (specialized pollinators of Orchidaceae and other tropical families; Roubik 1989), as sampled at chemical baits in central Amazonian wet forest, declined monotonically from continuous forest through fragments of 100, 10, and 1 ha embedded in a recently clear-cut matrix (Powell and Powell 1987). Failure to duplicate those results in a repeat study 6 years later (Becker et al. 1991) may have arisen in part from the use of a different sampling protocol, and also from changes in the nature of the matrix, by then a robust second-growth scrub. To this group of long-distance pollinators (Janzen 1971, 1974; Raw 1989), second growth might be a much less serious barrier than barren cleared areas.

Like their tropical neighbors, subtropical forests in the Americas are experiencing high rates of fragmentation, and in addition many have been seriously affected by overgrazing and selective logging since the time of European colonization (Bucher 1987; Adamoli et al. 1990; Lerdau et al. 1991). In a dry subtropical "chaco serrano" forest, we found fragmentation to be associated with a steep decrease in diversity and abundance of native insect pollinators as sampled by two polyphilic treelet species and by yellow pan traps (Aizen and Feinsinger 1994a). The study involved replicated comparisons among tracts of continuous forest, large forest fragments (>2 ha), small forest fragments (<1 ha), and the surrounding matrix (either cattle pasture or corn fields), with fragments isolated from forest by 40–700 m.

Of course, in terms of regional biodiversity the number of species sustained by a single fragment is less important than the cumulative number sustained by an entire set of fragments, in comparison with the same area of continuous forest (Simberloff and Abele 1982). Theoretically, total species counts might be similar in fragmented and unfragmented landscapes if the different fragments contained random subsets of the species pool of the intact habitat. This reasoning was not supported by our data. We observed similar declines in abundance and diversity of bees, the most important pollinators in this habitat, whether results were expressed on a per fragment basis or whether species lists were combined over replicates and sampling periods (Fig. 6.1). In pan-trap samples, numbers of individuals of two important families of mostly solitary bee pollinators, Anthophoridae and Megachilidae, declined dramatically (Aizen and Feinsinger 1994a).

The effect of fragmentation on pollinator faunas is not exclusive to the tropics and subtropics. In Sweden, Jennersten (1988) recorded numbers of bumblebee and butterfly species found in continuous habitat (a mosaic of forest and meadows) and two small habitat fragments of 1 ha, surrounded by

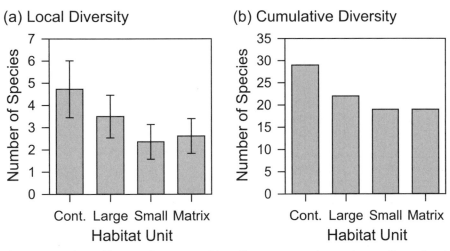

Fig. 6.1. Number of bee species captured in yellow pan traps in continuous forest, large forest fragments (>2 ha), small fragments (<1 ha), and surrounding agricultural matrix in a fragmented "chaco serrano" forest in northwestern Argentina. **a** Values are means ± SE of four replicates of each habitat unit type and three sampling periods (see Aizen and Feinsinger 1994a, for details and statistical analysis), **b** values represent cumulative numbers of species summed over replicates and sampling periods. (Redrawn from Aizen and Feinsinger 1994a)

barley and oat fields. Continuous habitat supported eight bumblebees and ten butterfly species, the fragments only two and four species, respectively.

When assessing the effects of fragmentation on species assemblages, though, we should ask not only "how many?" but also "who?" (cf. Patterson 1987). If the pollinator fauna of fragments consists mainly of the "weedy" species that abound in the converted matrix, then the number of species is irrelevant: it is quite improbable that fragmented landscapes could sustain robust regional faunas of the original, native flower-visitors. In the Argentine chaco, we found that bee assemblages in small fragments converged in species composition on those of the surrounding agricultural matrix (Aizen and Feinsinger 1994a). In particular, in flowers and in pan traps, the frequency of feral Africanized honeybees *(Apis mellifera)* increased with fragmentation. The spread of "weedy" insect flower-visitors may be enhanced not only by the mosaic-like nature of fragmented landscapes, but also by overall anthropogenic changes in habitat characteristics. In a very different Argentine habitat, temperate *Nothofagus* forests of the southern Andes, the senior author found that foragers of the only native bumblebee species, *Bombus dahlbomii*, declined in relative abundance while foragers of *Bombus ruderatus*, a recent invader from Europe (Roig-Alsina and Aizen 1996), increased along a gradi-

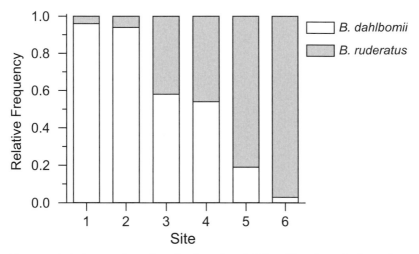

Fig. 6.2. Relative number of native *(Bombus dahlbomii)* and exotic *(B. ruderatus)* bumblebees observed visiting flowering patches of the native herb *Alstroemeria aurea* in six sites differing in forest type and disturbance level, Nahuel Huapi N.P., Argentina. Bumblebees made >70 % of visits to flowers of this species. Of the 462 visits recorded during a total of 360 10-min observation periods distributed across sites and throughout the 1996 flowering season, *B. dahlbomii* made 274 and *B. ruderatus* 188. Sites: *1* Old-growth *Nothofagus* forest (Challhuaco); *2* creek in lightly disturbed mixed *Austrocedrus* forest (Co. Otto); *3* moderately disturbed *Austrocedrus* forest (Co. Runge); *4* highly disturbed post-fire matorral (Casa de Piedra); *5* highly disturbed suburban *Austrocedrus* forest (Co. Otto); *6* urban *Pinus/Austrocedrus* forest (Co. Runge)

ent of anthropogenic habitat alteration (Fig. 6.2). Likewise, monotonic changes in butterfly community diversity and composition occurred along an urban-forest gradient in Porto Alegre, southeastern Brazil (Ruszczyk and de Araujo 1992).

6.3 Mechanisms and Processes Behind Changes in Pollinator Faunas

Many recent studies dealing with animals in habitat mosaics suggest that the absence of a species from small or isolated habitat patches may best be explained by metapopulation processes (Hanski 1994). For example, high extinction rates associated with small population size and reduced colonization imposed by habitat barriers, two characteristics of a species displaying metapopulation dynamics, may explain patterns of patch occupancy by butterflies in fragmented landscapes (Harrison et al. 1988; Thomas et al. 1992; Hanski et al. 1994). We do not know of any comparable study for species of other major groups of insect pollinators such as bees or flies.

In theory, the fate of each small, isolated population simply depends on stochastic demographic processes (as emphasized by metapopulation theory, Gilpin and Soulé 1986). In reality, though, more deterministic mechanisms related to species' natural history may be responsible for rapid disappearance of particular flower-visitors from, and for low recruitment to, habitat fragments. To date, no study dealing with patterns in insect diversity and abundance in fragmented landscapes has clearly linked the patterns observed with underlying biological mechanisms – which leaves us free to speculate.

Simple behavioral constraints may inhibit some insects from crossing habitat barriers even as narrow as a few tens of meters wide. This could explain the loss from fragments of many habitat specialists, for example certain forest-dwelling bees and butterflies, which are never found in highly modified habitats such as cattle pastures and crop fields. Restriction of these species to the forest interior might be related to specific microenvironmental requirements as well (Herrera 1995).

Habitat fragmentation may also decrease the density and quality of nesting sites for some insects, for example, many bees. Physical changes associated with edge effects (Lovejoy et al. 1986; Saunders et al. 1991; Murcia 1995), such as desiccation, might render a large proportion of a fragment's area unsuitable for nesting. If habitat fragmentation is associated with a change in disturbance regime, such as an increase in fire frequency, overgrazing, and/or soil compaction, then nesting conditions for many ground-dwelling solitary bees are likely to be affected (Vinson and Frankie 1991; Kearns and Inouye 1997). Firewood extraction, a common practice in forest remnants (cf. Schelhas and Greenberg 1996), may also affect the availability of nesting sites for trunk-nesting flower visitors such as Xylocopa bees (Roubik 1989).

Abundance of insect visitors at flowering plants is ultimately limited by the food resources available. The high metabolic rates of flying insects create a high demand for food, and overall the spatial and temporal patterns of occurrence of flower-visitors tend to track the spatial and temporal patterns of presentation of their food resources: nectar and pollen in flowers (Fleming 1992; Bronstein 1995). Furthermore, both social and solitary pollinators are able to distinguish between rich and poor food sources at scales ranging from individual flowers to whole flowering patches (Thomson 1981; Feinsinger 1987; Rathcke and Jules 1993; Murcia 1996). Therefore, the absence of some flower visitors from small habitat fragments might simply be a direct consequence of the low absolute abundance of floral resources available there. Independent of fragmentation, many studies have reported changes in diversity, abundance, and composition of flower visitors related to flower density and patch size (e.g., Silander 1978; Thomson 1981; Schmitt 1983; Kwak 1987; Sih and Baltus 1987; Schmitt et al. 1987; Sowig 1989; Klinkhamer and de Jong 1990; Widén and Widén 1990; Kunin 1992, 1993, 1997; Conner and Neumeier 1995).

Thus, some changes in pollinator faunas with fragmentation may result directly from behavioral responses to changed patch size or plant isolation, rather than to demographic stochasticity or, more generally, metapopulation dynamics. It follows that fragmentation may have especially strong effects on the occurrence of oligolectic flower-visitors (those insects depending exclusively on one or a few plant taxa for pollen food), who are likely to perceive their food availability decreasing more steeply than are generalist, opportunistic flower-visitors (Kunin 1993). Nevertheless, insects that are generalist nectar-feeders as adults may also be restricted by scarcities of food sources for specialized larvae, as in the case of many Lepidoptera.

The arguments above may seem to imply that habitat fragments are immersed within a lifeless matrix. On the contrary, the matrix surrounding habitat fragments may provide an insect species pool that may greatly influence the composition of the fragments' pollinator assemblages (Janzen 1983; Murcia 1996). For example, in the small fragments of "chaco serrano" we studied (Aizen and Feinsinger 1994a), the frequency of exotic Africanized honeybees relative to native insect pollinators converged on that in the surrounding agricultural matrix. The capacity of an insect species to thrive in the altered matrix and persist in, or invade, native habitats may relate to life-history traits. Africanized honeybees, whose invasion abilities continent-wide are phenomenal, nest under a great variety of conditions, exploit an astoundingly wide variety of flowers, and can exploit food sources >10 km away from the colony (Michener 1973, 1975; Taylor 1977; Roubik 1989, 1991; Smith 1991; Huryn 1995).

Exploitative competition between exotic and native flower visitors may exacerbate fragmentation-related declines in the latter, particularly if exotic or ruderal flower visitors recruit differentially to habitat fragments. Even native vertebrate pollinators may suffer. In Australia, exotic honeybees heavily exploit nectar and pollen from flowers normally visited by honeyeaters (Ramsey 1988; Paton 1993; Vaughton 1996). Generally speaking, native pollinator assemblages in tropical and subtropical environments may be nearer the limits of their nectar and pollen food resources than are assemblages in high-altitude or -latitude environments (e.g., Arroyo et al. 1985), making the former particularly sensitive to resource usurpation by invading species. Studies designed to investigate whether honeybees actually reduce population levels of native flower-visitors give mixed results (Roubik 1978, 1980, 1983, 1988, 1991; Schaffer et al. 1979, 1983; Ginsberg 1983; Roubik et al. 1986; Markwell et al. 1993). Clearly, though, high numbers of honeybees in habitat fragments depress the levels of nectar and pollen potentially available to native flower-visitors, thus reducing the maximum numbers of native flower-visitors a fragment could sustain. Roubik (1992) cites resource preemption by dominant bee species to explain the shifting nature, in space and time, of bee flower-visitor assemblages in Neotropical forests. We propose, however, that exploitative competition by honeybees or other weedy species luxuriating in altered habitats might have the opposite effect leading to the patterning of flower-visitor assemblages, when forests are fragmented.

6.4 Scale Considerations

Some landscape-level patterns in assemblages of flower visitors may be the cumulative result of processes operating at the level of populations and mechanisms operating at the level of individuals (Feinsinger 1976; Frankie et al. 1983; Bronstein 1995). Ideally, for conservation purposes we should be able to identify the key scale responsible for broad patterns of "habitat fragmentation effects". How much of the fragmentation-related change in pollinator faunas is due to small-scale and how much to landscape-level factors? If small-scale phenomena are most important, then conservation efforts should focus on management of individual habitat fragments and their immediate matrix, whereas if regional factors play the leading role then entire landscapes should be targeted (Pulliam 1988; Fahrig and Merriam 1994).

Obviously, the response of insect pollinator faunas to landscape alteration results from interplay between landscape- and local-scale phenomena (Murcia 1996; cf. Forman 1995), and any conservation efforts must embrace both scales. We emphasize, though, that the smaller scale should not be neglected. For instance, Herrera (1995) reported that pollinator assemblages differed greatly between neighboring lavender shrubs growing a few meters apart and attributed this to differences in plant micro-environments, rather than intrinsic plant features. Herrera (1995) proposed that the change in species composition was the cumulative result of specific thermal requirements of different insect species (i.e., shade- vs. sun-loving species). In the "chaco serrano" of Argentina, we observed that neighboring trees of the same species often supported quite different assemblages of flower visitors: for example, native insects prevailed in some trees while honeybees dominated numerically in their neighbors (Aizen and Feinsinger 1994a). In this case, the mechanism may have involved resource preemption by honeybees rather than differences in light environment. Another mechanism affecting the exact make-up of pollinator faunas may be small-scale disturbance. As Fig. 6.2 shows, some relatively undisturbed sites occurring under different forest types >20 km apart in south temperate Argentina (e.g., sites 1 and 2) had similar proportions of native and exotic bumblebees, while other sites quite near to one another (e.g., 2 and 5, 3 and 6, which were respectively 700 and 300 m apart), but with different grades of intervention presented contrasting proportions.

Neither fragments nor matrix should be viewed as internally homogeneous habitats. Micro-environmental conditions and the extent of floral resource usurpation by matrix-dwelling flower visitors undoubtedly change from borders to the interior of habitat fragments and from fragment borders outwards into the matrix as well. For example, in forest ecosystems fragmentation might induce changes in pollinator assemblages simply by exposing a large fraction of the vegetation to insolated edges (Murcia 1993, 1995, 1996). Thus, changes in the composition of pollinator assemblages may reflect "edge effects" rather than simply "habitat fragmentation effects". To our knowledge, no study has examined edge effects on insect pollinator assemblages (Murcia 1993, 1995, 1996; cf. Didham et al. 1996).

Central to scale considerations is the question of how different sorts of flower visitors perceive their "world". In general, vertebrate pollinators assess their foraging options at a broader spatial scale than invertebrates. However, tremendous variation exists in foraging ranges within any pollinator group (e.g., Feinsinger 1976; Herrera 1987). According to Bronstein (1995), generalist pollinators, which have a larger availability of food items locally, appear to make their foraging decisions at smaller scales than more specialized ones, which have to track the phenology of preferred food items usually scattered over much larger spatial scales. Extreme examples of large-scale landscape perception are provided by migratory butterflies, hummingbirds, and bats that can move over distances of even several thousands of kilometers following nectar corridors formed by regional gradients in flowering times (e.g., Fleming 1992).

Recent behavioral studies in honeybees demonstrate that pollinators may forage with some cognitive spatial picture of the distribution of their flower resources (reviewed in Menzel et al. 1997). Social bees, in particular, are central-place foragers (i.e., a foraging flight always starts and ends at the nest) whose capacity to locate, memorize, and relate close and distant "landmarks" apparently influences greatly the spatial scale of their foraging. We know little to nothing, however, about the cognitive capacities and, more basically, the foraging flight ranges of the myriad solitary bees that represent the most important component of pollinator diversity in many biomes (Kearns and Inouye 1997).

6.5 Pollination and Habitat Fragmentation

If composition, abundance, and foraging behavior of insect flower-visitors change with habitat fragmentation, then sexual reproduction of the plants involved may change as well (Murcia 1993, 1996; Rathcke and Jules 1993; Bond 1994; Kearns and Inouye 1997). For example, changes in pollinator abundance and behavior may affect the number of pollen grains deposited on flower stigmas. Pollen deposition may decrease so much as to diminish the number of seeds produced (Bierzychudek 1981; Burd 1994) or even impair seedling vigor (Mulcahy 1979; Lee 1984; Marshall and Folsom 1991; Walsh and Charlesworth 1992). Likewise, changes in species composition of pollinator assemblages may affect pollen deposition if efficient pollinators are replaced by sloppy ones (e.g., Ramsey 1988; Keys et al. 1995; Vaughton 1996). In dioecious plant species (with separate female and male individuals), by chance a fragment may end up with many more of one sex than the other producing a dramatic change in pollen deposition patterns independent of any changes in the animal pollinators themselves (House 1992, 1993; Cunningham 1995). Changes in pollen deposition and seed output associated with shifts in pollinator abundance or identity have been reported for habitat islands (Jennersten 1988; Lamont et al. 1993; Aizen and Feinsinger 1994b) and for true islands (Linhart and Feinsinger 1980; Spears 1987; Ågren 1996). In addition, as populations become smaller and sparser, pollination may be impaired due to increasing pollen losses to interspecific flowers (e.g., Feinsinger et al. 1991; Kunin 1993), or to interference associated with increased foreign pollen deposition (e.g., Waser and Fugate 1986; Murphy and Aarsen 1995; McLernon et al. 1996).

Not only quantity, but also quality of pollen may shift with habitat fragmentation. Where plant populations had been genetically structured pre-fragmentation (Loveless and Hamrick 1984), restricted pollen flow among a small number of related individuals now trapped in habitat fragments will increase the deposition of inbred pollen and decrease deposition of true outcross pollen (Levin and Kerster 1974; Coles and Fowler 1976; Handel 1983; Levin 1984, 1989; Sobrevila 1988; Hall et al. 1996; Murcia 1996; Young et al. 1996). Self-pollination may also increase if pollinators restrict most flights to plants within fragments, resulting in a higher frequency of revisits. Over time, plants in fragmented populations may suffer reduced fitness as a result of increasing genetic load (Menges 1991; Heschel and Paige 1995). Negative consequences of inbreeding may be expressed prezygotically as an increased proportion of pollen tubes aborted, or postzygotically as increased frequency of seed abortion, impaired seed germination, increased seedling mortality, or decreased adult performance (Ledig 1986; Waser and Price 1991; Barrett and Harder 1996).

Pollination quality may also change if habitat fragmentation results in replacement of foragers that move frequently among plants by restricted-area foragers such as honeybees (e.g., Kwak 1987). Although a honeybee colony as a whole exploits resources over a wide area, once recruited to a given source individual honeybee foragers tend to concentrate their foraging within a single flowering patch, the crown of a single shrub or tree, or even a restricted area within that crown (Visscher and Seeley 1982; Seeley 1985, 1989; Roubik 1989, 1992). In contrast, diverse taxa of native insects display a diversity of persistence times at a given plant and of flight distances between consecutive flowers or plants visited, in turn diversifying pollen flow distances (Herrera 1987). Thus, in general terms replacement of a diverse pollinator assemblage by honeybees or other area-restricted foragers might lead not only to increased deposition of self pollen, but also to a decreased diversity of parental genotypes represented in stigmatic pollen loads.

To our knowledge, no studies have directly assessed the consequences of turnover in pollinator assemblages to genetic constitution of seeds or to seed and seedling performance. Monitoring insect visits to two polyphilic, self-incompatible tree species in the "chaco serrano", *Prosopis nigra* and *Cercidium australe*, we found that the decreased number of visits made by native insects in small fragments (accompanied by a decrease in the number of flower-visiting taxa) was fully compensated by increased numbers of visits by Africanized honeybees. Consequently, the quantity of pollen deposited on either species' stigmas did not change, but apparently the quality did, as rates of pollen tube abortion increased with decreasing fragment size (Aizen and Feinsinger 1994b).

So far, we have stressed the negative effects of fragmentation on plant pollination and sexual reproduction. Sensitivity of plant reproduction to habitat fragmentation, though, undoubtedly varies greatly among species with different traits (Murcia 1993, 1996). These traits include breeding system (e.g., autogamous vs. self-incompatible or dioecious), degree of specialization relative to pollinator taxa (e.g., generalized vs. dependent on a single pollinator taxon), the identity of the pollinator taxa involved (e.g., social hymenoptera vs. fragmentation-sensitive bats or pesticide-sensitive hawkmoths), average abundance of the plant species (e.g., locally dense vs. very dispersed), life form (e.g., trees vs. annual herbs), and life history (e.g., semelparous vs. iteroparous). Nevertheless, the power of these traits to predict differential responses to habitat fragmentation has not yet been evaluated empirically. Case studies of single plant species are always valuable. Still, to evaluate effects of habitat fragmentation on ecosystem functioning via pollination, responses must also be assessed across whole plant assemblages.

To our knowledge, only one published study has evaluated the impact of forest fragmentation across several representatives from the plant species assemblage. Overall, most of the 16 plant species we studied in the Argentina "chaco serrano" (Aizen and Feinsinger 1994b) responded negatively to habitat fragmentation. Nevertheless, most changes in pollination and seed output between continuous forest and forest fragments were of low to moderate magnitude. With few exceptions, the effects of fragmentation at the pollination stage did not translate simply and directly into effects on fruit and seed production. These minor to moderate effects of forest fragmentation on plant reproduction contrasted greatly with the pronounced effects of fragmentation on the animal pollinator assemblage itself.

Perhaps it should not be surprising that pollination responses fail to reflect precisely pollinator responses to habitat fragmentation. Only a very few plant species engage in tight relationships with particular pollinators. Recent reviews on the evolution of plant-pollinator mutualisms stress the diffuse nature of most relationships (Feinsinger 1983; Jordano 1987; Herrera 1996; Waser et al. 1996; Kearns et al. 1998). The "chaco serrano" study (Aizen and Feinsinger 1994b) suggests that plant reproduction, at least in the proximate

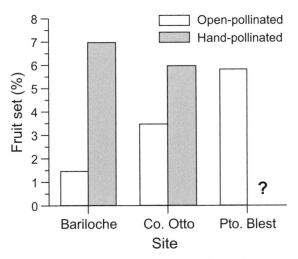

Fig. 6.3. Relative number of fruits set by the "hummingbird-type" flowers (Faegri and van der Pijl 1979) of *Embothrium coccineum* planted along streets of downtown San Carlos de Bariloche, Argentina (41°8′S, 71°19′W), and in two naturally occurring populations 5 km (Cerro Otto) and 40 km (Puerto Blest) west of the city. Fruit set (fruits produced per flower) of open-pollinated flowers was estimated from 3840 ($n=10$ trees), 605 ($n=8$), and 945 flowers ($n=6$) for the three respective populations. The overall difference in fruit set among populations was significant ($\chi^2=63.4$, df=2, $P<0.0001$). Results of the relative number of fruits set after unlimited, hand cross-pollination for the populations of Bariloche (4160 flowers, 10 trees) and Cerro Otto (620 flowers, 8 trees) are provided for comparison. Net bagging of a total of 610 flowers in Cerro Otto confirmed that fruits are not set in the absence of pollinators

sense, may be quite resilient in the face of considerable change in the pollinator assemblage. Likewise, numerous plant species continue to produce seeds when transplanted to urban settings, even half a world away from their original geographical range or when their presumed legitimate pollinators are "lost" (see references in Waser et al. 1996). For instance, the senior author found that self-incompatible *Embothrium coccineum*, a treelet with red showy flowers native to the south Andean temperate forest, still sets fruit in the streets of downtown San Carlos de Bariloche (a city of 80,000 inhabitants), where honeybees are the only visitors; although fruit set was higher in two natural populations 5 and 30 km distant, where hummingbirds and native insects are the pollinators (Fig. 6.3). Nevertheless, even though proximate effects of fragmentation appear to be less severe on plants than on pollinators, long-term effects may be quite severe. Furthermore, fragmentation apparently leads to simplification of plant-pollinator interactions, increasing the susceptibility of these relationships to further disruption (Waser et al. 1996).

6.6 Concluding Remarks and Research Needs

Throughout this review we have noted fundamental questions that merit further investigation, or investigation period. True, those few studies that exist suggest that habitat fragmentation strongly affects insect pollinator faunas. The mechanisms and processes behind those changes, however, are still largely unexplored. Do those mechanisms and processes operate at the local level, at the regional level, or both? Small-scale factors are apt to depend on distance from borders more than simply on fragment size; consequently, is much of the "fragmentation effect" really edge effect instead? Few of the many studies on edge effects have focused on insects and the ecological interactions in which they participate (Lovejoy et al. 1986; Murcia 1995; Didham et al. 1996).

Speaking specifically of the plant-pollinator mutualism, how might different types of matrix affect events within fragments? The contrasting results of the two different studies on euglossine bees near Manaus, Brazil suggests that a matrix of second-growth vegetation may ameliorate the negative consequences of fragmentation, but obviously better controlled studies in a great variety of sites are needed. Furthermore, we need information not only on how the matrix influences events within fragments, but also on how the fragment of original habitat influences events within the matrix. The latter theme is especially intriguing when the matrix supports an animal-pollinated crop. Seed production in many such crops is chronically pollinator-limited. Nearby chunks even of highly altered "wild" habitat may serve as sources of pollinators capable of increasing crop seed or fruit yields (Kevan 1975; Buchmann and Nabhan 1996; Kearns and Inouye 1997; Kearns et al. 1998).

Likewise, we need many more studies, in a great diversity of habitats, on responses to fragmentation of flower pollination and seed output. We suggest that high-resolution studies of single plant species (which, after all, represent an *n* of only 1 at the level of the question) should be complemented by studies of lower resolution, but broader scope, surveying an array of different species drawn from a given plant assemblage.

Several recent studies have examined the genetic consequences of fragmentation to plant populations (reviewed by Young et al. 1996). Nevertheless, we do not know how much genetic change has been driven by changes in pollinator assemblages and how much by the fragment's physical restriction on pollen flow. Furthermore, few data exist concerning effects on seed quality itself. Future studies might integrate effects of fragmentation on pollination, seed output, and the performance and demography of the seed progeny.

Considering the interaction between plants and insect pollinators, short-term conservation tactics might emphasize maintaining pollinator diversity (cf. Buchmann and Nabhan 1996) by recognizing and manipulating the processes that might decrease that diversity. From the plant perspective, though, in the short term pollination concerns may be less serious than problems associated with other life history stages (Aizen and Feinsinger 1994b). For example, plant populations in fragments may experience drastic changes in seedling recruitment due to overgrazing, soil compaction, and trampling. Only in those cases where so few plant individuals, of an obligately outcrossing species, exist that pollination and fertilization rates are severely reduced, will fragmentation's effects on pollination alone lead to a demographic bottleneck.

In the long term, though, fragmentation's effects on pollination may impinge upon the survival of plant taxa and thereby entire assemblages, if erosion of genetic variation compromises plant species' survival. Whether the simplification of pollinator assemblages in fragmented landscapes truly erodes the potential for evolutionary change in plants is an open question.

Acknowledgement. We thank Nick Waser for helpful comments on the manuscript. This work was supported by the International Foundation for Science (IFS Grant D/1700-1) and the National Research Council of Argentina (CONICET).

References

Adamoli J, Sennhauser E, Acero JM, Rescia A (1990) Stress and disturbance: vegetation dynamics in the dry Chaco region of Argentina. J Biogeogr 17:491–500

Ågren J (1996) Population size, pollinator limitation, and seed set in the self-incompatible herb *Lythrum salicaria*. Ecology 77:1779–1791

Aizen MA (1997) Influence of local floral density and sex ratio on pollen receipt and seed output: empirical and experimental results in dichogamous *Alstroemeria aurea* (Alstroemeriaceae). Oecologia 111:404–412

Aizen MA, Feinsinger P (1994a) Habitat fragmentation, native insect pollinators, and feral honeybees in Argentine "chaco serrano". Ecol Appl 4:378–392

Aizen MA, Feinsinger P (1994b) Habitat fragmentation, pollination, and plant reproduction in a chaco dry forest, Argentina. Ecology 75:330–351

Arroyo MTK, Arroyo JJ, Primack RB (1985) Community studies in pollination ecology in the high temperate Andes of central Chile. II. Effect of temperature on visitation rates and pollination possibilities. Plant Syst Evol 149:187–203

Barrett SCH, Harder LD (1996) Ecology and evolution of plant mating. Trends Ecol Evol 11:73–79

Barth FG (1991) Insects and flowers: the biology of a partnership. Princeton University Press, Princeton, NJ

Bawa KS (1974) Breeding systems of tree species of a lowland tropical community. Evolution 28:85–92

Bawa KS (1990) Plant-pollinator interactions in tropical rain forests. Annu Rev Ecol Syst 21:399–422

Bawa KS, Bullock SH, Perry DP, Colville RE, Grayum MH (1985) Reproductive biology of tropical trees. II. Pollination systems. Am J Bot 69:122–134

Becker P, Moure JS, Peralta FJA (1991) More about euglossine bees in Amazonian forest fragments. Biotropica 23:586–591

Bierzychudek P (1981) Pollinator limitation of plant reproductive effort. Am Nat 117:838–840

Bond WJ (1994) Do mutualisms matter? Assessing the impact of pollinator and disperser disruption on plant extinction. Philos Trans R Soc Lond B Biol Sci 344:83–90

Bronstein JL (1995) The plant-pollinator landscape. In: Hansson L, Fahrig L, Merriam G (eds) Mosaic landscapes and ecological processes. Chapman and Hall, London, pp 256–288

Bucher EH (1987) Herbivory in arid and semi-arid regions of Argentina. Rev Chil Hist Nat 60:265–273

Buchmann SL, Nabhan GP (1996) The forgotten pollinators. Island Press, Washington, DC

Bullock SH (1985) Breeding systems in the flora of a deciduous forest in Mexico. Biotropica 17:287–301

Burd M (1994) Bateman's principle and plant reproduction: the role of pollen limitation in fruit and seed set. Bot Rev 60:84–139

Coles JF, Fowler DP (1976) Inbreeding in neighboring trees in two white spruce populations. Silvae Genet 25:29–34

Conner JK, Neumeier R (1995) Effects of black mustard population size on the taxonomic composition of pollinators. Oecologia 104:218–224

Cunningham SA (1995) Ecological constraints on fruit initiation by *Calyptrogyne ghiesbreghtiana* (Arecaceae): floral herbivory, pollen availability and visitation by pollinating bats. Am J Bot 82:1527–1536

Didham RK, Ghazoul J, Stork NE, Davis AJ (1996) Insects in fragmented forests: a functional approach. Trends Ecol Evol 11:255–260

Eckhart VM (1992) Spatio-temporal variation in abundance and variation in foraging behavior of the pollinators of gynodioecious *Phacelia linearis* (Hydrophyllaceae). Oikos 64:573–586

Faegri K, van der Pijl L (1979) The principles of pollination ecology. Third edition, revised. Pergamon, New York

Fahrig L, Merriam G (1994) Conservation of fragmented populations. Conserv Biol 8:50–59

Feinsinger P (1976) Organization of a tropical guild of nectarivorous birds. Ecol Monogr 46:257–291

Feinsinger P (1983) Coevolution and pollination. In: Futuyma DJ, Slatkin M (eds) Coevolution. Sinauer, Sunderland, MA, pp 282–310

Feinsinger P (1987) Approaches to nectarivore-plant interactions in the New World. Rev Chil Hist Nat 60:285–319

Feinsinger P, Tiebout HM, Young BE (1991) Do tropical bird-pollinated plants exhibit density-dependent interactions? Field experiments. Ecology 72:1953–1963

Fleming TH (1992) How do fruit- and nectar-feeding birds and mammals track their food resources? In: Hunter MD, Ohgashi T, Price PW (eds) Effects of resource distribution on animal-plant interactions. Academic Press, New York, pp 355–391

Forman RTT (1995) Land mosaics: the ecology of landscapes and regions. Cambridge University Press, New York

Frankie GW, Haber WA, Opler PA, Bawa KS (1983) Characteristics and organization of the large bee pollination system in the Costa Rican dry forest. In: Jones CE, Little RJ (eds) Handbook of experimental pollination biology. Van Nostrand Reinhold, New York, pp 411–447

Gilbert LE (1980) Food web organization and conservation of neotropical diversity: an evolutionary-ecological perspective. In: Soulé ME, Wilcox BA (eds) Conservation biology: an evolutionary-ecological perspective. Sinauer, Sunderland, MA, pp 19–34

Gilpin ME, Soulé ME (1986) Minimum viable populations: processes of species extinctions. In: Soulé ME (ed) Conservation biology: the science of scarcity and diversity. Sinauer, Sunderland, MA, pp 11–34

Ginsberg HS (1983) Foraging ecology of bees on an old field. Ecology 64:165–175

Hall P, Walker S, Bawa K (1996) Effect of forest fragmentation on genetic diversity and mating system in a tropical tree, *Pithecellobium elegans*. Conserv Biol 10:757–768

Handel SN (1983) Pollination ecology, plant population structure, and gene flow. In: Real L (ed) Pollination biology. Academic Press, New York, pp 163–211

Hanski I (1994) Patch-occupancy dynamics in fragmented landscapes. Trends Ecol Evol 9:131–135

Hanski I, Kuussaari M, Nieminen M (1994) Metapopulation structure and migration in the butterfly *Melitaea cinxia*. Ecology 75:747–762

Harrison SM, Murphy DD, Ehrlich PR (1988) Distribution of the bay checkerspot butterfly, *Euphydryas editha bayensis*: evidence for a metapopulation model. Am Nat 132:360–382

Herrera CM (1987) Components of pollinator "quality": comparative analysis of a diverse insect assemblage. Oikos 50:79–90

Herrera CM (1988) Variation in mutualisms: the spatio-temporal mosaic of a pollinator assemblage. Biol J Linn Soc 35:95–125

Herrera CM (1995) Microclimate and individual variation in pollinators: flowering plants are more than their flowers. Ecology 76:1516–1524

Herrera CM (1996) Floral traits and plant adaptation to insect pollinators: a devil's advocate approach. In: Lloyd DG, Barrett SCH (eds) Floral biology. Chapman and Hall, New York, pp 65–87

Heschel MS, Paige KN (1995) Inbreeding depression, environmental stress, and population size variation in scarlet gilia *(Ipomopsis aggregata)*. Conserv Biol 9:126–133

Horvitz CC, Schemske DW (1990) Spatio-temporal variation in insect mutualists of a neotropical herb. Ecology 71:1085–1097

House SM (1992) Population density and fruit set in three dioecious tree species in Australian tropical rain forest. J Ecol 80:57–69

House SM (1993) Pollination success in a population of dioecious rain forest trees. Oecologia 96:555–561

Huryn VMB (1995) Use of native New Zealand plants by honey bees (Apis mellifera L.): a review. N Z J Bot 33:497–512

Janzen DH (1971) Euglossine bees as long-distance pollinators of tropical plants. Science 171:201–203

Janzen DH (1974) The deflowering of Central America. Nat Hist 83:48–53

Janzen DH (1983) No park is an island: increase in interference from outside as park size decreases. Oikos 41:402–410

Jennersten O (1988) Pollination in Dianthus deltoides (Caryophyllaceae): effects of habitat fragmentation on visitation and seed set. Conserv Biol 2:359–366

Jordano P (1987) Patterns of mutualistic interactions in pollination and seed dispersal: connectance, dependence asymmetries, and coevolution. Am Nat 129:657–677

Kearns CA, Inouye DW (1997) Pollinators, flowering plants, and conservation biology. Bioscience 47:297–306

Kearns CA, Inouye DW, Waser NM (1998) Endangered mutualisms: the conservation of plant-pollinator interactions. Annu Rev Ecol Syst 29:83–112

Kevan PG (1975) Pollination and environmental conservation. Environ Conserv 2:293–298

Keys RN, Buchmann SL, Smith SE (1995) Pollination effectiveness and pollination efficiency of insects foraging on Prosopis velutina in south-eastern Arizona. J Appl Ecol 32:519–527

Klinkhamer PGL, de Jong TJ (1990) Effects of plant size, plant density and sex differential reward on pollinator visitation in the protandrous Echium vulgare (Boraginaceae). Oikos 57:399–405

Kunin WE (1992) Density and reproductive success in wild populations of Diplotaxis erucoides (Brassicaceae). Oecologia 91:129–133

Kunin WE (1993) Sex and the single mustard: population density and pollinator behavior effects on seed-set. Ecology 74:2145–2160

Kunin WE (1997) Population size and density effects in pollination: pollinator foraging and plant reproductive success in experimental arrays of Brassica kaber. J Ecol 85:225–234

Kwak MM (1987) Pollination and pollen flow disturbed by honeybees in bumblebee-pollinated Rhinanthus populations? In: van Andel J (ed) Disturbance in grasslands. Dr W Junk, Dordrecht, pp 273–283

Lamont BB, Klinkhamer PGL, Vitkowski ETF (1993) Population fragmentation may reduce fertility to zero in Banksia goodii – a demonstration of the Allee effect. Oecologia 94:446–450

Ledig FT (1986) Heterozygosity, heterosis, and fitness in outbreeding plants. In: Soulé ME (ed) Conservation biology: the science of scarcity and diversity. Sinauer, Sunderland, MA, pp 77–104

Lee TD (1984) Patterns of fruit maturation: a gametophyte selection hypothesis. Am Nat 123:427–432

Lerdau M, Whitbeck J, Holbrook NM (1991) Tropical deciduous forest: death of a biome. Trends Ecol Evol 6:210–212

Levin DA (1984) Inbreeding depression and proximity-dependent crossing success in Phlox drummondii. Evolution 36:116–127

Levin DA (1989) Proximity-dependent cross-compatibility in Phlox. Evolution 43:1114–1116

Levin DA, Kerster HW (1974) Gene flow in seed plants. Evol Biol 7:139–220

Linhart YB, Feinsinger P (l980) Plant-hummingbird interactions: effects of island size and degree of specialization on pollination. J Ecol 68:745–760

Lovejoy TE, Bierregaard RO Jr, Rylands AB et al. (1986) Edge and other effects of isolation on Amazon forest fragments. In: Soulé ME (ed) Conservation biology: the science of scarcity and diversity. Sinauer, Sunderland, MA, pp 257–285

Loveless MD, Hamrick JL (1984) Ecological determinants of genetic structure in plant populations. Annu Rev Ecol Syst 15:65–95

Markwell TJ, Kelly D, Duncan KW (1993) Competition between honey bees (Apis mellifera) and wasps (Vespula spp.) in a honeydew beech (Nothofagus solandri var. solandri) forest. N Z J Ecol 17:85–93

Marshall DL, Folsom MW (1991) Mate choice in plants: an anatomical to population perspective. Annu Rev Ecol Syst 22:37–63

McLernon SM, Murphy SD, Aarsen LW (1996) Heterospecific pollen transfer between sympatric species in a midsuccessional old-field community. Am J Bot 83:1168–1174

Menges ES (1991) Seed germination percentage increases with population size in a fragmented prairie species. Conserv Biol 5:158–164

Menzel R, Gumbert A, Kunze J, Shmida A, Vorobyev M (1997) Pollinator's strategies in finding flowers. Isr J Plant Sci 41:141–156

Michener CD (1973) The Brazilian honeybee. Bioscience 23:523–527

Michener CD (1975) The Brazilian bee problem. Annu Rev Entomol 20:399–416

Mulcahy DL (1979) The rise of the angiosperms: a genecological factor. Science 206:20–23

Murcia C (1993) Edge effects on the pollination of tropical cloud forest plants. Unpublished Ph.D. dissertation, University of Florida, Gainesville, FL

Murcia C (1995) Edge effects in fragmented forests: implications for conservation. Trends Ecol Evol 10:58–62

Murcia C (1996) Forest fragmentation and the pollination of neotropical plants. In: Schelhas J, Greenberg R (eds) Forest patches in tropical landscapes. Island Press, Covelo, CA, pp 19–36

Murphy SD, Aarsen LW (1995) Reduced seed set in Elytrigia repens caused by allelopathic pollen from Phleum pratense. Can J Bot 73:1417–1422.

Nabhan GP, Fleming T (1993) The conservation of mutualisms. Conserv Biol 7:457–459

Paton DC (1993) Honeybees in the Australian environment. Bioscience 43:95–103

Patterson BD (1987) The principle of nested subsets and its implications for biological conservation. Conserv Biol 1:323–334

Powell AH, Powell GVN (1987) Population dynamics of male euglossine bees in Amazonian forest fragments. Biotropica 19:176–179

Pulliam HR (1988) Sources, sinks, and population regulation. Am Nat 132:652–661

Ramsey MW (1988) Differences in pollinator effectiveness of birds and insects visiting Banksia menziessi (Proteaceae). Oecologia 76:119–124

Rathcke BJ, Jules ES (1993) Habitat fragmentation and plant-pollinator interactions. Curr Sci 65:273–277

Raw A (1989) The dispersal of euglossine bees between isolated patches of eastern Brazilian wet forest (Hymenoptera, Apidae). Rev Brasil Entomol 33:103–107

Rodrigues JJS, Brown KS, Ruszczyk A (1993) Resources and conservation of neotropical butterflies in urban forest fragments. Biol Conserv 64:3–9

Roig-Alsina A, Aizen MA (1996) Bombus ruderatus, una nueva especies de Bombus para la Argentina (Hymenoptera: Apidae). Physis 51:120–121

Roubik DW (1978) Competitive interactions between neotropical pollinators and Africanized honey bees. Science 201:1030–1032

Roubik DW (1980) Foraging behavior of competing Africanized honeybees and stingless bees. Ecology 61:836–845

Roubik DW (1983) Experimental community studies: time-series tests of competition between African and neotropical bees. Ecology 64:971–978

Roubik DW (1988) An overview of Africanized honeybee populations: reproduction, diet, and competition. In: Needham G, Delfinado-Baker M, Page R, Bowman C (eds) Africanized honeybees and bee mites. Ellis Horwood, Chichester, pp 45–54

Roubik DW (1989) Ecology and natural history of tropical bees. Cambridge University Press, New York

Roubik DW (1991) Aspects of Africanized honeybee ecology in tropical America. In: Spivak M, Fletcher DJC, Breed MC (eds) The "African" honeybee. Westview Press, Boulder, CO, pp 259–281

Roubik DW (1992) Loose niches in tropical communities: why are there so few bees and so many trees? In: Hunter MD, Ohgushi T, Price PW (eds) Effects of resource distribution on animal-plant interactions. Academic Press, New York, pp 327–353

Roubik DW, Moreno JE, Vergara C, Wittman D (1986) Sporadic food competition with the African honeybee: projected impact on neotropical social bees. J Trop Ecol 2:97–111

Ruszczyk A, de Araujo AM (1992) Gradients in butterfly species diversity in an urban area in Brazil. J Lepid Soc 46:255–264

Saunders DA Jr, Hobbs RJ, Margules CR (1991) Biological consequences of ecosystem fragmentation: a review. Conserv Biol 5:18–32

Schaffer WM, Jensen DB, Hobbs DE et al. (1979) Competition, foraging energetics, and the cost of sociality in three species of bees. Ecology 60:976–987

Schaffer WM, Zeh DW, Buchmann SL et al. (1983) Competition for nectar between introduced honeybees and native North American bees and ants. Ecology 64:564–577

Schelhas J, Greenberg R (eds) (1996) Forest patches in tropical landscapes. Island Press, Covelo, CA

Schmitt J (1983) Density-dependent pollinator foraging, flowering phenology, and temporal dispersal patterns in *Linanthus bicolor*. Evolution 37:1247–1257

Schmitt J, Eccleston J, Erhardt DW (1987) Density-dependent flowering phenology, outcrossing, and reproduction in *Impatiens capensis*. Oecologia 72:341–347

Seeley TD (1985) Honeybee ecology. Princeton University Press, Princeton, NJ

Seeley TD (1989) The honeybee colony as a superorganism. Am Sci 77:546–553

Sih A, Baltus MS (1987) Patch size, pollinator behavior and pollinator limitation in catnip. Ecology 68:1679–1690

Silander JA (1978) Density-dependent control of reproductive success in *Cassia biflora*. Biotropica 10:292–296

Simberloff DS, Abele LG (1982) Refuge design and island biogeographic theory: effects of fragmentation. Am Nat 120:41–50

Smith DR (1991) African bees in the Americas: insights from biogeography and genetics. Trends Ecol Evol 6:17–21

Sobrevila C (1988) Effects of distance between pollen donor and pollen recipient on fitness components in *Espeletia schultzii*. Am J Bot 75:701–724

Sowig P (1989) Effects of flowering plant's patch size on species composition of pollinator communities, foraging strategies, and resource partitioning in bumblebees (Hymenoptera: Apidae). Oecologia 78:550–558

Spears EE (1987) Island and mainland pollination ecology of *Centrosema virginianum* and *Opuntia stricta*. J Ecol 75:351–362

Taylor OR (1977) The past and possible future spread of Africanized honeybees in the Americas. Bee World 58:19–30

Thomas CD, Thomas JA, Warren MS (1992) Distributions of occupied and vacant butterfly habitats in fragmented landscapes. Oecologia 92:563–567

Thomson JD (1981) Spatial and temporal components of resource assessment by flower-feeding insects. J Anim Ecol 50:49–59

Turner IM, Corlett RT (1996) The conservation value of small, isolated fragments of lowland tropical rain forest. Trends Ecol Evol 11:330–333

Vaughton G (1996) Pollination disruption by European honeybees in the Australian bird-pollinated *Grevillea barkluana* (Proteaceae). Plant Syst Evol 200:89–100

Vinson SB, Frankie GW (1991) Nest variability in *Centris aethyctera* (Hymenoptera: Anthophoridae) in response to nesting site conditions. J Kans Entomol Soc 64:156–162

Visscher PK, Seeley TD (1982) Foraging strategies of honeybee colonies in a temperate deciduous forest. Ecology 63:1790–1801

Walsh NE, Charlesworth D (1992) Evolutionary interpretations of differences in pollen tube growth rates. Q Rev Biol 67:19–37

Waser NM, Fugate ML (1986) Pollen precedence and stigma closure: a mechanism of competition for pollination between *Delphinium nelsonii* and *Ipomopsis aggregata*. Oecologia 70:573–577

Waser NM, Price MV (1991) Reproductive costs of self-pollination in *Ipomopsis aggregata* (Polemoniaceae): are ovules usurped? Am J Bot 73:1036–1043

Waser NM, Chittka L, Price MV, Williams NM, Ollerton J (1996) Generalization in pollination systems, and why it matters. Ecology 77:1043–1060

Widén B, Widén M (1990) Pollen limitation and distance-dependent fecundity in females of the clonal gynodioecious herb *Glechoma heredaceae* (Lamiaceae). Oecologia 83:191–196

Young A, Boyle T, Brown T (1996) The population genetic consequences of habitat fragmentation for plants. Trends Ecol Evol 11:414–418

7 Implications of Evolutionary and Ecological Dynamics to the Genetic Analysis of Fragmentation

L. JOSEPH, M. CUNNINGHAM, S. SARRE

7.1 Introduction

The expectation of reduced genetic diversity in fragmented environments is rooted in classical population genetics theory (Wright 1978). It can be formally expressed with the following genetic and demographic hypotheses: (1) genetic drift, the random fixation of alleles at a given locus, is increased; (2) inbreeding, the average level of relatedness within populations, is also increased; (3) gene flow between populations is reduced; and (4) the probability of local extinction of demes within a metapopulation is increased (Young et al. 1996). These hypotheses predict that erosion of genetic diversity should be manifest in two broad genetic outcomes. First, diversity within populations isolated in habitat fragments is expected to be reduced relative to that in similar sized areas in an unfragmented habitat. Second, divergence among populations isolated in fragments should increase relative to populations separated by the same distance(s) in an unfragmented habitat if initial allele frequencies are different or if there are multiple alleles in high frequencies (see McCauley 1991).

Theoretical and empirical approaches to these population genetic issues have been reviewed elsewhere (see Young et al. 1996 and references therein). Our primary aim is to explore the impact of long-term evolutionary history on how one designs a study of the population genetic consequences of anthropogenic fragmentation. To do this, we compare two projects set up with the same aims and methodologies for studying the population genetics of such fragmentation. Their outcomes were very different and comparing them will emphasize the importance of understanding the deeper evolutionary history and nonequilibrium demography of species in fragmented environments. We also briefly review three other recent studies of genetic structure in several species from three very different nonfragmented habitats. We suggest that they and many similar projects are windows to the range of pre-fragmentation genetic structures that exist (one hopes that post-fragmentation comparisons will never be needed!). Fragmentation is, after all, a system-level phe-

Ecological Studies, Vol. 162
G.A. Bradshaw and P.A. Marquet (Eds.)
How Landscapes Change
© Springer-Verlag Berlin Heidelberg 2003

nomenon and to assess its effect we must presuppose some understanding of how the intact system works (Crome 1993). In addition, these three studies offer an alternative approach to the point that we will make in the first part of the chapter. We hope that this chapter will, therefore, usefully complement more traditionally organized reviews. Unless otherwise indicated, the term fragmentation hereafter refers to anthropogenic fragmentation.

7.2 Post-Fragmentation. A Comparison of Fragmentation Genetics in the Western Australian Wheat Belt and the Rainforests of the Wet Tropics

For details of the work described in this section see Cunningham and Moritz (1998), Joseph et al. (1993, 1995), Moritz et al. (1993, 1997), Joseph and Moritz (1994) and Schneider et al. (1998) for the rainforest material, and Sarre (1995a,b) and Sarre et al. (1995) for the wheatbelt work.

7.2.1 Study Areas

The wheatbelt of Western Australia in the continent's temperate southwest (Fig. 7.1) is one of the most heavily fragmented landscapes in Australia. Hobbs and Saunders (1993, see also references therein) noted that clearing for agriculture has seen 93 % of the wheatbelt's formerly continuous, low *Eucalyptus* woodland reduced to many remnant islands in a sea of farmland dedicated to sheep and wheat production. They further noted that the remnants are not a representative sample of the pre-existing vegetation as the vegetation on richer soils was cleared before that on sandier soils. Even now, there remains a significantly greater proportion of sand plain shrub communities than woodland communities. The clearing has resulted in the known extinction of 24 plant species; 348 plant species are listed as rare or endangered and of these only 79 have populations within designated reserves, 135 are found in roadside reserves and 53 are on private land.

The tropical rainforests of the Wet Tropics in northeastern Queensland are comparable in extent to the wheatbelt (Fig. 7.1). They are mostly surrounded by *Eucalyptus* forests and woodlands, though in some places they extend right to the beaches of the Coral Sea. The boundary between rainforest and eucalypt is often sharp, *Eucalyptus* and the floristic elements of rainforests meeting only in ecotonal rainforests. Forming part of a chain of tropical and subtropical rainforests along Australia's eastern seaboard, these rainforests have had a long and dynamic history and are possibly the most ancient rainforests in the world (Adam 1992). The Wet Tropics rainforests are naturally fragmented into a number of different units or blocks (Nix and Switzer 1991). One

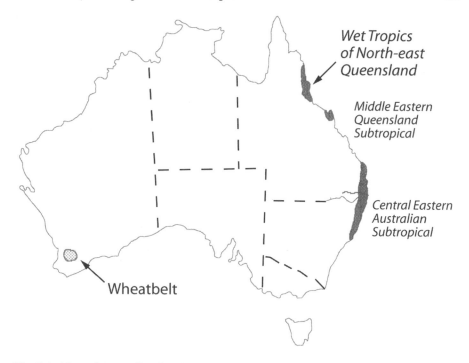

Fig. 7.1. Map of Australia showing approximate locations of the heavily fragmented areas discussed in the text, the Wet Tropics rainforests of northeastern Queensland and the wheatbelt of Western Australia. Also shown are other tracts of rainforest in eastern Australia referred to in the text

of these, that on the Atherton Tablelands, has been heavily fragmented during this century owing to its rich volcanic soils, which make the area suitable for agriculture. Most of the deforestation there occurred in the 1920s and 1930s and many of the present rainforest fragments can be identified on World War II maps. By 1983, >76,000 ha of forest had been removed (Frawley 1983; Winter 1987; Crome 1993). The rainforest fragments of the Atherton Tableland, surrounded as they are by uncleared rainforest on mountain slopes, appear to provide an ideal experimental setting in which to test the hypotheses concerning erosion of genetic diversity in fragmented habitats.

7.2.2 Study Species

In the wheatbelt, a single species was studied, *Oedura reticulata*, which is a small arboreal gecko distributed through southwestern Western Australia. In the Wet Tropics a suite of bird species and a skink were chosen for study. The birds chosen were insectivorous understorey passerines, because it has been shown that this guild is likely to be particularly susceptible to fragmentation effects (e.g., Willis 1974, 1979; Karr 1982). Among them were the scrubwrens *Sericornis* spp, which are known to have varying degrees of specialization to rainforest (Howe et al. 1981; Crome 1990). The most rainforest specialized (Atherton scrubwren *S. keri* and yellow-throated scrubwren *S. citreogularis*) were predicted to be least likely to be able to move between fragments and should, therefore, be most likely to meet the broad predictions outlined in the Introduction. *S. keri*, for example, is a strict rainforest specialist found only in the Wet Tropics. In contrast, the more habitat-generalized species, the large-billed scrubwren *S. magnirostris* and the white-browed scrubwren *S. frontalis* should be able to occur in nonrainforest habitats and would, therefore, be less likely to have been affected by fragmentation. Of these two species, *S. frontalis* is a widely distributed habitat generalist that occurs from eastern Australia's wet rainforests to semi-arid coastal shrublands in Western Australia. Also studied were the Chowchilla, *Orthonyx spaldingii,* and the grey-headed robin, *Poecilodryas albispecularis*, both of which are rainforest specialists, though the latter is common along rainforest edges. Finally, another Wet Tropics endemic, the prickly skink, *Gnypetoscincus queenslandiae*, was studied. This skink is abundant in suitable habitat, but has relatively low fecundity, making it potentially susceptible to demographic fluctuations (Cunningham 1993).

7.2.3 Methodology

In the wheatbelt, geckoes were sampled in 12 sites including remnants ranging from <0.25 to 5.4 ha and three larger nature reserves. In the Wet Tropics rainforests, birds and the skink were sampled in habitat fragments approximately 5–100 ha in area and separated by distances ranging from 0.5 to 10 km and also at sites in uncleared rainforest separated by similar distances. The two studies were directly comparable in the regional scales of the areas sampled. The wheatbelt study spanned approximately 30 km and the Atherton Tableland sites, including fragmented and uncleared sites, spanned a comparable area for the skink *G. queenslandiae* and a little less for the birds.

Both studies examined mitochondrial DNA because it is expected to be more sensitive to reductions in population size than the generally more slowly evolving DNA of the cell nucleus (Moritz et al. 1987; Avise 1991; but see below). Genetic diversity of mtDNA was measured with restriction enzymes in both studies, which cut DNA at particular short recognition sequences, each enzyme recognizing a unique sequence. If the DNA sequences of two individuals differ with respect to a recognition site for a given restriction enzyme (present in one, absent in the other), different patterns of DNA fragments will result and indices to genetic diversity can be developed. In addition, variable allozyme loci were examined in *G. queenslandiae* to contrast patterns of nuclear and mitochondrial differentiation.

7.2.4 Results

In the wheatbelt study of the gecko, there was considerable polymorphism within and among the study populations. In all, 22 haplotypes were observed among the samples from the 12 sites. There was little regional structure of the mtDNA haplotypes in the wheatbelt and some haplotypes were present in each of the three regions in which the 12 sites were distributed. However, generally higher mean numbers of haplotypes were present in the populations from nature reserves compared with those in the remnants (6.0 vs. 3.3, respectively, $t_{10}=2.42$, $P<0.05$) and haplotype diversity was also higher in the reserves (0.64 vs. 0.37, respectively, $t_{10}=1.63$, $P>0.05$). The data indicated that considerable genetic drift has occurred in some remnant populations with the concomitant loss of rare haplotypes, and a tendency towards fixation for one haplotype that is common in the nature reserves. Sarre (1995a) concluded that post-fragmentation populations of *O. reticulata* are unable to form a metapopulation structure and that stochastic extinction forces alone will be sufficient to severely reduce the gecko's regional distribution. The findings seem consistent with the expectations concerning reduced genetic diversity within fragments and increased diversity between fragments.

In contrast, in the rainforest study, restriction enzymes revealed extremely little mtDNA diversity in the birds either within and among fragments or between fragments and uncleared rainforest. Mitochondrial diversity in *G. queenslandiae*, which was studied in more detail, was extremely variable both within and among populations, and was largely determined by regional genetic structuring. Populations of *G. queenslandiae* in the southern and central Atherton Tableland showed low levels of mitochondrial diversity regardless of whether these were in fragments or continuous forest. Haplotype and nucleotide diversities were, respectively, 0.22±0.12 and 0.23 % for the southern sites and 0.70±0.08 and 0.53 % for the eastern sites. Higher levels of diversity, and a different mitochondrial lineage, were found in continuous forest populations from the eastern Atherton Tableland and this diversity was significantly structured (Gst=0.31, $P<0.001$). Both these eastern and southern lineages were present in two fragments on the east Atherton Tableland, resulting in high levels of diversity in these populations. Allozyme analysis also showed evidence of reduced diversity in the south and a cline in allele frequencies consistent with secondary contact among the east Atherton and south Atherton populations. As expected, given the larger effective population size of nuclear markers, there was greater diversity within sites and less differentiation among sites with allozymes than with mitochondrial DNA.

The rainforest study's scope soon expanded to include a larger geographical scale. The various species had also been sampled in other naturally defined rainforest blocks in the Wet Tropics to the north of Atherton and, where distributions permitted, in subtropical rainforest nearly 2000 km to the south. Coupled with more detailed sequence level data, clear macrogeographic patterns in the organization of genetic diversity emerged at these scales. Within the Wet Tropics, concordant breaks in the location of within-species patterns of genetic diversity emerged in unrelated rainforest specialists. The patterns differed principally in the magnitude of the divergence between populations on either side of the concordantly located genetic disjunctions. Thus, populations of these rainforest specialists in the fragmented Atherton Tableland rainforests were more closely related to populations several hundred kilometers to the south than they were to populations only a short distance to the north on the other side of the primary disjunction. This scale of structuring represents a much deeper level than that observed for *G. queenslandiae* across the Atherton Tableland.

The absence of detectable genetic effects from recent fragmentation in the Wet Tropics was attributed to the dynamic long-term history of the rainforests (see also Joseph et al. 1995). Specifically, repeated episodes of climatic fluctuation may have induced natural fragmentation and expansion of the rainforests of the Wet Tropics and so shaped patterns of genetic diversity at the level of the resolution detectable with restriction enzyme analysis of mtDNA. At that level of resolution in the birds, presumably not enough time had elapsed since the rainforests colonized the volcanic soils of the Atherton Tablelands for mutation in the pre-fragmentation populations to have generated the levels of diversity necessary for applying the classical theory of fragmentation effects. In *G. queenslandiae* diversity was sufficient only in the east Atherton sites, yet clear evidence of introgression was found in those sites making them unsuitable for comparison with the east Atherton continuous forest. In both cases historical perturbations to the distribution and amount of genetic diversity made it impossible to test for effects of recent fragmentation.

In summary, although the wheatbelt and rainforest studies were designed and conducted with similar aims and methodologies in similarly fragmented habitats, the patterning and regional structure of the genetic diversity in the two areas were very different. We suggest that this reflects differing long-term histories of the ecosystems involved (Fig. 7.2) with the wheatbelt system possibly having been stable for much longer than the rainforests of the Wet Tropics. In the wheatbelt, the diversity at loci accessible by restriction enzyme analysis of whole mtDNA enabled the classical fragmentation theory to be applied in understanding population structure. Either historical effects have not operated in the wheatbelt, or enough time has elapsed since historical effects may have occurred there for mutation to have generated adequate diversity for study. In contrast, the diversity accessed by the same laboratory methodology applied to the Wet Tropics situation was unsuitable for applying the theory. There, however, the existing samples could still be analyzed at loci having much higher mutation rates than mtDNA, such as microsatellites, and which therefore reveal greater diversity than restriction enzyme analysis of mtDNA.

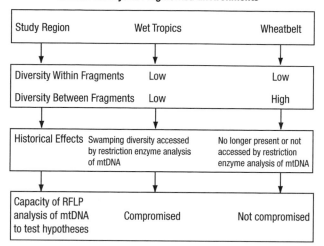

**Comparative Summary of Two
Genetic Surveys of Fragmented Environments**

Study Region	Wet Tropics	Wheatbelt

Diversity Within Fragments	Low	Low
Diversity Between Fragments	Low	High

Historical Effects	Swamping diversity accessed by restriction enzyme analysis of mtDNA	No longer present or not accessed by restriction enzyme analysis of mtDNA

Capacity of RFLP analysis of mtDNA to test hypotheses	Compromised	Not compromised

Fig. 7.2. Schematic comparison of two studies designed to examine the genetic effects of fragmentation, but which revealed differing magnitudes of historical effects on the study areas

The main messages we should like to draw from this comparison are those of study design. Before launching into programs of field sampling and laboratory study of the genetic effects of anthropogenic fragmentation, a pilot survey of macrogeographic genetic diversity of the species involved can improve study design in two critical ways. First, one will be better able to select appropriate and historically matched sites for study of anthropogenic fragmentation. Second, an understanding of the magnitude of historical effects on genetic diversity in the study area can guide the choice of a suitable genetic marker system (or, at least, of markers that will not be suitable; see also Bermingham and Avise 1986) and hence enable access to levels of diversity appropriate for studying the genetic consequences of anthropogenic fragmentation. A similar point has been made by Steinberg and Jordan (1997). Figure 7.3 summarizes the suggested relationship between the time since historical effects may have homogenized genetic diversity and various methodologies one could apply to the study of fragmentation.

The Wet Tropics study is perhaps sufficient to make these points generally. The comparison between the Wet Tropics and the wheatbelt studies, however, enables more specific conclusions. When studying the genetic effects of anthropogenic fragmentation, critical factors to determine within the proposed study area are the levels of diversity in populations and the degree of genetic structuring among them. These two factors demand differing sampling regimes. The former requires analysis of large samples from relatively few sites. For the latter, relatively few samples are required per site, but these sites must be scattered throughout the study area, the number of sites depending on the scale of genetic structuring encountered in a pilot macro-

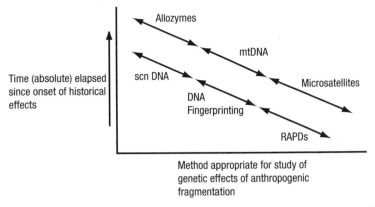

Fig. 7.3. Schematic summary of how the time elapsed since the onset of historical effects on genetic diversity can contribute to the choice of genetic markers one uses in studying the effects of fragmentation

geographic study. When the data are collected and analyzed, a further caveat applies. Whereas we can be confident about results indicating significant population subdivision, results indicating no subdivision may still be a function of the sensitivity of the molecular analysis employed and the number of individuals sampled in each subpopulation.

7.3 Pre-Fragmentation. An Alternative Perspective on Genetic Structure of Natural Populations

The comparison above of two published studies has arrived retrospectively at the conclusion that an initial macrogeographic perspective is a useful and important precursor to designing laboratory and field studies of the genetic effects of fragmentation. We should now like to approach the same point prospectively by considering three recent studies of population genetic structure as hypothetical surveys of pre-fragmentation diversity. Potentially, any number of similar studies of genetic structure in natural populations might be viewed in this way. Although none of the three studies was done in the context of fragmentation, we discuss them here, because together they give some insight into the range of geographical scales and methods with which pre-fragmentation structures can be addressed.

7.3.1 Three Sympatric Amazonian Rodents: Contrasting Genetic Structures

Patton et al. (1996) noted that risk assessments in conservation biology, such as assessing the genetic risks of fragmentation, require knowledge of population genetic structure. Accordingly, they used approximately 400 base pairs of sequence from the cytochrome b gene of mtDNA to investigate the population genetic structure of three species of Amazonian rodents (*Oligoryzomys microtis, Oryzomys capito* and *Mesomys hispidus*) sampled from the same set of localities along a 1,000-km stretch of the Rio Juruá in western Brazil. Sample sites were at the headwaters and upper and lower central sectors of the river and, in one species, from the river mouth.

Each of the three species exhibited distinctly different patterns of haplotype differentiation. In *M. hispidus* 74 % of the overall low total diversity was between regions, whereas only 11.2 % was within local populations within regions. It showed a strong pattern of isolation by distance throughout its range along the entire river indicating that gene flow is largely restricted to adjacent localities. In contrast, *Ol. microtis* showed weaker isolation by distance only over the largest geographical distances. Finally, in *Or. capito*, 90 % of the variation was within local populations within regions such that this species had no discernible genetic structure throughout its range along the river.

Should fragmentation occur along a given sector of Rio Juruá, and the prospect is not impossible, how might Patton et al.'s (1996) macrogeographic overview guide the design of study of its genetic effects? Recalling that their study was primarily designed to look at macrogeographic structure rather than within-population level diversities, we suggest the following. If sufficient time had elapsed post-fragmentation to expect pre-existing diversity to have become homogenized, a study with the same methods used by Patton et al. (1996) would be appropriate for *Or. capito* because diversity was equally distributed along the river. In the strongly structured *M. hispidus*, however, fragmentation in one region would be affecting a pool of genetic diversity different from that in another. From the outset, careful matching of sites would be necessary to avoid confusing high divergence among sites due to anthropogenic fragmentation with similarly high divergence due to the original isolation-by-distance structure. Further, it would probably be necessary to use more rapidly evolving loci to study the effects of anthropogenic fragmentation.

7.3.2 Yellow-Footed Rock-Wallabies:
a Naturally Patchily Distributed Species

Pope et al. (1996) studied the macro- and microgeographic genetic population structure of the yellow-footed rock-wallaby *Petrogale xanthopus*. Strictly confined to rocky outcrops scattered across inland southeastern Australia, the species is a superb example of a naturally patchily distributed species. As such, data on its genetic population structure should be of interest to anyone contemplating the effects of fragmentation on similarly patchy species. Although the authors' primary aim was to determine whether colonies in Queensland separated by between 10 and 70 km are temporally stable or transient, they began with a macrogeographic overview comparing the two recognized subspecies, *P. x. xanthopus* and *P. x. celeris*, sampled from localities some 1000 km apart. Measured with the relatively rapidly evolving control region of mtDNA, divergence between the two subspecies was generally an order of magnitude greater than that between samples from different colonies within the main study area. This initial macrogeographic survey supported the view that the two subspecies were evolutionarily distinct units, but also indicated that the study's main aims would ideally require a genetic marker with a rate of evolution more rapid than the control region. Accordingly, the main aims were then addressed with nuclear microsatellites, but also with the mtDNA control region. Populations separated by 70 km of unsuitable habitat differed significantly for mtDNA and at microsatellite loci. Populations separated by 10 km of apparently suitable habitat, however, showed a significant difference in allele frequency at one microsatellite locus, but were statistically homogeneous for mtDNA.

This case amply demonstrates the value of a macrogeographic pilot survey in aiding one's choice of a marker with which to study fragmentation effects at a microgeographic scale. Specifically, the pilot study demonstrated the worth of making the substantial investment of time and money that can be involved in developing microsatellite primers for new study species.

7.3.3 *Eucalyptus argutifolia*: Clonal Reproduction and Fragmentation

Clonal reproduction challenges study of the effects of fragmentation because pre-fragmentation effective population sizes are likely considerably less than observed census sizes. Selection of a genetic marker system is therefore especially critical. In this context a pilot macrogeographic study might appropriately be scaled at orders of magnitude less than the areas considered in all the previous examples in this chapter.

Kennington et al. (1996) reported on a genetic study of *Eucalyptus argutifolia*, a rare multistemmed (mallee) eucalypt confined to Western Australia. Like the preceding species *E. argutifolia* is naturally patchy and has 15 widely disjunct populations in which most mallees cluster and form single or several distinct groups. Three putative clones of the species from two populations 20 km apart had earlier been identified by allozymes and these consisted of 6, 15 and 21 mallees, covering areas of 58, 75 and 160 m², respectively. The question addressed by Kennington et al. (1996) was whether the clustering of identical genotypes in this species resulted from recruitment of allozymically uniform siblings around a maternal plant or from clonal growth. They used M13 DNA fingerprinting of nuclear DNA to test for genetic variation in these clones. Different seedlings from the same maternal plant could be discriminated, but there was no variation within each of the putative clones. The clustering of mallees with identical multilocus allozyme genotypes within populations, therefore most likely reflects the clonal habit rather than limited seed dispersal clustering allozymically uniform individuals. Aside from highlighting the great potential vulnerability of stands with few individuals, this case shows that under certain circumstances the geographical scale for a pilot macrogeographic survey can appropriately be set at the order of magnitude of the two post-fragmentation studies compared earlier in this chapter.

7.4 A Final Theoretical Consideration

Young et al. (1996) discussed several theoretical difficulties that have arisen in the interpretation of genetic data from fragmented species. We close these remarks on the genetics of fragmentation with brief comments on a theoretical issue, the use of statistics of population structure in fragmented environments (i.e., Fst – Wright (1965); Gst – Takahata and Palumbi (1985); Nst – Lynch and Crease (1990); $\emptyset st$ – Excoffier et al. (1992); Rst – Slatkin1995).

As noted above, Sarre (1995a) concluded that the gecko O. *reticulata* could not maintain a metapopulation structure in the Western Australian wheatbelt. Against this, however, he further noted the paradoxical conclusion that estimates of the number of individuals being exchanged between habitat fragments were greater than one per generation. This finding flags a theoretical concern in studying the population genetic consequences of fragmentation involving (1) the inter-relationships between the correlations of genes within local populations and those among populations and (2) Nm, the number of individuals migrating between populations. Nm is inversely related to the above population structure statistics, which for simplicity, we will represent below only with Gst.

Varvio et al. (1986) theoretically demonstrated that in fragmented or subdivided populations Gst approaches equilibrium much more quickly than heterozygosity. Caution is needed when applying this result. Sarre (1995a) noted that his calculation of Nm being greater than 1 for O. *reticulata* between wheatbelt fragments probably indicates that the fragmentation of the wheatbelt has been so recent that even Gst has not yet equilibrated and its values are being underestimated, therefore leading to overestimates of Nm. This in turn demonstrates a phenomenon seen in a number of recent studies, i.e., a time lag between the onset of demographic changes in a species and the response to those changes that can be measured with population genetic parameters (see also Barton and Wilson 1995; Lavery et al. 1996a,b). Where such a time lag applies, temporally staggered genetic surveys would give useful insights into estimates of parameters such as Nm.

7.5 Conclusion

We have stressed two elements of study design that need careful attention when wishing to apply classical population genetics theory to population structure in anthropogenically fragmented environments: the importance of selecting study sites that are appropriately matched and loci that will offer suitable levels of diversity. They can be assessed through pilot macrogeographic study ranging beyond the area of interest. There may always be some merit in going direct to microsatellites or AFLP markers (Mueller and Wolfenbarger 1999), which can be expected to reveal high within-population level diversity if it is present. One has to weigh this, however, against the investments of time and money that may be necessary to develop new primers for a microsatellite study. We have also briefly mentioned the time lag that can occur between the onset of changes in a population's demography and the ability to detect signs of those changes in genetic parameters. In conclusion, historical and analytical factors are such that applying the classical theory of eroded genetic diversity in fragmented environments can be far from simple. Appreciation of deeper evolutionary history, especially as it is reflected in the regional structure of genetic diversity, and the demography of a nonequilibrium system can be critical to the success with which theory is applied and tested in anthropogenically fragmented environments.

Acknowledgement. L.J. thanks Pablo Marquet and Gay Bradshaw for the invitation to attend the IAI Workshop at which this work was outlined and his co-authors for their later collaboration. L.J. and M.C. conducted their work in the Wet Tropics under the supervision of Craig Moritz to whom we both, once again, express our thanks. We are once again happy to thank all those mentioned in earlier papers for their help with our fieldwork. The Wet Tropics work was funded by grants from the Australian Research Council, the Commonwealth Scientific and Industrial Research Organization and the Queensland Department of Environment. For comments, critical reading of the manuscript and discussion we thank L. Belasco, E. Lessa, J. Patton, L. Pope and J. Sumner. Writing of the present paper was supported by the Consejo Nacional de Investigacion Científica y Técnologica (CONICYT), Programa de Desarollo de las Ciencias Basicas (PEDECIBA), and CSIC, all of Montevideo, Uruguay and the Academy of Natural Sciences, Philadelphia, Pennsylvania, USA.

References

Adam P (1992) Australian rainforests. Clarendon Press, Oxford

Avise JC (1991) Ten unorthodox perspectives on evolution prompted by comparative population genetic findings on mitochondrial DNA. Annu Rev Genet 25:45–69

Barton NH, Wilson I (1995) Genealogies and geography. Philos Trans R Soc Lond Ser B 349:49–59

Bermingham E, Avise JC (1986) Molecular zoogeography of freshwater fishes in the southeastern United States. Genetics 113:939–965

Crome FHJ (1990) Vertebrates and succession. In: Webb LJ, Kikkawa J (eds) Australian tropical rainforests: science, values, meaning. CSIRO, Melbourne, Australia, pp 53–64

Crome FHJ (1993) Tropical forest fragmentation: some conceptual and methodological issues. In: Moritz C, Kikkawa J (eds) Conservation biology in Australia and Oceania. Surrey Beatty and Sons, Chipping Norton, Australia, pp 61–76

Cunningham M (1993) Reproductive biology of the Prickly Forest Skink, *Gnypetoscincus queenslandiae*, an endemic species from northern Queensland. Mem Queensl Mus 34:131–138

Cunningham M, Moritz C (1998) Genetic effects of forest fragmentation in a rainforest restricted lizard (Scincidae: *Gnypetoscincus queenslandiae*). Biol Conserv 83:19–30

Excoffier L, Smouse PE, Quattro JM (1992) Analysis of molecular variance inferred from metric distances among DNA haplotypes: application to human mitochondrial DNA restriction data. Genetics 131:479–491

Frawley KJ (1983) A history of forest and land management in Queensland, with particular reference to the north Queensland rainforest. Rainforest Conservation Society of Queensland, Brisbane, Australia

Hobbs RJ, Saunders DA (1993) Effects of landscape fragmentation in agricultural areas. In: Moritz C, Kikkawa J (eds) Conservation biology in Australia and Oceania. Surrey Beatty and Sons, Chipping Norton, Australia, pp 77–95

Howe RW, Howe TD, Ford HA (1981) Bird distributions in small rainforest fragments in New South Wales. Aust Wildl Res 8:637–651

Joseph L, Moritz C (1993) Phylogeny and historical aspects of the ecology of eastern Australian scrubwrens *Sericornis* spp: evidence from mitochondrial DNA. Mol Ecol 2:161–170

Joseph L, Moritz C (1994) Mitochondrial DNA phylogeography of birds in eastern Australian rainforests: first fragments. Aust J Zool 42:385–403

Joseph L, Moritz C, Hugall A (1993) A mitochondrial DNA perspective on the historical biogeography of middle eastern Queensland rainforest birds. Mem Queensl Mus 34:201–214

Joseph L, Moritz C, Hugall A (1995) Molecular support for vicariance as a source of diversity in rainforests. Proc R Soc Lond B 260:177–182

Karr J (1982) Avian extinction on Barro Colorado Island, Panama: a reassessment. Am Nat 119:220–239

Kennington WJ, Waycott M, James SH (1996) DNA fingerprinting supports notion of clonality in a rare mallee, *Eucalyptus argutifolia*. Mol Ecol 5:693–696

Lavery S, Moritz C, Fielder DR (1996a) Indo-Pacific population structure and evolutionary history of the coconut crab *Birgus latro*. Mol Ecol 5:557–570

Lavery S, Moritz C, Fielder DR (1996b) Genetic patterns suggest exponential population growth in a declining species. Mol Biol Evol 13:1106–1113

Lynch M, Crease TJ (1990) The analysis of population survey data on DNA sequence variations. Mol Biol Evol 7:377–394

McCauley DE (1991) Genetic consequences of local population extinction and recolonization. Trends Ecol Evol 6:5–8

Moritz C, Dowling T, Brown W (1987) Evolution of animal mitochondrial DNA: relevance for population biology and systematics. Annu Rev Ecol Syst 18:269–292

Moritz C, Joseph L, Adams M (1993) Cryptic diversity in an endemic rainforest skink (*Gnypetoscincus queenslandiae*). Biodiv Conserv 2:412–425

Moritz C, Joseph L, Cunningham M, Schneider C (1997) Molecular perspectives on historical fragmentation of Australian tropical and subtropical rainforest: implications for conservation. In: Laurance WF, Bierregaard RO Jr (eds) Tropical forest remnants: ecology, management and conservation of fragmented communities. University of Chicago Press, Chicago, pp 442–454

Mueller U, Wolfenbarger L (1999) AFLP genotyping and fingerprinting. Trends Ecol Evol 14:389–394

Nix HA, Switzer M (1991) Rainforest animals. Atlas of vertebrates endemic to Australia's wet tropics. Kowari 1. Australian National Parks and Wildlife Service, Canberra

Patton J, da Silva M, Malcolm JR (1996) Hierarchical genetic structure and gene flow in three sympatric species of Amazonian rodents. Mol Ecol 5:229–238

Pope L, Sharp A, Moritz C (1996) Population structure of the yellow-footed rock-wallaby *Petrogale xanthopus* (Gray, 1854) inferred from mtDNA sequences and microsatellite loci. Mol Ecol 5:629–640

Sarre S (1995a) Mitochondrial DNA variation among populations of *Oedura reticulata* (Gekkonidae) in remnant vegetation: implications for metapopulation structure and population decline. Mol Ecol 4:395–405

Sarre S (1995b) Size and structure of populations of *Oedura reticulata* (Gekkonidae) in woodland remnants: implications for the future regional distribution of a currently common species. Aust J Ecol 20:288–298

Sarre S, Smith GT, Meyers JA (1995) Persistence of two species of gecko (*Oedura reticulata* and *Gehyra variegata*) in remnant habitat. Biol Conserv 71:25–33

Schneider CJ, Cunningham M, Moritz C (1998) Comparative phylogeography and the history of endemic vertebrates in the Wet Tropics rainforests of Australia. Mol Ecol 7:487–498

Slatkin M (1995) A measure of population subdivision based on microsatellite allele frequencies. Genetics 139:457–462

Steinberg EK, Jordan CE (1997) Using molecular genetics to learn about the ecology of threatened species: the allure and illusion of measuring genetic structure in natural populations. In: Feidler PF, Kareiva P (eds) Conservation for the coming decade. Chapman and Hall, New York, pp 438–460

Takahata N, Palumbi SR (1985) Extranuclear differentiation and gene flow in the finite island model. Genetics 109:441–457

Varvio S-L, Chakraborty R, Nei M (1986) Genetic variation in subdivided populations and conservation genetics. Heredity 57:189–198

Willis E (1974) The composition of avian communities in remanescent woodlots in southern Brazil. Papeis Avulsos Zool, Sao Paulo 33:1–25

Willis E (1979) Populations and local extinctions of birds on Barro Colorado Island, Panama. Ecol Monogr 44:153–169

Winter JW (1987) Rainforest clearfelling in northeastern Australia. Proc R Soc Queensl 98:41–57

Wright S (1965) The interpretation of population structure by F-statistics with special regard to systems of mating. Evolution 19:395–420

Wright S (1978) Evolution and the genetics of natural populations. Volume 4. Variability within and among natural populations. University of Chicago Press, Chicago

Young A, Boyle T, Brown T (1996) The population genetic consequences of habitat fragmentation for plants. Trends Ecol Evol 11:413–418

8 Forest Fragmentation, Plant Regeneration and Invasion Processes Across Edges in Central Chile

R.O. Bustamante, I.A. Serey, S.T.A. Pickett

8.1 Introduction

Forest degradation is a worldwide phenomenon (Myers 1988; Groom and Schumaker 1993). Annually, millions of hectares of tropical and temperate forests are deforested and fragmented for agriculture, farming and forestry (Schelhas and Greenberg 1996), creating a new landscape that differs notably in pattern and process relative to the original forested one, producing a significant decrease in biodiversity (Saunders et al. 1991; Viana and Tabanez 1996).

The effect of the matrix on remnant fragments (the edge effect) has received the majority of attention in forest fragmentation research (Chen et al. 1992; Ranney et al. 1981; Saunders et al. 1991; Santos and Tellerías 1992; Murcia 1995; Didham 1997; Kapos et al. 1997; Laurance 1997; Turton and Freiburger 1997). The matrix may induce several types of drastic abiotic changes in fragments. For example, a gradient of photosynthetically active radiation (PAR) decreases and soil moisture increases from the edge to the interior forest (Kapos 1989). The new abiotic conditions induced by the matrix affect plant establishment and recruitment (Chen et al. 1992) and favor the invasion of alien plants from the matrix, modifying biotic interactions such as plant competition (Saunders et al. 1991) and pollination (Jennersten 1988; Aizen and Feinsinger 1994a,b).

Edges have a great impact on plant regeneration. Generally, shade-intolerant plants are favored near the forest-matrix edge, in contrast to shade-tolerant plants that will be favored only in the interior (Ranney et al. 1981; Chen et al. 1992). Similarly, edge creation modifies seed predation (Santos and Tellerías 1992; Murcia 1995). Though limited in scope, variation in seed consumption as a function of forest-type has been reported (Sork 1983; Burkey 1993). For instance, in temperate forest fragments seed consumption decreases from the edge (Sork 1983), in contrast to tropical forests where seed consumption tends to increase toward the interior (Burkey 1993). Seed con-

Ecological Studies, Vol. 162
G.A. Bradshaw and P.A. Marquet (Eds.)
How Landscapes Change
© Springer-Verlag Berlin Heidelberg 2003

sumption also varies depending on plant species. In temperate forest fragments (northeast New York, USA), seeds of native species such as *Quercus rubra* and *Prunus serotina* are heavily consumed in edges, in contrast to the alien plant *Ailanthus altissima* whose seeds are not consumed at all (Bustamante et al., in prep.).

Studies dealing with the invasion of plants across edges are scarce. However, some generalizations have emerged from tropical forests. First, fragments are more or less prone to plant invasion depending on structural aspects of fragments such as cover, disturbance regime or life attributes of invaders (Brothers and Spingarn 1992). Second, weeds, herbs and lianas seem to be the most successful invaders occupying edges, gaps or disturbed sites within fragments (Laurance 1997; Viana et al. 1997). Third, fragments of evergreen forests are less prone to invasion than deciduous or semi-deciduous forests (Janzen 1983; Laurence 1997; Viana et al. 1997). To what extent these generalizations apply to temperate forests is an issue that deserves further research.

Studies of edge effects in both tropical and temperate forests have tended to be descriptive, lacking sufficient investigation of the underlying mechanisms that link patterns and processes at population, community or ecosystem levels (Murcia 1995). The absence of such a mechanistic approach poses difficulties in dealing with more complex edge effects, where a hierarchy of factors is responsible for the observed regeneration patterns for native and exotic plants. For example, abiotic requirements for the regeneration and/or the invasion of a plant may be modulated by biotic interactions (e.g., seed predation, plant competition) across edges. The inclusion of other factors, in addition to abiotic requirements of plants, will give us a more realistic picture of forest fragmentation and its effect on plant population process.

In this chapter, we focus on the mechanisms underlying edge effects on exotic and native plant recruitment. First, we document the invasion of *Pinus radiata* across edges of fragments of the Maulino forest (sensu San Martín and Donoso 1996) in central Chile. Second, we develop a demographic model to explore recruitment probability for shade-intolerant and shade-tolerant plant species across forest edges, specifically considering the effect of seed predation. Third, we use the model to explain some plant regeneration patterns documented for exotic (pines) and native trees in temperate forests.

8.2 Edge Effect and the Invasion of *Pinus radiata* into Temperate Forests of Central Chile

During the 20th century, Chilean native forests have suffered massive degradation due to human activities (Donoso and Lara 1995). Large areas of land, formerly covered by forests, have been converted to agricultural lands and

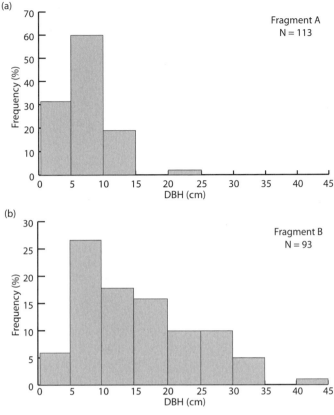

Fig. 8.1. Size distribution (evaluated by the DBH in cm) of *Nothofagus glauca* (hualo) in a *fragment A* (disturbed) and *fragment B* (less disturbed), Pantanillo Forest Station, Rio Maule Region, Chile

plantations of exotic trees (Lara and Veblen 1993). *Pinus radiata* (Monterey pine) is a colonizer and shade-intolerant plant, native to California, with wing-dispersed pollen and seeds (Lewis and Ferguson 1993). This species was introduced to Chile in 1885 for ornamental purposes only. However, this exotic plant is now the base of the Chilean forestry industry. By 1943, there were 143,500 ha of Monterey pine plantations. By 1993, this area amounted to 1.24 million ha, the largest area of Monterey pine plantations of the world (Lewis and Ferguson 1993). This expansion has come at the expense of native forests. Up to 18 % of the native forest of the coastal range in the Rio Maule Region, central Chile (35°–36°S lat.), was converted to plantations in the period 1978–1987 (Lara and Veblen 1993). Currently, the landscape of this region is a mosaic where the original forest remains in a highly fragmented state relative to its former condition, immersed in an "ocean" of pine plantations (San Martín and Donoso 1995; Bustamante and Castor 1998; Grez et al. 1998).

The hualo forest is a deciduous forest located between 35° and 37°S lat. in the Coastal and Andean range of the Rio Maule Region. The dominant species is the endemic tree *Nothofagus glauca* (hualo). These trees reach a height of 20 m, with a mean diameter at breast height (DBH) of 24 cm, and a density of about 4,000 ind./ha (San Martín and Donoso 1995). The altitudinal range of *Nothofagus* forest varies between 300 to 1200 m a.s.l. (Donoso 1975). The forest interior includes diverse species of vines (e.g., *Lapageria rosea*) and shrubs (e.g., *Escalonia pulverulenta*) as understory components. Fires are common in the Río Maule Region, a product of agriculture and forestry practices (Sáiz 1990). Hualo trees are cut for fuel and regenerate very fast by vegetative reproduction. Currently, there are few localities with old-growth forest. Most of the fragments are composed mainly of trees regenerating from stumps.

Pinus radiata has been an extremely successful invader in diverse ecosystems of the Southern hemisphere (Richardson et al. 1994), probably as a consequence of its rapid maturation, serotony, resistance to fires, and the high ability of its seeds to disperse by wind (Richardson et al. 1990; Rejmánek 1996). In Chile, this species has been considered noninvasive (Kruger et al. 1989; but see San Martín et al. 1984). However, the evidence supporting this assertion is anecdotal without quantification of seed dispersal, germination, survival, establishment and spread of this species. In the next section, we present preliminary evidence documenting the active invasion of the hualo forest by this exotic species.

The study was conducted at Pantanillo Forest Station, property of the University of Chile (35°30′S; 72°20′W). The station includes approximately 200 ha of fragmented hualo forest, surrounded by extensive plantations of Monterey pines. In this area, the forest was severely disturbed in the past by cutting, logging and fire. The forest consists primarily of secondary growth from stumps (pers. obs.). Fragments tend to be small (mode 5 ha) and range from 0.5 to 315 ha (for more details, see Estades and Temple 1999).

Two fragments of hualo forest were studied, one disturbed (fragment A) and one less disturbed (fragment B), located approximately 5 km apart. They are similar in size (25–26 ha), but different in structure. In fragment A, hualo trees average 6 m in height, with evidence of recent human disturbances (cutting, burning) and a canopy cover of about 45 %. In fragment B, hualo trees average 15 m in height, with a canopy cover of about 95 %. There is no evidence of recent human disturbance. Furthermore, fragment B has more abundant lichens, mosses and vines (*Lapageria rosea*, pers. observ.), suggesting a more humid environment. Based on the general assertion that more structured and closed fragments become more resistant to the invasion by alien plants (Brothers and Spingarn 1992), we tested the prediction that fragment A will be more prone to invasion than fragment B.

In each fragment, we evaluated the distribution and abundance of Monterey pines and hualo trees. We counted the number of individuals using a rectangle of 30 × 3 m², each 10 m, along a transect 80 m in length, running from the edge to the interior of the fragment. We also recorded the DBH of individuals (in cm). Individuals less than 1.5 m height were evaluated at the base of the shoot. We recorded the frequency of pine trees with cones, in order to determine the reproductive stage of individuals. As we have not yet found adequate replicates for fragments A and B, the results should be taken as pre-liminary.

8.3 Results

In fragment A, the hualo trees were smaller (in DBH) than trees in fragment B (fragment A: 7.03±0.68 cm; average±2 SE; fragment B: 15.28±1.72 cm, aver-age±2 SE; Wilcoxon, Z=8.38, $P<0.001$). In fragment A the majority of individ-uals were concentrated into the small-size classes (Fig. 8.1a), whereas in frag-ment B, the size of individuals was more evenly distributed (Fig. 8.1b).

The density and size of pines were clearly higher in fragment A than in fragment B (Table 8.1). Nevertheless, the proportion of reproductive individ-uals was similar in both fragments (Table 8.1). Additionally, in fragment A, pines were distributed along the whole fragment, whereas in fragment B, they were concentrated in the edges of the fragment (Fig. 8.2).

Table 8.1. Comparison of Monterrey pine attributes between two fragments of hualo forest, differing in the degree of disturbance, Río Maule Region, central Chile. Figures represent average ±2 SE

Attributes	Fragment A (more disturbed)	Fragment B (less disturbed)	P
Pines density	5±1.5	1.09±0.6	0.004[a]
Pines size (DBH in cm)	5.8±2.0	2.19±0.6	0.003[a]
Proportion of reproductive individuals	0.24	0.22	0.42[b]

[a] Wilcoxon test.
[b] χ^2 test.

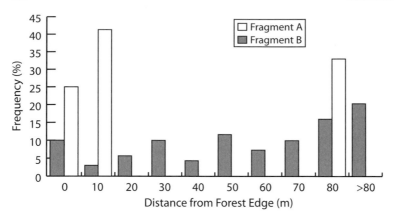

Fig. 8.2. Relative distribution of Monterrey pine across edges in fragments A and B, Pantanillo, Forest Station, Rio Maule Region, Chile. *Open bars* Pines of fragment A; *shaded bars* pines of fragment B

8.4 Discussion

Our results suggest that Monterey pine is invading the fragments of hualo forest, but the magnitude of this process depends on fragment structure. In fragments with smaller hualo trees and with a more sparse canopy (like fragment A), invasion occurs across the whole fragment and the established invaders are likely to reproduce. In contrast, in fragments with large individuals and with a dense canopy (fragment B), invasion occurs at the edges only, although successful invaders are equally likely to reproduce. Thus, we corroborated the prediction advanced by Brothers and Spingarn (1992). These drastic differences in invasiveness may be explained by comparing regeneration requirements of the pines with the structural attributes of the two types of fragments. Monterey pine is a colonizing, shade-intolerant plant (Lewis and Ferguson 1993). Moreover, although the hualo forest is deciduous, the leaves of the dominant tree *Nothofagus glauca*, remain in the canopy about 240 days (Mancilla 1987) generating a shaded habitat almost all year around. Obviously, this effect is more important in fragments with low human disturbance. Therefore, although seeds of pines can arrive in the interior of well-developed forest fragments (pers. observ.), abiotic conditions impose constraints to germination and establishment that prevent seedling recruitment.

Current landscape-level anthropogenic disturbance (cutting, intentional fires) that occurs in the Río Maule region makes native forest more susceptible to invasion by Monterey pine. First, this species prefers open and disturbed habitats for colonization (Richardson and Bond 1991; Richardson and et al. 1994). Second, because fragments are embedded in a matrix of pine plantations, they are exposed to a massive seed rain every year. In fact, *Pinus radiata* trees become reproductive at 5 years of age (Richardson and Bond 1991) and the rotation period of a plantation is about 25 years (Lara and Veblen 1993). Therefore, a reproductive individual can be producing and dispersing seeds for 20 years before cutting. This huge amount of seed coming from plantations every year increases the invasiveness of pines into the forest by a "mass effect" (Richardson and Cowling 1992). In this context, the small number of less disturbed and well-structured forest fragments is critical for conservation issues as they can withstand invasion by Monterey pine.

In summary, the conditions for a successful invasion of Monterey pine into hualo forest are met by existing conditions. First, the profuse seed rain assures seeds will arrive inside fragments. Second, once individuals are established they grow vigorously. Third, a substantial fraction of established individuals are reproductive (22–24 %), with the potential to produce and disperse offspring inside the forest, independent of the external subsidy of seeds. The fragments that pose resistance to this invasion (e.g., those with a dense canopy like fragment B), are scarce in the area and they are being replaced by Monterey pine plantations. We predict that the invasion of this exotic species will continue occurring, except in those fragments with a dense canopy. We have a unique opportunity to monitor this process, and to evaluate to what extent this exotic species will affect ecological processes at population, community and landscape levels. This knowledge will help to forest managers to conduct conservation and/or restoration practices so as to assure the persistence of the hualo forest.

8.5 A Graphic Model

The dominance of one mechanism (e.g., seed predation) or the interaction of mechanisms (e.g., seed predation and changes in light intensity) in explaining regeneration across forest edges have not been addressed in forest fragmentation research. For example, the regeneration of a shade-intolerant plant may be depressed in the edges if seed predation counterbalances it in that habitat. We propose a demographic model that predicts the shape of plant recruitment probability, P(R), across a spatial gradient x, from the edge to the interior of a fragment considering both abiotic requirements for regeneration and seed predation. This model will be centered on seeds and seedlings, stages considered critical for plant population dynamics and regeneration processes (Chambers and MacMahon 1994).

P(R) is a function of both seed survival probability P(S) and the seedling establishment probability P(E). P(S) is estimated by dividing the number of unconsumed seeds by the total number of dispersed seeds. P(E) is obtained dividing the number of established seedlings by the number of seeds that escaped seed predation. For the present discussion, we consider only the extremes of the regeneration niche of plants, that is, shade-tolerant vs. shade-intolerant plant species and assume that P(S) and P(E) are considered linear functions of x.

As forest fragmentation creates a gradient of PAR from edge to interior (Kapos 1989), two opposing linear functions of P(E) are used: a decreasing function (b<0) in shade-intolerant plants (Fig. 8.3a) and an increasing function (b>0) in shade-tolerant plants (Fig. 8.3b). Moreover, three curves are defined to represent P(S), each differing with respect to slope: positive slope (b>0; Fig. 8.4a), zero slope (b=0; Fig. 8.4b) and negative slope (b<0; Fig. 8.4 c), corresponding to a situation of increasing, equal and decreasing seed predation across the spatial gradient x.

P(R) is derived from the multiplication of P(S) by P(E) (Price and Jenkins 1986; Chambers and McMahon 1994). In the model, each P(E) (Fig. 8.3a, b)

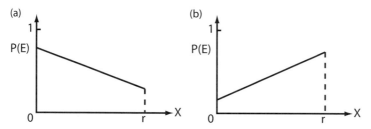

Fig. 8.3. Establishment probability P(E) across edges in fragmented forests: **a** shade-intolerant, **b** shade-tolerant plants

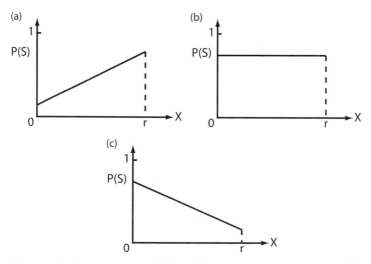

Fig. 8.4. Seed survival probability P(S) across edges in fragmented forests. Three curves are presented representing curves with **a** positive slope, **b** zero slope, and **c** negative slope

function is multiplied with each of the three linear functions P(S) (Fig. 8.4a–c). The resultant P(R) curves are derived for shade-intolerant (Fig. 8.5a–c) and shade-tolerant plants (Fig. 8.6a–c). If P(S)=1 (i.e., there is no seed predation across edges) then P(R)=P(E); if P(E)=1 (i.e., 100 % of the surviving seeds establish as seedlings) then P(R)=P(S). For mathematical details about the resulting P(R) functions, see Appendix 8.1)

The number of seedlings that recruit across edges, N_R, is obtained by multiplying the curve P(R) with N_S, the number of surviving seeds across edges. Because the seed shadows (the seed distribution curve from parent plants after dispersal) of adjacent parent trees tend to overlap each other in the interior of forests (Condit et al. 1992; Houle 1992; Bustamante and Simonetti 2000), it is reasonable to assume that N_S remains constant along x, thus resulting in an N_R shape identical to P(R).

Some interesting results emerge from the model. In the case of shade-intolerant plants, when P(S) is constant P(R) decreases across edges (Fig. 8.5b). When P(S) decreases from the edge, then P(R) increases from the edge (Fig. 8.5c) This result emerges from mathematical analysis (see Appendix 8.1), but we have no biological explanation for it. In the case of shade-tolerant plants, when P(S) is constant or when it increases from edges to the interior of a fragment, P(R) increases as well (Fig. 8.6a, b). Therefore, in both cases, seed predation does not change the recruitment pattern predicted by abiotic requirements alone. However, the picture changes drastically when P(S) is negatively correlated with P(E). In these cases, the shape of P(R) is identical for both shade-intolerant and shade-tolerant plants (Figs. 8.5a, 8.6c), with P(R) being maximized at some intermediate distance x′ from the edge of

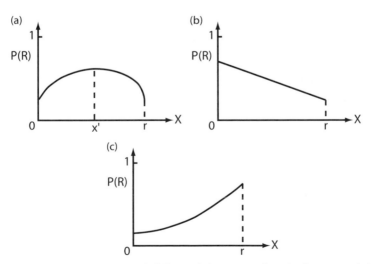

Fig. 8.5. Recruitment probability P(R) across edges in fragmented forests for shade-intolerant plant species (**a, b, c**). These functions were obtained multiplying the functions P(S) and P(E) from Figs. 8.3 and 8.4. x′ Distance where the function is maximal

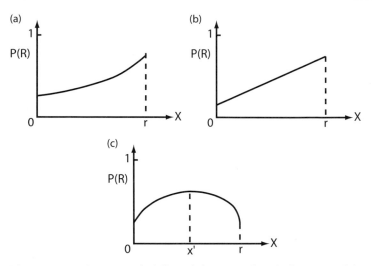

Fig. 8.6. Recruitment probability P(R) across edges in fragmented forests for shade-tolerant plants (**a, b, c**). These functions were obtained in the same manner as in Fig. 8.5. x' Distance where the function is maximal

the fragment to the interior. That is, when P(S) and P(E) are negatively associated, we are not able to discern if the predicted P(R) comes from shade-intolerant or shade-tolerant plant species.

More complex recruitment curves may be generated simply by modifying the slope of P(E) or P(S). In the next section some of these modifications will be examined in more detail in order to explain some recruitment curves observed in the field for exotic and native trees.

8.6 Model Application

8.6.1 Recruitment of Native Trees

Some studies dealing with regeneration of native plants across edges show patterns that cannot be explained by abiotic causes alone. Whitney and Runkle (1981), working in a hardwood forest (Ohio, USA), noted that the abundance of two shade-tolerant trees (*Carpinus caroliana* and *Ostrya virginiana*) was higher at the edges of the forests than in the interior. For instance, *Ostrya* was at least six times more abundant at the edges than in the interior of forests. Chen et al. (1992) described similar results working in Douglas-fir forests (Cascade Range, Oregon, USA). They observed that the density of seedlings and saplings of *Tsuga heterophylla*, a shade-tolerant tree, decreased significantly from the edge to the interior of fragments in the same manner as did *Pseudotsuga menziesii*, a shade-intolerant tree species in the same forest. These studies did not address these "anomalous" results or discuss why a shade-tolerant plant might decrease recruitment to the interior of forest, presumably more favorable for its germination and establishment.

The regeneration pattern observed by Whitney and Runkle (1981) and Chen et al. (1992) may be explained by the interaction of biotic and abiotic factors as our model suggests. It is reasonable to assume that x' (the distance where the recruitment curve is maximal) is a function of the slope of P(S), such b that when seed predation is more intense in the interior than at the edge of a forest at a level where seed abundance falls below a critical threshold, then recruitment may be severely depressed in that microhabitat even if abiotic conditions for regeneration are optimal (Crawley 1992). In terms of the model, P(S) has a strong negative slope and the resulting x' will move along the spatial gradient toward the edge (see x'$_1$, Fig. 8.7a). It is clear that as the slope of P(S) becomes more gentle, x' will tend to be located toward the interior of the fragment, which means that seed predation becomes lower toward the interior of the fragment (Fig. 8.7a). In sum, intense seed predation in the interior of fragments can generate the recruitment curves described for shade-tolerant trees in temperate forests of the USA.

8.6.2 Recruitment of Monterey Pine

The invasion pattern observed for the Monterey pine in the hualo forest (with pines concentrated at the edges of fragments), may be explained by our model as follows. Fragments with a dense canopy and with no evidence of human disturbance (like fragment B) are closed relative to fragments with a sparse

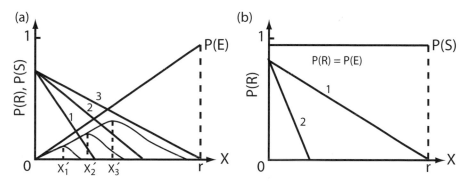

Fig. 8.7. a Expected recruitment probability P(R) across edges for shade-tolerant plants. These curves were obtained by varying the slope of P(S) curves *1, 2* and *3*. Note that x', the distance where the function is maximal, moves from the edge to the interior of forest; **b** expected recruitment probability P(R) across edges for *Pinus radiata* (Monterrey pine) in closed fragment (*curve 2*) and opened fragments (*curve 1*). We assumed that P(S)=1, therefore P(R)=P(E)

canopy and signs of intense human disturbance (fragment A). Consequently, because the Monterey pine is shade-intolerant, the slope of P(E) will be more negative in fragment A than in fragment B. No seed consumption has been reported for *Pinus radiata* (Muñoz-Pedreros et al. 1990), therefore P(S)=1 and P(R)=P(E). That is, the recruitment curve will depend on P(E) exclusively. In consequence, P(R) will be maximal near the edges in both kinds of fragments, but its slope will be less negative in fragment A, which means that recruitment will penetrate farther into the fragment, relative to more closed fragments such as fragment B (Fig. 8.7b).

8.7 General Conclusions

We have presented a graphical demographic model which examined the effects of abiotic/biotic conditions prevailing across edges on two population processes often considered independently in forest fragmentation research: regeneration of native trees and the invasion of exotic trees. First, our model explains that some temperate shade-tolerant trees increase recruitment at the edges, when severe seed predation occurs in the interior of fragments. Second, our model explains how fragments that maintain a dense canopy and reduced human disturbance discourage the invasion of exotic colonizers, such as the Monterey pine in the Maulino forest. These denser fragments may play a key role in the persistence of the native species they shelter, particularly if they are rare phenomena in the landscape (Schelhas and Greenberg 1996). Although fragmentation clearly predisposes forests to plant invasion (Laurance et al. 1997), the magnitude of this process may be moderated by the structure of fragments rather than their size or shape (Turner and Corlett 1996).

Acknowledgement. This work is a contribution of the Programa Bosques Nativos, D.I.D., Universidad de Chile. We acknowledge the 1994 Cary Summer Fellowship to Ramiro Bustamante, funded by the Mary Flager Cary Charitable Trust and the Institute of Ecosystem Studies (IES), Milbrook, New York, USA. This fellowship gave the senior author the opportunities to discuss many of the issues advanced in this study.

8.8 Appendix

$P(E), P(S)$ and $P(R)$ are function of x, the distance from the edge to the interior of a fragment. For simplicity, we assume that fragments are circular, therefore $0 \leq x \leq r$, 0 being located at the edge and r, the radius of the circular fragment.

Shade-tolerant plant strategies are represented by two linear functions: $P(E)=-m_1x+b_1$ for shade-intolerant species (Fig. 8.3a) and $P(E)=m_2x+b_2$ for shade-tolerant species (Fig. 8.3b). Seed survival is represented by three linear functions: with positive slope $P(S)=m_3x+b_3$, (Fig. 8.4a) with zero slope $P(S)=K$ (Fig. 8.4b) and with negative slope $P(S)=-m_4x+b_4$ (Fig. 8.4c). $P(R)$ results for the multiplication of both $P(E)$ and $P(S)$.

In the case of shade-intolerant plants, three $P(R)$ curves are obtained:

1. $P(E) = m_1x+b_1$
$$P(S) = m_3x+b_3$$
$$P(R) = -Ax^2-Bx + C \tag{1}$$

$A=m_1m_3$, $B=(m_1b_3-m_3b_1)$ and $C=b_1b_3$.

$P(R)$ is a convex function with has a maximal at an intermediate distance x' between 0 and r (Fig. 8.5a).

2. $P(E) = -m_1x+b_1$
$$P(S) = K$$
$$P(R) = -Dx + E \tag{2}$$

$D=Km_1$, $E=Kb_1$.

In this case, $P(R)$ is a linear function with a negative slope that decreases from 0 to r (Fig. 8.5b).

3. $P(E) =-m_1x+b_1$
$$P(S) =-m_4x+b_4$$
$$P(R) =Fx^2-Gx+H \tag{3}$$

$F=m_1m_4$, $G=(m_1b_4+m_4b_1)$ and $H=b_1b_4$.

$P(R)$ is a concave function that increases from 0 to r (Fig. 8.5c).

In the case of shade-tolerant plants, three $P(R)$ curves are obtained as well:

1. $P(E) = m_2x+b_2$
$$P(S) = m_3x+b_3$$
$$P(R) = Ix^2+Jx+K \tag{4}$$

$I=m_2m_3$, $J=(m_2b_3+b_2m_3)$ and $K=b_2b_3$.
$P(R)$ is a concave function that increases from 0 to r (Fig. 8.6a).

2. $P(E) = m_2x + b_2$
      $P(S) = K$
      $P(R) = Lx + M$ (5)

$L=Km_2, M=Kb_2$.
In this case, $P(R)$ is a linear function that increases from 0 to r (Fig. 8.6b):

3. $P(E) = m_2x + b_2$
      $P(S) = -m_4x + b_4$
      $P(R) = -Nx^2 + Ox + P$ (6)

$N=-m_2m_4, O=(m_2b_4-m_4b_1)$ and $P=b_2b_4$.
$P(R)$ is a convex function with a maximal at an intermediate distance between 0 and r (Fig. 8.7)

References

Aizen MA, Feinsinger P (1994a) Habitat fragmentation, native insect pollinators and feral honeybees in Argentine "Chaco Serrano". Ecol Appl 4:4478–4492
Aizen MA, Feinsinger P (1994b) Forest fragmentation, pollination and plant reproduction in a Chaco dry forest, Argentina. Ecology 75:330–351
Brothers TS, Spingarn A (1992) Forest fragmentation and alien plant invasion of central Indiana old-growth forest. Conserv Biol 6:91–100
Burkey TV (1993) Edge effects in seed and egg predation at two neotropical rainforest sites. Biol Conserv 66:139–143
Bustamante RO, Castor C (1998) The decline of an endangered ecosystem: the case of the ruil forest in Central Chile. Biodiv Conserv 7:1607–1626
Bustamante RO, Simonetti JA (2000) Seed predation and seedling recruitment in plants: the effect of the distance between parents. Plant Ecol 147:173–183
Chambers JC, MacMahon JA (1994) A day in the life of a seed: movements and fates of seeds and their implications for natural and managed systems. Annu Rev Ecol Syst 25:263–292
Chen J, Franklin JF, Spies TA (1992) Vegetation responses to edge environment in old-growth. Ecol Appl 2:387–396
Condit R, Hubbell SP, Foster RB (1992) Recruitment near conspecific adults and the maintenance of tree and shrub diversity in a Neotropical forest. Am Nat 140:261–288
Crawley MJ (1992) Seed predators and plant population dynamics. In: Fenner M (ed) Seeds. The ecology of regeneration in plant communities. Redwood Press, Melksham, UK, pp 157–191
Didham RK (1997) The influence of edge effects and forest fragmentation on leaf litter invertebrates in Central Amazonia. In: Laurance WF, Bierregaard RO Jr (eds) Tropical forest remnants: ecology, management and conservation of fragmented communities. University of Chicago Press, Chicago, pp 55–70
Donoso C (1975) Distribución ecológica de las especies de Nothofagus en la zona mesomórfica. Boletín Técnico 33. Facultad de Ciencias Forestales Universidad de Chile, Santiago, Chile
Donoso C, Lara A (1995) Utilización de los bosques nativos en Chile: pasado, presente y futuro. In: Armesto JJ, Villagrán C, Arroyo MK (eds) Ecología de los bosques nativos de Chile. Editorial Universitaria, Santiago, Chile, pp 363–387

Estades CF, Temple SA (1999) Deciduous-forest bird communities in a fragmented landscape dominated by exotic pine plantations. Ecol Appl 9:573–585

Grez AA, Bustamante RO, Simonetti JA, Fahrig L (1998) Landscape ecology, deforestation and habitat fragmentation: the case of the ruil forest in Chile. In: Salinas-Chávez E, Middleton J (eds) Landscape ecology as a tool for sustainable development in Latin America. (http://www.brocku.ca/epi/lebk/grez.htlm)

Groom MJ, Schumaker N (1993) Evaluating landscape change: patterns of worldwide deforestation and local fragmentation. In: Kareiva PM, Kingsolver JG, Huey RB (eds) Biotic interactions and global change. Sinauer, Sunderland, MA, pp 24–44

Houle G (1992) Spatial relationships between seed and seedling abundance and mortality in a deciduous forest of northeastern North America. J Ecol 80:99–108

Janzen DH (1983) No park is an island: increase in interference from outside as park size decreases. Oikos 41:402–410

Jennersten O (1988) Pollination in *Dianthus deltoides* (Caryophyllaceae): effects of habitat fragmentation on visitation and seed set. Conserv Biol 2:359–366

Kapos V (1989) Effects of isolation on the water status of forest patches in the Brazilian Amazonas. J Trop Ecol 5:173–185

Kapos V, Wandelli E, Camrago JL, Ganade G (1997) Edge related changes in environment and plant responses due to fragmentation in Central Amazonia. In: Laurance WF, Bierregaard RO Jr (eds) Tropical forest remnants: ecology, management and conservation of fragmented communities. University of Chicago Press, Chicago, pp 33–44

Kruger FJ, Breitenbach GJ, Macdonald IAW, Richardson DM (1989) The characteristics of invaded mediterranean-climate regions. In: Drake JA, Mooney HA, Di Castri F, Groves RH, Kruger FJ, Rejmnanek M, Williamson M (eds) Biological invasions: a global perspective. Wiley, Chichester, pp 181–213

Lara A, Veblen TT (1993) Forest plantations in Chile: a successful model? In: Mather A (ed) Afforestation: policies, planning and progress. Belhaven Press, London, pp 118–139

Laurance WF (1997) Hyper-disturbed parks: edge effects and the ecology of isolated rainforest reserves in tropical Australia. In: Laurance WF, Bierregaard RO Jr (eds) Tropical forest remnants: ecology, management and conservation of fragmented communities. University of Chicago Press, Chicago, pp 71–84

Laurance WF, Bierregaard RO Jr, Gascon C et al. (1997) Tropical forest fragmentation: synthesis of a diverse and dynamics discipline. In: Laurance WF, Bierregaard RO Jr (eds) Tropical forest remnants: ecology, management and conservation of fragmented communities. University of Chicago Press, Chicago, pp 502–514

Lewis NB, Ferguson IS (1993) Management of radiata pine. Inkata Press, Melbourne, Australia

Mancilla E (1987) Fenología del follaje de *Nothofagus glauca* (Phil) Krasser y su relación con sus características sucesionales. Masters Thesis, University of Chile, Santiago, Chile

Muñoz-Pedreros A, Murúa R, González L (1990) Nicho ecológico de micromamíferos en un agroecosistema forestal de Chile central. Rev Chil Hist Nat 63:267–277

Murcia C (1995) Edge effect in fragmented forests: implications for conservation. Trends Ecol Evol 10:58–62

Myers N (1988) Tropical forests and their species: going, going...? In: Wilson EO, Peter FM (eds) Biodiversity. National Academy Press, Washington, DC, pp 28–35

Price MV, Jenkins SH (1986) Rodents as seed disperser and consumers. In: Estrada A, Fleming TH (eds) Frugivores and seed dispersers. Dr Junk Publishers, Dordrecht, , pp 191–233

Ranney JW, Brunner MC, Levenson JB (1981) The importance of edge in the structure and dynamics of forest stands. In: Burgess RL, Sharpe DM (eds) Forest island dynamics in man-dominated landscapes. Springer, Berlin Heidelberg New York, pp 67–95

Rejmánek M (1996) A theory of seed plant invasiveness: the first sketch. Biol Conserv 78:171–181

Richardson DM, Bond WJ (1991) Determinants of plant distribution: evidence from pine invasions. Am Nat 137:639–668

Richardson DM, Cowling RM (1992) Why is mountain fynbos invasible and which species invade? In: van Wilgen BW, Richardson DM, Kruger FJ, Hensbergen HJ (eds) Fire in South African mountain fynbos: ecosystem, community and species responses at Swartboskloof. Springer, Berlin Heidelberg New York, pp 161–189

Richardson DM, Cowling RM, Le Maitre DC (1990) Assessing the risk of invasive success in *Pinus* and *Banksia* in South African mountain fynbos. J Veg Sci 1:629–642

Richardson DM, Williams PA, Hobbs RJ (1994) Pines invasions in the southern hemisphere: determinants of spread and invadability. J Biogeogr 21:511–527

Sáiz F (1990) Incendios forestales en el Parque Nacional La Campana, sector Ocoa, V Region, Chile. Y Problema e incidencia de incendios forestales en Chile. An Mus Hist Nat Valparaiso (Chile) 21:5–13

San Martin J, Donoso C (1996) Estructura florística e impacto antrópico en el bosque Maulino de Chile. In: Armesto JJ, Villagrán C, Arroyo MK (eds) Ecología de los bosques nativos de Chile. Editorial Universitaria, Santiago, Chile, pp 153–168

San Martín J, Figueroa H, Ramírez C (1984) Fitosociología de los bosques de ruil (*Nothofagus alessandri* Espinosa) en Chile Central. Rev Chil Hist Nat 57:171–200

Santos T, Tellerías JL (1992) Effects of fragmentation on a guild of wintering passerines: the role of habitat selection. Biol Conserv 71:61–67

Saunders DA, Hobbs RJ, Margules CR (1991) Biological consequences of ecosystem fragmentation: a review. Conserv Biol 5:18–32

Schelhas J, Greenberg R (1996) The value of forest patches. In: Schelhas J, Greenberg R (eds) Forest patches in tropical landscapes. Island Press, Covelo, CA, pp xv-xxxvi

Sork VL (1983) Mast fruiting in hickories and availability of nuts. Am Nat 109:81–88

Turner M, Corlett RT (1996) The conservation value of small, isolated fragments of lowland tropical rain forest. Trends Ecol Evol 11:330–333

Turton SM, Freiburger HJ (1997) Edge and aspect effects on the microclimate of a small tropical forest remnant on the Atherton Tableland, Northeastern Australia. In: Laurance WF, Bierregaard RO Jr (eds) Tropical forest remnants: ecology, management and conservation of fragmented communities. University of Chicago Press, Chicago, pp 45–54

Viana VM, Tabanez AAJ (1996) Biology and conservation of forest fragments in the Brazilian Atlantic moist forest. Schelhas J, Greenberg R (eds) Forest patches in tropical landscapes. Island Press, Covelo, CA, pp 151–167

Viana VM, Tabanez AAJ, Batista JL (1997) Dynamics and restoration of forest fragments in Brazilian Atlantic moist forest. In: Laurance WF, Bierregaard RO Jr (eds) Tropical forest remnants: ecology, management and conservation of fragmented communities. University of Chicago Press, Chicago, pp 351–365

Whitney GG, Runkle JL (1981) Edge versus age effects in the development of a beech-maple forest. Oikos 37:377–381

9 The Ecological Consequences of a Fragmentation-Mediated Invasion: The Argentine Ant, *Linepithema humile*, in Southern California

A. V. Suarez and T.J. Case

9.1 Introduction

Habitat fragmentation can facilitate species loss through a number of processes. The remaining habitat may only sample a subset of the existing fauna (Preston 1962). Alternatively, species may become locally extinct within the remaining habitat after fragmentation (Bolger et al 1991). Small populations associated with the reduction in total area may become sensitive to stochastic processes, both demographic (Shaffer 1981; Gilpin and Soulé 1986; Goodman 1987) and environmental (Shaffer 1981). Fragmentation may reduce colonization, or immigration into the habitat remnants, especially by less mobile species (Turner and Corlett 1996). Deterministic processes associated with detrimental edge effects (Wilcove 1985; Yahner 1988; Robinson et al. 1995) and the loss of landscape level processes, such as fire regimes (Leach and Givnish 1996), also cause local extinctions post fragmentation. Fragmentation may also facilitate the immigration or invasion of exotic species that may directly compete with, prey upon, parasitize or otherwise indirectly affect native species (Diamond and Case 1986). Some of these processes are correlated. For example, an increase in edge around a habitat fragment will facilitate the penetration of exotic species that may be prominent along those edges. This is particularly a problem in urban landscapes because many successful exotic species are associated with human-mediated disturbance (Elton 1958; Fox and Fox 1986; Orians 1986; Petren and Case 1996).

Exotic ants have been inadvertently introduced worldwide (Williams 1994). Many species, such as the big headed ant (*Pheidole megacephala*), the little fire ant (*Wasmannia auropunctata*), the imported fire ant (*Solenopsis invicta*) and the Argentine ant (*Linepithema humile*, formerly *Iridomyrmex humilis*) have wreaked ecological havoc where they have become established (Williams 1994). Most of the impact of these species has been measured in loss of native ant diversity, which is greatly reduced or completely eliminated

Ecological Studies, Vol. 162
G.A. Bradshaw and P.A. Marquet (Eds.)
How Landscapes Change

in areas where they have invaded. Ants are important components of many ecosystems and changes in the native ant community may have repercussions on other taxa and trophic levels. Ants compete with rodents for seeds in desert and montane systems (Brown et al. 1975) and may help shape the plant community. In South African fynbos shrublands, native ant displacement by the Argentine ant has reduced the establishment of a native shrub whose seeds are ant-dispersed (Bond and Slingsby 1984). Some islands, like Hawaii, have no native ant species, and there is evidence that the introduction of the Argentine ant has negatively impacted the native arthropod fauna (Cole et al. 1992). There is also evidence that exotic ants detrimentally impact vertebrates (Mount et al. 1981; Allen et al. 1995).

The San Diego coastal horned lizard (*Phrynosoma coronatum blainvillei*) is listed by the California Department of Fish and Game as an animal of special concern, and is also a candidate for listing by the US Fish and Wildlife Service (Jennings and Hayes 1994). Like most horned lizards, this species specializes on ants (Pianka and Parker 1975; Montanucci 1989), particularly harvester ants of the genera *Messor* and *Pogonomyrmex* (Rissing 1981; Munger 1984; Suarez et al. 2000). The upland scrub habitats where they live have been subject to massive fragmentation as a result of urbanization over the past 100 years (Westman 1981; Alberts et al. 1993). In addition to habitat loss, the native ants which horned lizards eat are being displaced by an exotic invader, the Argentine ant, which has been invading the remaining suitable habitat in southern California (Suarez et al. 1998).

The Argentine ant was inadvertently introduced to the United States around 1891 in New Orleans (Titus 1905; Newell 1908). It was first detected in California in 1905 and spread rapidly throughout much of the state (Woodworth 1908; Smith 1936; Mallis 1941). Argentine ants have general and opportunistic nesting and dietary requirements (Newell and Barber 1913; Mallis 1942; Flanders 1943; Markin 1970a). Colonies have multiple queens which are inseminated within the nest prior to dispersal and do not undergo nuptial flights (Markin 1970b). The Argentine ant has been shown to displace native ant species throughout California (Erickson 1971; Tremper 1976; Ward 1987; Holway 1995, 1998a; Human and Gordon 1996; Suarez et al. 1998) through a combination of exploitative and interference competition (Human and Gordon 1996; Holway 1999).

In this study we provide evidence for the facilitation of spread of an exotic species, the Argentine ant (*Linepithema humile*), into natural areas through increased urban edges associated with habitat fragmentation. We demonstrate the negative impact of Argentine ants on native ground-foraging ants, and also provide evidence that these changes in ant communities may affect a specialist ant predator, the coastal horned lizard (*Phrynosoma coronatum*).

9.2 Methods

9.2.1 Ant Communities of Coastal Scrub Fragments in Southern California

Habitat fragments of native scrubland were isolated from the surrounding continuous vegetation throughout the last 100 years as a result of urban and suburban development (Soulé et al. 1988). Suarez et al. (1998) sampled the ant communities in 40 habitat fragments throughout coastal San Diego County, and one area of continuous scrub habitat, to determine how fragmentation promotes species loss. The habitat fragments vary in size (ranging in size from 0.4 to 101.6 ha), the number of years since isolated from the surrounding vegetation (age ranging from 3 to 95 years), the remaining percentage of native vegetation, the degree of isolation from areas of continuous vegetation, and the relative amount of urban edge surrounding the fragment. Detailed methods of how these parameters were obtained can be found in Suarez et al. (1998). Ant communities within the habitat fragments were compared to communities in plots of similar size, vegetation and topography within a continuous habitat in the University of California's Elliot Chaparral Reserve and the Miramar Naval Air Station (Fig. 9.1). These plots were at least 500 m from the nearest developed edge and represented areas of 1, 4, 10, 20, 30, and 50 ha. Additional point sampling throughout the 88-ha Elliot reserve was used to estimate the ant fauna for the entire reserve.

Ant communities were sampled with a combination of pitfall trapping and visual surveys. Pitfall sampling consisted of placing an array of five pitfall traps every 100 m along a transect through the longest axis of the habitat fragment or plot within the continuous area. Each trap consisted of a 60-mm-wide, 250-ml glass jar and was filled halfway with a mixture of water and nontoxic Sierra brand antifreeze. The jars were collected after 5 days. This method provides an estimate of ant activity for each array by counting the number of workers falling into the jars in each 5-day sample period. Pitfall trapping was repeated three times seasonally, with sample periods in the fall, winter, and spring/summer. At each array, we measured the distance to the nearest urban edge and relative amount of exotic vegetation (Suarez et al. 1998). Extensive visual surveys, consisting of walking throughout the area and overturning objects, were conducted in each fragment and control plot to compliment the pitfall trapping. Each fragment was visited and inspected visually for ants six times. Only ground-foraging ants were included in analyses (Suarez et al. 1998).

To examine within-fragment differences in ant species distributions, the average number of Argentine ant workers per jar and native ant species per array were calculated for each five-jar pitfall array. This average is compared to the percentage of exotic vegetation and the distance to the nearest devel-

Fig. 9.1. Map of southwestern San Diego County, California, USA. Extensive urbanization (the *light gray "background"* color indicates developed areas and the *white intersecting lines* indicate highways) has eliminated most native vegetation, leaving behind a system of habitat fragments (*black*). Some of the larger fragments used to measure the distance to potential source populations include the UC Elliot Reserve, the southeastern corner of the Miramar Naval Air Station, the Cabrillo National Monument, Proctor Valley, and Otay Valley. In addition, 38 of 40 fragments surveyed in Suarez et al. (1998) are indicated with *numbers* in decreasing order of size

oped edge. All arrays, regardless of the fragment they occurred in, were grouped together for these analyses.

To examine between-fragment distributions of native ant species, an average number of Argentine ants per jar (later referred to as Argentine ant activity) was determined for each fragment by pooling the information from each pitfall array in the fragment and all sampling dates. Linear and stepwise regressions were used to test for correlations between various fragment descriptors (age, area, vegetation, amount of edge and degree of isolation), the relative abundance of the Argentine ant, and the number of native ant species (Suarez et al. 1998).

Native ant species may vary in their sensitivity to habitat fragmentation and the subsequent invasion of the Argentine ant. Logistic regressions were used to construct incidence functions for the more common ant species and genera in relation to fragment size: *Solenopsis molesta*, *Leptothorax andrei*, *Prenolepis imparis*, *Dorymyrmex insanus*, *Solenopsis xyloni* and *Cremato-gaster californica*. Genera were used rather than species for some uncommon groups in order to increase the power of the tests. These include the genera *Camponotus*, *Pheidole*, and *Neivamyrmex*. The genera *Messor* and *Pogono-myrmex* are both commonly referred to as harvester ants because of their diets and behaviors (Davidson 1977; Holldobler and Wilson 1990). For this reason, and their importance as food for ant predators such as the coastal horned lizard (Pianka and Parker 1975; Rissing 1981; Munger 1984), they were also combined into one group.

9.2.2 The Effects of Argentine Ants on Coastal Horned Lizard Diet

Coastal horned lizard diet information was obtained through the examination of fecal pellets in three reserves in southern California, USA. Lizards were collected from the University of California's Elliot Chaparral Reserve (Elliot), Torrey Pines State Park (Torrey Pines), and the Southwestern Riverside County Multispecies Reserve (Riverside). Lizards were brought into the lab until they deposited a fecal pellet and then returned to the field. These areas were also visually searched for horned lizards fecal pellets, which are very distinctive (Rissing 1981, pers. observ.). Fecal pellets were dissected and ants were identified to species when possible using a reference collection and a key to the head capsules of ants from fecal pellets (Snelling and George 1979). Other insects were identified to order. While Argentine ants are absent from our study sites in Riverside, they are invading Elliot and Torrey Pines (Suarez et al. 2000).

To compliment data collected on horned lizard diet in the wild, we performed laboratory experiments consisting of paired presentations of ants to lizards (Suarez et al. 2000). Lizards were collected from Elliot and Riverside and brought into the lab. Each lizard was marked and placed into its own 25 ×

50-cm terrarium. Each terrarium contained sand, a water dish and some sparse vegetation. A heat lamp was placed over the terrarium on a timer set from 6 a.m. to 6 p.m. The temperature was maintained at 92–96 °F during mid-day. The lizards were kept in the lab only for the duration of the trials and then were returned to the field at the site of capture.

The lizards were starved for 2 days before the start of their feeding experiments. The experiments consisted of a prey preference trial repeated every other day. Trials took place in a different 25 × 50-cm terrarium which was coated with limousine grade tinting, allowing an observer to peer into the tank and watch the lizard's behavior without being seen. The lizard was allowed to become accustomed to the tank for 5 min before the trial began. Five individuals each of two species of ants were then placed into the tank with lizard. Ants were replaced as they were eaten by the lizard so there were five individuals of each species of ant at all times during the trial. The trial lasted for 45 min or until the lizard buried itself in the sand. The following information was recorded during the trial: each time the lizard fixed on an ant, whether the ant was eaten, and how many steps the lizard took to reach the ant.

The percentage of total ants eaten was calculated for each prey species after every trial (Suarez et al. 2000). By using the percentage of total ants eaten during the trial instead of the absolute number of ants eaten, we can reduce the effects of hunger or other random variation that may result in more or less total numbers of ants being eaten in any trial. These values summarize the prey preference of the lizards including any effect the behavior of the ants may have had. Each time a lizard turned its head and noticed an ant we recorded a fix. The ratio of fixed to eaten ants then gives us an approximation of how many ants were eaten after detection by the lizard. Finally, by measuring the number of steps taken by the lizard to reach the prey item we are quantifying effort or cost (in terms of energy expenditure or increased predation risk) the lizard is willing to take for that particular prey type.

The following native ants were paired with the exotic Argentine ant: *Pogonomyrmex rugosus*, *Pogonomyrmex californicus*, *Messor andrei*, and *Crematogaster californica*. The order the ant pairs were presented to the lizards was chosen at random. The three harvester ants vary in size, *P. rugosus* (6.5–7.5 mm) is the largest, then *P. californicus* (5.5–6.8 mm), and *Messor andrei* (4.5–6.0 mm) is the smallest, although there is considerable overlap in size between *P. californicus* and *M. andrei*. *Crematogaster californica* (3–4 mm) and the Argentine ant (2.5–3.0 mm) are considerably smaller than the harvester ants (Wheeler and Wheeler 1973). The native ant species were chosen because they are common prey items for coastal horned lizards in southern California (Suarez et al. 2000).

9.3 Results

9.3.1 Ant Communities of Coastal Scrub Fragments in Southern California

Within fragments, the number of native ant species detected at any sample array decreased in the presence of the Argentine ant from an average of almost seven to two (Fig. 9.2). The average number of Argentine ants per jar was negatively correlated with the distance to the nearest urban edge (Fig. 9.3a). Only at points greater than 100 m from the nearest urban edge were Argentine ants found at densities of less than five workers per jar. Subsequently, the number of native ant species at any point was positively correlated with the distance from the nearest urban edge (Fig. 9.3b). The amount of exotic vegetation also decreased with distance from an urban edge (Fig. 9.3 c).

The direct role of urbanization on the success of Argentine ants (and the subsequent loss of native ant species) within remaining habitat patches is illustrated in one of the fragments, Rice Canyon. Rice Canyon was isolated approximately 3 years before sampling from a larger area of continuous scrub habitat. Urbanization began at the east and south end of the canyon and has continued until the present, slowly working westward. This has resulted in a gradient of disturbance from the east to west end of the fragment. While native ants are abundant in the west end of the fragment, the east end is dom-

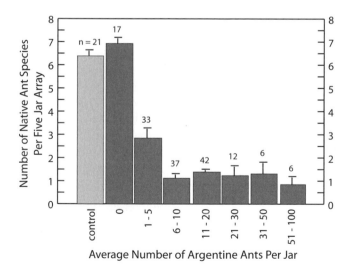

Fig. 9.2. The number of native ant species detected at an array vs. the average number of Argentine ants per trap at that array. Numbers of Argentine ants were averaged across the three sample seasons for each array

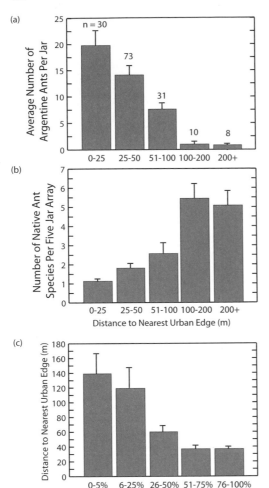

Fig. 9.3. Associations between the distance of an array to the nearest urban edge and the average number of Argentine ants per jar at that array (**a**), the number of native ant species detected at the array (**b**), and the approximate cover of exotic vegetation surrounding the array (**c**). *Error bars* indicate one standard error. These graphs include all points sampled within the 40 habitat fragments

inated by Argentine ants. The transition between the decline in native ant species and the penetration of Argentine ants corresponds spatially with the onset of development (Fig. 9.4).

Among fragments, bivariate regressions reveal that log area ($R^2=0.441$, $P<0.0001$) and percent native vegetation ($R^2=0.349$, $P<0.0001$) were positively correlated, while log age ($R^2=0.152$, $P<0.02$) and log Argentine ant activity ($R^2=0.483$, $P<0.0001$) were negatively correlated with the remaining number of native ants species within a fragment (Fig. 9.5). The amount of edge and degree of isolation were not correlated with the number of native ant species. In a stepwise multiple regression with the number of native ant species as the dependent variable, however, only log area ($P<0.0001$), log age ($P<0.0005$), and log Argentine ant activity ($P<0.0002$) significantly explained any variance in the number of native ant species among the fragments (total model:

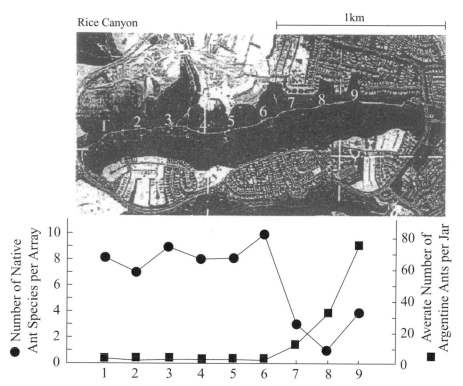

Fig. 9.4. A photograph of Rice Canyon (San Diego County, California, USA) and a transect of sampling points which runs the length of the canyon. The developed area at the northern end of the canyon begins at *point 7* and continues east through *point 9*. The number of native ant species detected at each sample array along with the average number of Argentine ants per jar at each array are plotted below the photograph (from Suarez et al. 1998)

d.f.=36, R^2=0.747, $P<0.001$). A stepwise multiple regression with Argentine ant activity as the dependent variable, revealed log area as the only significant predictor of log Argentine ant activity (total model: d.f.=38, R^2=0.195, $P<0.005$; Suarez et al. 1998).

Sample plots within continuous vegetation averaged more native ant species, 16.4±4.6 (mean ± SD), than the isolated fragments that only averaged 5.9±4.9 species. In addition, the regression slope of log native ant species on log area was significantly lower in plots within continuous vegetation than in isolated canyons (t=2.52; v=40; $P<0.01$; Fig. 9.6).

Native ant species/genera varied in their probability of going locally extinct due to habitat fragmentation and the subsequent invasion of Argentine ants. Incidence functions of presence/absence data versus fragment size found *Prenolepis imparis* and *Dorymyrmex insanus* were the least sensitive species, and the genera *Neivamyrmex*, the army ants, and *Pogonomyrmex* and *Messor*, the harvester ants, were the most sensitive (Fig. 9.7). *Solenopsis molesta* and

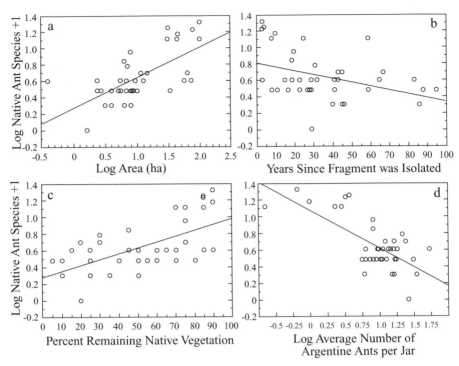

Fig. 9.5a–d. Linear regressions of various fragment descriptors on log native ant species detected within the 40 isolated habitat fragments surveyed. Only independent variables that were significantly correlated with log native ant species number are presented including: **a** log area of habitat fragment (Y=0.271+0.372 × X, R^2=0.441, P<0.0001); **b** age, in years, since isolation (Y=0.814−0.005 × X; R^2=0.152, P<0.02); **c** percent cover of native vegetation remaining within the fragment (Y=0.283+0.007 × X, R^2=0.349, P<0.0001); and **d** log Argentine ant activity, measured in average number of workers per jar, within the fragment (Y=1.076−0.451 × X, R^2=0.483, P<0.0001). Fragment descriptors that did not significantly correlate with the number of native ant species within the fragments include the distance to the nearest area of continuous vegetation and the relative amount of urban edge

Leptothorax andrei occurred in nearly every habitat fragment sampled, 34 and 32 of 40 fragments respectively, and were not significantly associated with fragment area or the presence of the Argentine ant. Species lists for the fragments and plots within continuous habitat can be found in Suarez et al. (1998).

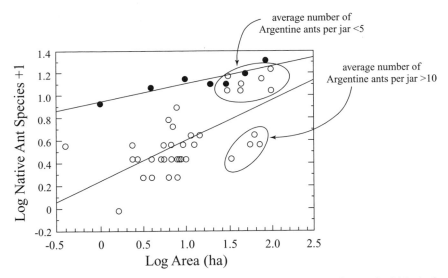

average number of
Argentine ants per jar <5

average number of
Argentine ants per jar >10

Fig. 9.6. Regression of the log number of native ant species +1 detected within isolated fragments (*open circles*, Y=0.271+0.372 × X; R^2=0.441) and similarly sized plots within continuous habitat at the UC Elliot reserve (*closed circles*, Y=1.013+0.164 × X; R^2=0.806) vs. their area in hectares (from Suarez et al. 1998)

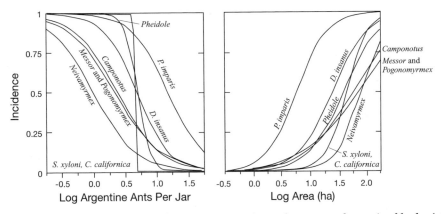

Fig. 9.7. Incidence functions of various ant species and genera as determined by logistic regressions of presence/absence data within a fragment vs. log fragment area. Only species with significant association ($P<0.05$) are shown. Species tested for which the logistic regression showed no significant correlation ($P>0.05$) include *Solenopsis molesta*, and *Leptothorax andrei* (from Suarez et al. 1998)

9.3.2 The Effects of Argentine Ants on Coastal Horned Lizard Diet

At the three reserves where diet information was gathered, native ants made up over 98 % of the prey items identified from all fecal pellets combined in areas never occupied by the exotic Argentine ant. In contrast, ants averaged only 11 % of total prey items in fecal pellets collected from lizards in areas that were occupied by Argentine ants at Elliot. Harvester ants (genera *Messor* and *Pogonomyrmex*) were the most common prey item in all three reserves. The proportion of total diet consisting of harvester ants ranged from 49 % (by prey item) at the Riverside Multispecies Reserve to 69 % at the Elliot Reserve. In areas occupied by Argentine ants, however, harvester ants only made up from 0 to 6 % of their diet (Table 9.1). At Torrey Pines, the native ant *Dorymyrmex insanus* persisted in areas that were invaded by Argentine ants resulting in a diet shift from harvester ants to *D. insanus* in those areas. Argen-

Table 9.1. Summary of diet composition for coastal horned lizards (*Phrynosoma coronatum*) obtained from fecal pellets at three reserves. The data from the UC Elliot Reserve and Torrey Pines State Park have been separated into areas with and without Argentine ants (*Linepithema humile*). The numbers represent the percentage of total diet for all fecal pellets combined. The numbers in parentheses next to present and absent represent the number of fecal pellets examined

	Elliot *L. humile*		Torrey pines *L. humile*		Riverside *L. humile*
	Absent (67)	Present (10)	Absent (4)	Present (3)	Absent (98)
Hymenoptera: Formicidae					
Messor andrei	69.19	6.19	a	a	2.21
Pogonomyrmex subnitidus	a	a	62.83	0	a
Pogonomyrmex rugosus	a	a	a	a	28.73
Pogonomyrmex californicus	a	a	a	a	18.52
Crematogaster californica	23.46	3.26	0	0	7.95
Dorymyrmex insanus	0.31	0.98	36.43	77.78	0.16
Pheidole vistana	2.25	0	0	2.22	0.04
Formica spp.	0.06	0	0	0	14.74
Myrmecocystus spp.	0.06	0	0	0	14.45
Camponotus spp.	1.86	0.65	0.74	6.67	11.86
Forelius spp.	1.31	0	a	a	0
Solenopsis xyloni	0.03	0	0	0	0
Neivamyrmex californicus	0.05	0	0	0	0
Other ant	0.09	0	0	0	0.02
Total harvester ants	69.19	6.19	62.83	0	49.46
Total ants	98.62	11.07	100	86.67	98.69
Total non-ants	1.38	88.93	0	13.33	1.3

[a] Ant species does not occur in the reserve.

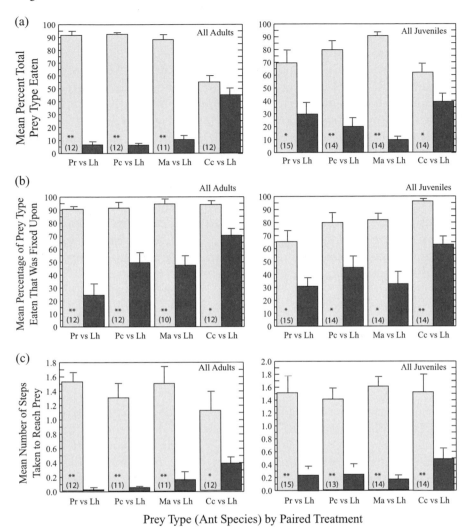

Fig. 9.8a–c. Summary of results of the laboratory prey preference experiments. Average percent of prey items eaten during prey preference trials (**a**), average percent of detected (fixed upon) prey items that were consumed during prey preference trials (**b**), and the mean number of steps taken to reach prey items (**c**) by horned lizards during paired prey preference trials of various native ant species vs. the exotic Argentine ant (*Linepithema humile*). Between 25 and 27 lizards were tested in each category. *Error bars* represent one standard error. The abbreviated ant species are Pr = *Pogonomyrmex rugosus*; Pc = *Pogonomyrmex californicus*; Ma = *Messor andrei*; Cc = *Crematogaster californica*; and Lh = *Linepithema humile*. Significant differences in values between the native ant species tested and the Argentine ant are represented by *asterisks* within the columns (*P<0.05; **P<0.001; Mann-Whitney U-Test) (from Suarez et al. 1998)

tine ants were never detected in their diet, even in areas where they had invaded and displaced native ant species.

In laboratory prey preference experiments, horned lizards always preferred native ant species to the Argentine ant in paired trials (Fig. 9.8). Lizards always ate more of the three harvester ants than the Argentine ant, averaging 87 % of prey items eaten. The preference for *C. californica* over the Argentine ant was not as strong, averaging about 60 % of prey eaten (Fig. 9.8a). For ants that were clearly fixed upon visually by the horned lizards, and then either eaten or not, lizards always significantly preferred the native ant species tested to the Argentine ant (Fig. 9.8b). Lizards ate between 78 % (*Pogonomyrmex rugosus*) and 95 % (*Crematogaster californica*) of the native ants they fixed upon, while only eating 45 %, on average, of the Argentine ants they fixed on. Horned lizards also took significantly more steps to reach the native ant species than the exotic Argentine ant (Fig. 9.8 c). For all native ant species combined, lizards averaged 1.44 steps to reach prey, while only taking 0.21 steps to reach Argentine ants when they were eaten.

9.4 Discussion

9.4.1 Local Extinction of Native Ground-Foraging Ants

The native ground-foraging ant communities in the scrub habitats of coastal southern California are prone to local extinction due to fragmentation and the subsequent presence of the invasive Argentine ant (Suarez et al. 1998). Argentine ants penetrate into habitat fragments from surrounding urban edges where they are more abundant. This results in smaller fragments, which have a higher edge to interior ratio, to become saturated with Argentine ants. Native ant species, which may already be vulnerable to extinction in smaller fragments due to stochastic processes, are lost in these fragments due to interference and exploitative competition from the invading Argentine ants (Human and Gordon 1996; Holway 1999). In larger fragments, the effective size of the remaining habitat is reduced by the Argentine ant, which can displace native species up to 200 m in from a developed edge. In addition, internal edges may be as detrimental as external, boundary edges, and preventing the penetration of exotic species such as the Argentine ant into reserves may be difficult. In South African fynbos shrublands, Argentine ants were more common in human disturbed areas, and penetrated into the reserve along roads, particularly paved roads, regardless of the presence of exotic vegetation (De Kock and Giliomee 1989).

The role of urban development in facilitating the spread of Argentine ants is seen in Rice Canyon which is completely isolated from other vegetation, but has only been developed on the east side (Fig. 9.4). Argentine ants have spread into the canyon from housing developments in place on the east end of the fragment and native species are only persisting in the west end where Argentine ant have not yet penetrated. Argentine ants appear to be limited by water availability (Smith 1936; Holway 1998b). It is likely that Argentine ants are benefiting from water runoff into the fragment from the landscaped areas above the canyon. Like Argentine ant density, the abundance of exotic vegetation is also correlated with distance to the nearest urban edge. However, in Rice Canyon (Fig. 9.4) the vegetation in the east end is predominately native, implying that the spread of Argentine ants into the habitat fragment and the subsequent loss of native species is not dependent on exotic vegetation. This is also supported at the University of California's Elliot Reserve and Torrey Pines State Park where Argentine ants have penetrated over 400 and 1000 m, respectively, into the reserves in areas dominated by native scrub vegetation (Suarez et al. 1998; J. King, unpubl.). This also highlights that the degree to which Argentine ants can penetrate into natural habitat varies depending upon the topography and abiotic conditions of the landscape. For example, in more xeric sites in Riverside County, California, Argentine ants appear only able to penetrate up to 50 m into native vegetation from neighboring urban developments (Suarez and Case, unpubl. data).

Within fragments, the number of native ants declines from an average of seven species per sample array to one or two species in areas occupied by the Argentine ant. The species that persist with Argentine ants include *Solenopsis molesta, Leptothorax andrei*, and *Prenolepis imparis* (Suarez et al. 1998). Both *S. molesta* and *L. andrei* are small species (1–2 mm) that are often categorized as hypogeic, or belowground foraging (Ward 1987). *Prenolepis imparis*, however, is an aboveground foraging ant similar in size and biology to the Argentine ant. The coexistence between *P. imparis* and Argentine ants is likely due to temporal niche partitioning between these species. While Argentine ants forage most abundantly during summer and spring months, *P. imparis* is exclusively active in the cold and wet winter months (Tschinkel 1987; Ward 1987; Suarez et al. 1998). The most vulnerable species to fragmentation and Argentine ant invasion, as detected by logistic regressions on presence/absence data, included *Neivamyrmex* (the army ants), and *Messor* and Pogonomyrmex (the harvester ants; Fig. 9.8) both of which are important to ecosystem level processes (Brown et al. 1975; Gotwald 1995). The loss of harvester ants, in particular, can have drastic effects on both plant species, which depend on them for seed dispersal (see Bond and Slingsby 1984), and on horned lizards, whose diets are composed primarily of harvester ants (Pianka and Parker 1975; Rissing 1981; Munger 1984; Suarez et al. 2000).

9.4.2 Diet and Prey Preference in Coastal Horned Lizards

Horned lizards have been declining in California (Jennings 1988) and their decline in some areas has been correlated with the presence of Argentine ants (Fisher et al. 2002). We provide a potential mechanism behind the decline of horned lizards in these areas. The loss of native ground-foraging ants due to interactions with Argentine ants directly affects the preferred diet of coastal horned lizards. Native ground-foraging ants made up at least 98 % of the total diet based on prey items in all three reserves examined (Table 9.1). Harvester ants are the most common prey item found in horned lizard diets, making up between 50 and 70 % of their diet in undisturbed areas. These ants are very sensitive to habitat fragmentation and the subsequent presence of Argentine ants. Harvester ants were only found in the largest of the fragments surveyed and then only in areas where Argentine ants had not invaded (Suarez et al. 1998).

In areas where Argentine ants have invaded, the diet of coastal horned lizards changed from primarily ants to non-ant arthropods, and Argentine ants were never eaten. The small sample sizes of fecal pellets within areas occupied by Argentine ants is due to the fact that horned lizards were rarely seen in invaded areas and will actively avoid them (Suarez et al. 2000; Fisher et al. 2002). In addition, recent experiments suggest that hatchling horned lizards are unable to maintain positive growth rates on diets of Argentine ants or arthropods typical of an invaded community (Suarez and Case 2002).

Argentine ants were never preferred in paired prey preference tests with native ant species. In context of total percentage of ants eaten in any trial, native ants were always preferred over Argentine ants (Fig. 9.8). The lack of a strong preference for *C. californica* over Argentine ants in total ants eaten, is probably due to the behavior of *C. californica* which will frequently freeze, and remain motionless when a predator is nearby (pers. observ.). Therefore, *C. californica* may not have been detected as often during the trials. This is reflected in the percentage of ants that were eaten after being fixed upon. When *C. californica* was detected, lizards did eat them much more frequently than Argentine ants (Fig. 9.8b).

Horned lizards tend to pick a spot near a foraging trail or concentration of ants (i.e., the colony entrance) and consume ants as they approach the lizard (Munger 1984; J. Richmond and A. Suarez, pers. observ.). This "sit and wait" behavior may minimize detection of the lizard by potential predators as well as by the prey, preventing a colony response or attack. Lizards in both age classes took significantly fewer steps to predate Argentine ants then any native ant tested (Fig. 9.8 c). This suggests that when they do consume an Argentine ant they do so at a minimal "cost" compared to native ant species, which they are willing to pursue or even seek out (Shafir and Roughgarden 1998).

Argentine ants were never eaten by horned lizards in the wild, even though at Torrey Pines horned lizards ate the similarly sized *Dorymyrmex insanus*.

This may reflect an avoidance of Argentine ants due to their aggressive, swarming behavior (pers. observ., see Rissing 1981). It is also possible that Argentine ants differ chemically from the native ants typically eaten, and are either not recognized as potential prey by the lizards or are avoided outright (Suarez et al. 2000). Horned lizards ate Argentine ants in the lab, although rarely, but this reflects the fact that lizards were starved every other day and then limited in the choice and availability of prey. Statistical differences in prey choice, as seen under these conditions, probably reflects strong preferences against eating Argentine ants, as seen in the wild.

9.5 Conclusions and Implications for Reserve Management

Among other detrimental consequences of habitat fragmentation, the subdivision of natural areas facilitates the invasion of exotic species into the remaining habitat. We demonstrate that native ground-foraging ants are prone to local extinction due to habitat fragmentation through the subsequent invasion of Argentine ants from urban edges. Native ants vary in their vulnerability to these processes, the most sensitive include army ants and harvester ants, both of which are important to ecosystem level processes. Harvester ants, in particular, are important seed dispersers, seed predators, and food sources for certain native species such as horned lizards.

Coastal horned lizards have a diet dominated by ants, particularly harvester ants of the genera *Messor* and *Pogonomyrmex*. In areas where the Argentine ant has invaded and native ants are displaced, horned lizards are forced to change their diet to incorporate more non-ant arthropods. While habitat loss, in the form of fragmentation due to urbanization, is likely a major cause of the coastal horned lizard's decline, the invasion of Argentine ants into the remaining natural areas in southern California compounds the detrimental effects of habitat loss on horned lizards.

Changes in non-ant arthropod communities associated with Argentine ant invasion (e.g., Cole et al. 1991; Bolger et al. 2000) may have detrimental effects on many insectivorous vertebrates, and still needs to be quantified. To preserve viable populations of sensitive species, such as coastal horned lizards, we not only need to save an adequate amount of habitat, but also manage the remaining areas to prevent the penetration of exotic species. Some practical guidelines include minimizing the penetration of roads and landscaping into the reserve. If any type of replanting or landscaping is to be done, care must be taken to ensure that Argentine ants are not imported with the plants.

Acknowledgement. This project was made possible through financial support from the Metropolitan Water District of Southern California, Southwestern Riverside County Multi-species Reserve Management Committee – RFP NO. 128 (T. Case), United States National Science Foundation grant DEB-9610306 (T. Case), a NIH CMG training grant – GMO 7240 (A. Suarez), and a fellowship from the Canon National Parks Science Scholars Program (A. Suarez). Matt Blankenbiller, Kevin Crooks, Alex Goreman, Kevin Haight, Michelle Hollenbeck, Katherine Howard, Mike Jefferson, Jason Lahmani, Milan Mitrovich, Stephanie Widmann, Ben Williams and particularly Jon Richmond helped with various components of the fieldwork. Melissa Bennett and Phil Ward provided their expertise in ant identification. We would also like to thank Allison Alberts, Doug Bolger, Paul Griffen, David Holway, Trevor Price, Ray Radtkey, Adam Richman, Jon Richmond, Neil Tsutsui, and Phil Ward for comments and insight throughout the course of this work.

References

Alberts AC, Richman AD, Tran D et al. (1993) Effects of habitat fragmentation on native and exotic plants in southern California coastal scrub. In: Keeley JE (ed) Interface between ecology and land development in southern California. Southern California Academy of Sciences, Los Angeles, USA, pp 103–110

Allen CR, Lutz RS, Demarais S (1995) Red imported fire ant impacts on northern bobwhite populations. Ecol Appl 5:632–638

Bolger DT, Alberts A, Soulé M (1991) Occurrence patterns of bird species in habitat fragments: sampling, extinction, and nested species subsets. Am Nat 137:155–166

Bolger DT, Suarez AV, Crooks KR, Morrison SA, Case TJ (2000) Effects of habitat fragmentation and Argentine ants on arthropods in southern California. Ecol Appl 10:1230–1248

Bond W, Slingsby P (1984) Collapse of an ant-plant mutualism: the Argentine ant (*Iridomyrmex humilis*) and myrmecochorous Proteaceae. Ecology 65:1031–1037

Brown JH, Grover JJ, Davidson DW, Lieberman GA (1975) A preliminary study of seed predation in desert and montane habitats. Ecology 56:987–992

Cole FR, Medeiros AC, Loope LL, Zuehlke WW (1992) Effects of the Argentine ant on arthropod fauna of Hawaiian high-elevation shrubland. Ecology 73:1313–1322

Davidson DW (1977) Species diversity and community organization in desert seed-eating ants. Ecology 58:711–724

De Kock AE, Giliomee JH (1989) A survey of the Argentine ant, *Iridomyrmex humilis* (Mayr), (Hymenoptera, Formicidae) in south African fynbos. J Entomol Soc S Afr 52:157–164

Diamond J, Case TJ (1986) Overview: introductions, extinctions, exterminations, and invasions. In: Diamond J, Case TJ (eds) Community ecology. Harper and Row, New York, pp 65–79

Elton CS (1958) The ecology of invasions. Wiley, New York

Erickson JM (1971) The displacement of native ant species by the introduced Argentine ant *Iridomyrmex humilis* (Mayr). Psyche 78:257–266

Fisher RN, Suarez AV, Case TJ (2002) Spatial patterns in the abundance of a declining species: the coastal horned lizard. Conserv Biol 16:205–215

Flanders S (1943) The Argentine ant versus the parasites of the black scale. Citrograph 28:117

Fox MD, Fox BJ (1986) The susceptibility of natural communities to invasion. In: Groves RH, Burdon JJ (eds) Ecology of biological invasions. Cambridge University Press, Cambridge, pp 57–66

Gilpin ME, Soulé ME (1986) Minimum viable populations: process of species extinctions. In: Soulé ME (ed) Conservation biology: the science of scarcity and diversity. Sinauer Associates, Sunderland, MA, pp 19–34

Goodman D (1987) The demography of chance extinction. In: Soulé ME (ed) Viable populations for conservation. Cambridge University Press, Cambridge, pp 11–34

Gotwald WH Jr (1995) Army ants: the biology of social predation. Cornell University Press, Ithaca, NY

Holldobler B, Wilson EO (1990) The ants. Belknap Press/Harvard University Press, Cambridge, MA

Holway DA (1995) The distribution of the Argentine ant (*Linepithema humile*) in central California: a twenty-year record of invasion. Conserv Biol 9:1634–1637

Holway DA (1998a) Effects of Argentine ant invasions on ground-dwelling arthropods in northern California riparian woodlands. Oecologia 116:252–258

Holway DA (1998b) Factors governing the rate of invasion: a natural experiment using Argentine ants. Oecologia 115:206–212

Holway DA (1999) Competitive mechanisms underlying the displacement of native ants by the invasive Argentine ant. Ecology 80:238–251

Human KG, Gordon DM (1996) Exploitation and interference competition between the invasive Argentine ant, *Linepithema humile*, and native ant species. Oecologia 105:405–412

Jennings MR (1988) *Phrynosoma coronatum*. Cat Am Amphib Reptiles 428:1–5

Jennings MR, Hayes MP (1994) Amphibian and reptile species of special concern in California. Final report. California Department of Fish and Game Inland Fisheries Division, Contract Number 8023. 1701 Nimbus Rd, Rancho Cordova, CA 95701, USA

Leach MK, Givnish TJ (1996) Ecological determinants of species loss in remnant prairies. Science 273:1555–1558

Mallis A (1941) A list of the ants of California with notes on their habits and distribution. Bull South Calif Acad Sci 40:61–100

Mallis A (1942) Half a century with the Argentine ant. J Sci Monthly 55:536–545

Markin GP (1970a) The seasonal life cycle of the Argentine ant, *Iridomyrmex humilis* (Hymenoptera: Formicidae) in southern California. Ann Entomol Soc Am 63:1238–1242

Markin GP (1970b) Food distribution within laboratory colonies of the Argentine ant, *Iridomyrmex humilis* (Mayr). Insectes Soc 17:127–158

Montanucci RR (1989) The relationship of morphology to diet in the horned lizard genus *Phrynosoma*. Herpetologica 45:208–216

Mount RH, Trauth SE, Mason WH (1981) Predation by red imported fire ant, *Solenopsis invicta* (Hymenoptera: Formicidae), on eggs of the lizard *Cnemidophorous sexlineatus* (Squamata: Teiidae). J Ala Acad Sci 52:66–70

Munger JC (1984) Optimal foraging? Patch use by horned lizards (Iguanidae: *Phrynosoma*). Am Nat 123:654–680

Newell W (1908) Notes on the habits of the Argentine ant or "New Orleans ant", *Iridomyrmex humilis*. J Econ Entomol 1:21–34

Newell W, Barber TC (1913) The Argentine ant. USDA Bureau of Entomology Bulletin 122, USDA, Washington, DC

Orians GH (1986) Site characteristics favoring invasions. In: Mooney HA, Drake JA (eds) Ecology of biological invasions of North America and Hawaii. Ecological studies 58. Springer, Berlin Heidelberg New York, pp 133–145

Petren K, Case TJ (1996) An experimental demonstration of exploitation competition in an ongoing invasion. Ecology 77:118–132

Pianka ER, Parker WS (1975) Ecology of horned lizards: a review with special reference to *Phrynosoma platyrhinos*. Copeia 1975:141–162

Preston FW (1962) The canonical distribution of commonness and rarity: part I. Ecology 43:185–215

Rissing SW (1981) Prey preferences in the desert horned lizard: influence of prey foraging method and aggressive behavior. Ecology 62:1031–1040

Robinson SK, Thompson FR III, Donovan TM, Whitehead DR, Faaborg J (1995) Regional forest fragmentation and the success of migratory birds. Science 267:1987–1990

Shaffer ML (1981) Minimum population sizes for species conservation. Bioscience 31:131–134

Shafir S, Roughgarden J (1998) Testing predictions of foraging theory for a sit and wait forager, *Anolis gingivinus*. Behav Ecol 9:74–84

Skaiffe SH (1961) The study of ants. Longman, Green and Co, London

Smith MR (1936) Distribution of the Argentine ant in the United States and suggestions for its control or eradication. United States Department of Agriculture Circular Number 387, USDA, Washington, DC

Snelling RR, George CD (1979) The taxonomy, distribution and ecology of California desert ants. Report to Bureau of Land Management, United States Department of the Interior, Riverside, CA

Soulé ME, Bolger DT, Alberts AC et al. (1988) Reconstructed dynamics of rapid extinctions of chaparral-requiring birds in urban habitat islands. Conserv Biol 2:75–92

Suarez AV, Case TJ (2002) Bottom-up effects on the persistence of a specialist predator: ant invasions and coastal horned lizards. Ecol Appl 12:291–298

Suarez AV, Bolger DT, Case TJ (1998) The effects of fragmentation and invasion on the native ant community in coastal southern California. Ecology 79:2041–2056

Suarez AV, Richmond JQ, Case TJ (2000) Prey selection in horned lizards following the invasion of Argentine ants in southern California. Ecol Appl 10:711–725

Titus ESG (1905) Report on the "New Orleans" ant (*Iridomyrmex humilis* Mayr). US Bur Entomol Bull 52:79–84

Tremper BD (1976) Distribution of the Argentine ant, *Iridomyrmex humilis* Mayr, in relation to certain native ants of California: ecological, physiological, and behavioral aspects. PhD Thesis, University of California, Berkeley

Tschinkel WR (1987) Seasonal life history and nest architecture of a winter active ant, *Prenolepis imparis*. Insectes Soc 34:143–164

Turner IM, Corlett RT (1996) The conservation value of small, isolated fragments of lowland tropical rain forest. Trends Ecol Evol 11:330–333

Ward PS (1987) Distribution of the introduced Argentine ant (*Iridomyrmex humilis*) in natural habitats of the lower Sacramento Valley and its effects on the indigenous ant fauna. Hilgardia 55:1–16

Westman WE (1981) Diversity relations and succession in Californian coastal sage scrub. Ecology 62:170–184

Wheeler GC, Wheeler J (1973) Ants of Deep Canyon. University of California Press, Berkeley, CA

Wilcove DS (1985) Nest predation in forest tracts and the decline of migratory songbirds. Ecology 66:1211–1214

Williams DF (ed) (1994) Exotic ants: biology, impact, and control of introduced species. Westview Press, Boulder, CO

Woodworth CW (1908) The Argentine ant in California. Univ Calif Agric Exp Stn 38:1–192

Yahner R (1988) Changes in wildlife communities near edges. Conserv Biol 2:333–339

Part III
Ecosystem Fragmentation: Theory, Methods, and Implications for Conservation

10 A Review and Synthesis of Conceptual Frameworks for the Study of Forest Fragmentation

G.H. KATTAN and C. MURCIA

10.1 Introduction

Forest fragmentation is the large-scale transformation of a forested landscape to one in which remnant forest patches are isolated in a matrix of anthropogenic habitats. Forest fragmentation is a major cause of loss of biological diversity, in particular in the species-rich wet tropics, where landscape transformation is an ongoing process (Whitmore 1997; Viña and Cavelier 1999). A large body of literature gives evidence of the negative effects of fragmentation, which include changes in the physical environment, and regional and local extirpation of populations of many species of plants and animals (Saunders et al. 1991; Schelhas and Greenberg 1996; Turner 1996; Laurance and Bierregaard 1997).

Research on the effects of forest fragmentation has traditionally concentrated on exploring the consequences of isolating populations and communities in small habitat patches. This research was first inspired by the theory of island biogeography, which predicted reductions of diversity as the size of the habitat patches decreased and isolation increased (Simberloff 1988, 1997). This paradigm generated an enormous amount of research that provided important lessons, but it only addressed a limited part of a complex problem. Some of the shortcomings of the application of the island biogeography framework to terrestrial habitat fragmentation were that: (1) the main parameter of this theory is the number of species, but for conservation purposes the identity of species and their interactions are more important than the simple species count, and (2) in terrestrial ecosystems the matrix may be composed of a diversity of habitats that may be used by some forest organisms, and the matrix itself may support a variety of organisms that interact with the fragments. While the empirical evidence indicates that fragmentation has the general effect of reducing local diversity of forest organisms, there is variation in the response of specific taxonomic groups, and the specific mechanisms and attributes of organisms that determine susceptibility to

Ecological Studies, Vol. 162
G.A. Bradshaw and P.A. Marquet (Eds.)
How Landscapes Change
© Springer-Verlag Berlin Heidelberg 2003

extinction are poorly understood (Turner 1996; Harrison and Bruna 1999; Debinski and Holt 2000). The difficulties in understanding the varied responses to fragmentation are exacerbated by the lack of an appropriate conceptual framework.

Two theoretical frameworks developed rapidly during the last decade, and have made important contributions to our understanding of wildlife responses to fragmentation. These are landscape ecology (Pickett and Cadenasso 1995; Wiens 1995) and metapopulation biology (McCullough 1996; Hanski and Gilpin 1997). Landscape ecology provides a large-scale perspective, exploring the interactions among habitats and wildlife in heterogeneous landscapes. Metapopulation biology, in contrast, concentrates on the population dynamics of species in systems of small patches that have low levels of flow of individuals among them. Metapopulation biology is emerging as the favored frame of reference for studying fragmentation issues (Hanski and Simberloff 1997; Harrison and Bruna 1999). Metapopulation models, however, apply to situations in which habitat patches are embedded in an inert matrix. In contrast, landscape ecology views matrices as active parts of heterogeneous habitat mosaics. While a synthesis of the two theories has the potential for providing a comprehensive framework for understanding the effects of fragmentation (Wiens 1996, 1997a,b), they apply to very particular situations. We believe the framework needs to be expanded to encompass the great variety of landscape configurations that exist in reality.

In this paper, we review a series of concepts and theories relevant to the question of how a particular organism will respond to habitat alteration, in an attempt to weave them together in a broad conceptual framework. The connecting thread for this integration is the concept of scale, and we stress three main points: (1) there is no simple theory and no single mechanism that will explain *all* the effects of forest fragmentation; (2) how organisms respond to fragmentation depends on the interaction of the scale at which the organisms function and the scale of fragmentation; and (3) responses at a local scale may depend on what is going on at larger scales. In the following sections, we first briefly review the evolution of fragmentation studies in terms of the empirical evidence and the conceptual frameworks in which they have been embedded. Then we review the theory that would be required for producing a comprehensive conceptual framework, and finally we examine the potential of this framework to explain the evidence. Our emphasis is on animals in humid tropical forests, but we expect the framework to have general applicability to other organisms and ecosystems.

10.2 Evolution of Studies on the Effects of Forest Fragmentation: Empirical Evidence and Conceptual Frameworks

Early research on the effects of habitat fragmentation concentrated on an empirical search for patterns: what are the patterns of diversity of different types of organisms in differently sized fragments? What are the patterns of extinction of species in different taxonomic or functional groups? The expectations of these studies were that forest-dwelling taxa would exhibit a decrease in species diversity as size of fragments decreased, and that some functional groups of organisms would prove to be particularly vulnerable (e.g., Willis 1974, 1979; Leck 1979; Newmark 1991). The existence of repetitive patterns of extinction of ecologically similar species at different sites would indicate that fundamental processes are being altered, which would point to possible mechanisms of extinction. In contrast, the absence of patterns would suggest that extinction is random (Terborgh and Winter 1980).

The search for patterns of extinction requires knowing which species have gone extinct. A direct approach to this problem requires knowledge of the species content of an area before and after fragmentation, which has been rarely available (e.g., Lovejoy et al. 1986; Kattan et al. 1994; Bierregaard and Stouffer 1997; Tocher et al. 1997; Renjifo 1999). For example, the availability of historical data allowed documentation of the extinction of one third of the forest bird species at two sites in the Colombian Andes (Kattan et al. 1994; Renjifo 1999). In the absence of historical data, an indirect approach is to compare the species content of fragments of different sizes, under the assumption that absence of species in smaller fragments would indicate that the species have gone locally extinct. A reduction in diversity concomitant with a reduction in area has indeed been found for a variety of taxa (review in Turner 1996), but results have not been consistent across taxonomic groups, in that a decreased diversity is not necessarily observed in smaller fragments (e.g., Stouffer and Bierregaard 1995 for hummingbirds; Brown and Hutchings 1997 for butterflies; Malcolm 1997 for small mammals).

On the other hand, the search for patterns of vulnerability has produced some important results (Turner 1996). Some patterns have been predicted based on theoretical considerations. One of the earliest predictions was that large animals and top predators would be particularly vulnerable, due to their large area requirements (Terborgh 1974; Terborgh and Winter 1980). This pattern has been supported by studies on mammals and birds (e.g., Thiollay 1989; Newmark 1995, 1996; Jullien and Thiollay 1996; Chiarello 1999). Other patterns have been inferred from the empirical evidence. For example, large-bodied fruit-eating birds and mammals consistently disappear from fragmented landscapes (Turner 1996). Because fruit is a highly variable resource in space and time, frugivores need to move over large areas in the wake of fruiting waves (Levey and Stiles 1992; Powell and Bjork 1995). Therefore, these animals are vulnerable to perturbations that disrupt their movement patterns.

One important development was the realization that landscape context, and in particular the interaction with the matrix, was important in determining the dynamics of fragments. Birds recolonized forest fragments in central Amazonia after surrounding fields were abandoned and second growth allowed interpatch movement (Bierregaard and Stouffer 1997). A study of cloud forest fragments surrounded by different types of matrix in the Colombian Andes showed that in comparison with pastures, conifer plantations buffered population declines of some species of birds (Renjifo 2001). Therefore, some matrix habitats are more benevolent than others, and a heterogeneous matrix is likely to facilitate movement for a wide variety of organisms (Kattan and Alvarez-López 1996; Power 1996; Wiens 1996; Estrada et al. 1998; Medina et al. 2002).

An important interaction between fragments and the matrix is represented by edge effects. Forest fragments usually present abrupt borders with the surrounding matrix. Juxtaposition with an exotic ecosystem often causes changes in species composition, vegetation structure and ecological interactions on the perimeter of the fragment (Murcia 1995). All these changes may negatively affect the populations of most forest dwelling species, and may favor an increase of abundance of the matrix species, which may invade the fragment (Andrén 1994; Murcia 1995).

While the accumulation of studies on the effects of fragmentation has provided some lessons, the study of fragmentation is far from being a predictive science (Leigh 1997). One of the main lessons derived so far is that effects are varied, site- and organism-specific, and conclusions can be applied only at a very general level (Turner 1996; Harrison and Bruna 1999; Debinski and Holt 2000). A variety of general mechanisms of extinction have been identified, but the relative importance of each mechanism is unclear (Terborgh 1992; Turner 1996; Laurance et al. 1997). Is this a situation that can be remedied by framing these studies in a different conceptual framework? Two limitations of some of these studies were that they (1) tended to ignore the context in which the fragments were embedded, and (2) did not consider whether the scale of study (centered on a fragment) was appropriate for the scale of the focal organisms.

Both of these aspects are central to the theory of landscape ecology. The essence of a landscape is heterogeneity, that is, the presence of a mosaic of patches or multiple habitat types, and the purpose of landscape ecology is to understand how the configuration of the landscape (i.e., patch size, geometry and arrangement with respect to other patches) affects ecological patterns and processes (Wiens 1995). Both natural and anthropogenic landscapes can be viewed as mosaics of patches at different scales (Wiens 1976, 1997a). Thus, whereas the classical view of fragmentation is that of a dichotomous landscape, with patches of habitat isolated in an ocean of nonhabitat, and the emphasis is on the populations and communities contained in the isolated fragment, the view from the perspective of landscape ecology is that of a mosaic of habitat types, and the emphasis is on the species assemblages and population dynamics at the regional scale. The spatial structure of the mosaic (geometry of patches) determines the distribution and abundance of organisms in the landscape. From this point of view, a particular patch is simply part of a regional system of patches, and what happens in the patch must be considered in the landscape context.

Central to the theory of landscape ecology are the concepts of habitat use and animal movement (Wiens 1994; Kozakiewicz 1995). Habitat use is not necessarily an all-or-none phenomenon; instead, there is a gradation of intensity of use of different habitats (Block and Brennan 1993; Kozakiewicz 1995). Likewise, even though animals may not settle in some habitats, they may move through them to reach patches of appropriate habitats across the landscape. Thus, depending on the spatial configuration of patches, and the patterns of movement and habitat use, different animals will have particular patterns of distribution and abundance in the landscape.

Animal movement is the central theme in neutral landscape models, which are computer simulations that attempt to elucidate how animals will respond to different spatial configurations of patches, depending on movement rules and other life history characteristics of the organisms. One particular focus of these models is to understand how landscape connectivity will be ruptured, that is, at what proportion of habitat vs. nonhabitat will different kinds of animals be unable to move across the landscape (With and King 1997). These models suggest that landscape connectivity is a threshold phenomenon, that is, there is a critical amount of percent habitat cover below which a particular animal will not be able to move through the landscape. What this critical value is depends on dispersal and other life history characteristics of the particular organism (Andrén 1994; With et al. 1997).

The patterns of movement of animals in the landscape link landscape ecology with metapopulation theory (Wiens 1996, 1997b). A metapopulation is a collection of subpopulations, each confined to a discrete habitat patch. Each subpopulation has a certain probability of extinction, but there is some flux of individuals between patches, such that patches may be recolonized. Thus, the metapopulation will persist at the regional scale, although some patches may be unoccupied at times. Metapopulation theory applies to situations where flow rates of individuals between patches are relatively low; if flow rates are high, then we have a panmictic, although patchily distributed population. Metapopulation models are being applied to real life situations, with a great potential for providing management possibilities (McCullough 1996; Hanski and Gilpin 1997).

In summary, the study of fragmentation has been approached from three different perspectives, relevant to particular landscape configurations. First, what we call the empirical approach has emphasized the patterns of diversity in habitat fragments, and the patterns of vulnerability of different taxonomic and functional groups. The interaction with the matrix has been considered in the context of how it will affect the dynamics and species content of the fragments (e.g., edge effects), but without being considered as important in the dynamics of focal organisms. Thus, this approach emphasizes extinction and focuses on the individual fragment. The second approach, landscape ecology, addresses these issues by considering the large scale patterns of interaction among all habitats, allowing for the possibility that organisms may use different habitats with different intensities. Thus, organisms may flow through the landscape depending on its spatial configuration, or even maintain populations in matrix habitats. The spatial structure of the population in the landscape would depend on the patterns of habitat use and movement of particular organisms. Finally, metapopulation models concentrate on the dynamics of populations that are subdivided by a matrix that may be crossed, but not used. Thus, these models approach the problem from a regional perspective, but apply only to archipelago-type landscapes.

In the next section we bring these theories together by providing a broad view of fragmentation as part of a process of landscape transformation that may follow many different trajectories (process of fragmentation) and reach a diversity of endpoints (pattern of fragmentation; Wiens 1994). In this context, fragmentation is just one particular configuration in a transformed landscape.

10.3 A Comprehensive Framework

10.3.1 The Process of Fragmentation

A conceptual framework for explaining the effects of forest fragmentation requires understanding the sequence of events that occur as a landscape is transformed, and the relevant scales at which different events operate. We begin with a tropical forest landscape, defined on a scale of tens to hundreds of square kilometers; this landscape is composed of a mosaic of habitat types, caused by irregularities in soil type, topography, and climate at a local scale, which result in different vegetation types and associated species assemblages. The local patterns of natural disturbance (tree-fall gaps, landslides, river flooding) add to habitat heterogeneity, resulting in a dynamic mosaic of habitats at local and regional scales. Initial clearings made by humans at a small scale may mimic patterns of natural disturbance, and add to the regional habitat heterogeneity. These disturbances have very local effects, with little impacts on populations at regional scales (e.g., shifting agriculture systems in Amazonia; Andrade and Rubio 1994).

As the size of clearings increases, disturbances begin to have a major impact, both at local and regional scales. Deforestation is rarely spatially random; instead, it may be concentrated on certain areas, depending on factors such as topography and soil types. For example, deforestation may first concentrate on gentle slopes and high ground along rivers, while sparing steep slopes or flooded terrain. This may result in the elimination of entire habitats and their associated species assemblages, as well as species that depend on these habitats for some stages of their life cycle (e.g., local migrations, or species that require two different habitats for completion of life cycle). This pattern of deforestation may have a regional impact, as it reduces habitat diversity and simplifies the regional mosaic, hence reducing gamma diversity (Rosenzweig 1996), as well as reducing the diversity of local communities by mesoscale effects (Holt 1993). Another source of disturbance at this stage is that human activities begin to radiate from centers of colonization; for example, hunting may affect populations of game animals in a large area around centers of human habitation (Thiollay 1989; Redford 1992).

As deforestation continues, anthropogenic habitats begin to dominate in areal extent, becoming the matrix. At this stage, the original forest habitats may be shredded rather than fragmented, as habitat patches may adopt many different geometric shapes and interdigitate in complex patterns (Feinsinger 1994). Instead of neat, discrete islands, natural and human-created habitats may form mosaics with blurred boundaries (Wiens 1995). Even though forests may still occur in relatively large blocks and the landscape may be relatively heterogeneous, landscape changes may be having a major impact, as the spatial and temporal dynamics of processes that operate at regional levels may be disrupted. One such impact is caused by rupture of connectivity at the landscape level, as happens in the case of fruit-eating birds that move at regional scales.

Further transformation may produce a fragmented landscape, with the typical arrangement of forest patches isolated in intensely managed and homogeneous habitats (pastures, croplands, suburban habitats). In this type of landscape arrangement, forest patches tend to be small, internally homogeneous, and with sharp boundaries. Forest organisms that can disperse across the matrix may persist in this landscape, even though individual fragments may not be large enough to support a viable population. Many organisms, however, will be confined to forest patches and vulnerable to effects such as demographic accidents, catastrophes, and edge effects.

Probably no landscape is totally static, at least in a long enough time frame. When exploring the effects of fragmentation, one should consider the history of landscape change in the region, the current configuration, and if possible, future trends. The matrix surrounding fragments can change in time towards a more heterogeneous ecosystem that is more favorable for the fragments, or can change towards an extreme ecosystem that proves hostile to the forest species. For example, large-scale replacement of pasturelands with scattered trees and hedges by highly mechanized sugar cane plantations has resulted in the regional decline and extinction of many raptors in the Cauca Valley, in southwestern Colombia (Alvarez-López and Kattan 1995).

10.3.2 Effects of Fragmentation on Animal Populations

The effect that habitat alteration will have on wildlife in any of the above scenarios depends on the interaction between the scale at which the organism functions in the landscape, and the spatial configuration of the habitat mosaic; in other words, it depends on the patterns of distribution and abundance of organisms in the landscape, and how they are altered by fragmentation. Responses of organisms to fragmentation have been modeled as a function of mainly three factors: habitat use or specialization, mobility or gap-crossing abilities, and population density or area requirements (Dale et al. 1994; MacNally and Bennett 1997; With and King 1997). These three variables can be put together with other factors into a comprehensive model that explains the distribution and abundance of animals in a landscape, and how fragmentation alters the spatial structure of the population (Fig. 10.1). The key to understanding how a species will respond to habitat alteration is knowing which factors affect the spatial structure of the population. It is important to recognize two different spatial and temporal scales that determine the distribution of the species in the landscape (Fig. 10.1):

1. Scale of individuals: how an individual uses space is determined by its home range (daily and seasonal patterns of movement), which is a function of factors such as body size and physiological requirements, and distribution and abundance of resources (food, nest sites, refuges); resources may in turn be affected by intra- and interspecific interactions (Stern 1998). The interaction of these factors translates into patterns of habitat use of individuals (Kozakiewicz 1995) and their spatial dispersion.
2. Scale of populations: the spatial extent of a population depends on its dispersal and dispersion patterns (Goodwin and Fahrig 1998), which are a function of factors such as demography, interspecific interactions, life cycle, and phenology of the species and its resources.

Some factors, such as interspecific interactions and patch dynamics, operate across scales, affecting the spatial distribution of individuals, populations, and communities (Fig. 10.1), and it is critical to understand how these factors will be affected by fragmentation. For example, the dynamics of resource patches at small scales will determine patterns of individual behavior (e.g., an animal establishing a territory or entering a particular habitat), while at intermediate scales they will determine the dispersion of the population, and at larger scales they will determine the structure of the patch mosaic and, hence, the spatial structure of the population. Scale is critical for patch dynamics:

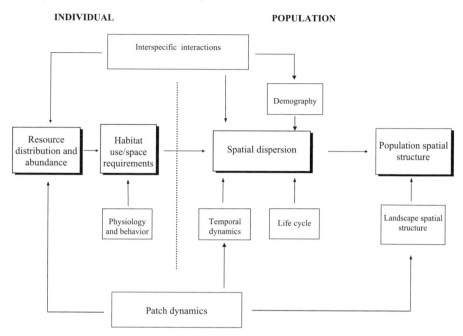

INDIVIDUAL   **POPULATION**

Fig. 10.1. Relationships between factors that determine the patterns of habitat use, dispersion, and spatial structure of a population. These factors may operate at the scale of the individual, the population, or across scales. For example, patch dynamics at a small scale determines the distribution and abundance of resources (e.g., patterns of fruit production in different habitats), while at a large scale it determines the distribution of habitat patches in the landscape and, therefore, the spatial distribution of the population. Understanding these factors is the key to predicting the response of a species to landscape alteration, depending on how these alterations interact with the spatial and temporal dynamics of the species

fragmentation that mimics the natural patch dynamics of a species would have no major effect. In contrast, fragmentation that occurs at a scale (both extent and grain) larger than the natural patch dynamics of a species is likely to disrupt cycles and movements.

The patterns of movement and dispersion of individuals interact with the structure of the landscape to determine the spatial structure of the population (Goodwin and Fahrig 1998). In a natural setting, populations may be continuous, but with variations in density. Density variations may be the result of a heterogeneous landscape, in which the pattern of spatial distribution of the population depends on its patterns of habitat use. Habitats differ in their suitability for a given species, depending on the distribution and abundance of resources within the patch; thus, individuals of a species may occupy a patch permanently or only temporarily (Block and Brennan 1993; Kozakiewicz 1995). Here, it is important to keep in mind how the size of the patch compares with the spatial scale of the individual and the population. Some species are restricted to one habitat type. For these species the spatial distribution and size of habitat patches will determine the spatial structure of the population. Other species are habitat generalists, using many or all of the different types of habitats in the landscape; for these species, the spatial structure of the population is determined by their patterns of habitat use. Habitat suitability may be reflected in population density and population dynamics within and among patches (Block and Brennan 1993; Kozakiewicz 1995). Regional populations of species that use more than one habitat type may be driven by source-sink dynamics, with subpopulations in optimal habitat producing excess individuals that maintain subpopulations in marginal habitat (Pulliam 1996).

Habitat specialists and habitat generalists are extremes in a continuum of patterns of habitat use. Many species may be able to use some habitats in the landscape, but not others. One effect of fragmentation is to make populations discontinuous, but not necessarily disconnected. Species responses to habitat fragmentation will largely depend on their capacity to move across and use matrix habitats. Animal movement is a central theme in landscape ecology and pivotal in understanding and predicting responses to fragmentation (Wiens 1996, 1997b). It is important to recognize two forms of movements. First is the movement of individuals during their normal activities, such as foraging within a home range. Some animals may spend their entire lifetimes in a small patch while others may wander widely. A particularly sensitive case in the context of habitat fragmentation is that of animals that perform seasonal migrations, such as many frugivorous birds (Levey and Stiles 1992; Powell and Bjork 1995). The normal patterns of movement define the scale of the animal and its perception of habitat patchiness. An animal that spends its life in a small patch (e.g., a leaf-litter ant) is probably very sensitive to small-scale patchiness. In contrast, a highly mobile organism probably has a coarse-grained perception of patchiness and moves across many habitat patches (Stern 1998).

The second type of movement is the post-natal dispersal movement that is usually made once in a lifetime. Each species may have a characteristic dispersal function, defined by the mean and shape of the distribution of dispersal distances (Fig. 10.2). Although it may be assumed that dispersal functions in general taper off with large distances, dispersal distances are difficult to measure and field studies may underestimate the frequency of long-distance dispersal events (Koenig et al. 1996). It is precisely these long-distance dispersal events that may be critical in determining the possibility of recolonizing habitat patches in a fragmented landscape (Peacock and Smith 1997). Dispersal functions may be altered by factors such as landscape configuration. In a fragmented population, the distribution of dispersal distances may be different from that in an unbounded population in continuous habitat. For example, nuthatches (*Sitta europaea*) in a highly fragmented landscape in Belgium had a much larger mean dispersal distance than in a forested landscape,

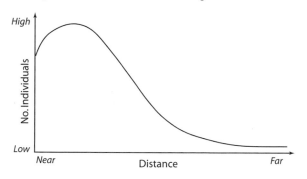

Fig. 10.2. Hypothetical dispersal function of a species, showing the numbers of individuals that disperse to different distances from the natal site

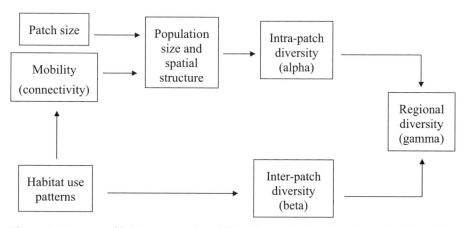

Fig. 10.3. Patterns of habitat use and mobility of an animal are key factors in determining the spatial structure of a population, and within- and between-habitat patterns of diversity. Depending on body size and other ecological factors (see Fig. 10.1), the number of individuals in a given patch will be determined by patch size; patterns of mobility and habitat use will determine rates of flow of individuals among patches and the spatial structure of the population. Adding these patterns for an assemblage of species would determine intra-patch diversity. On the other hand, patterns of habitat use will also determine how different the species composition is in the patches forming the landscape mosaic (inter-patch component of diversity). Finally, within- and between-habitat patterns of diversity will determine regional diversity

although the proportion of juveniles settling near their natal area did not differ between fragmented and nonfragmented landscapes (Matthysen et al. 1995).

Habitat use and movement are critical factors in determining the response of a species to a particular landscape configuration, and within- and between-habitat patterns of diversity (Fig. 10.3). Figure 10.4 shows how the interaction between habitat use, movement, and landscape configuration leads to different situations, and the relevant frameworks that apply to each situation. Habitat specialists that cannot cross the matrix (neither by individual movement nor by dispersal) are the species most likely to be severely affected by fragmentation, as individuals will be confined to habitat patches (Figs. 10.3 and 10.4). Whether a particular habitat patch will support a viable population or not depends on the size of the patch in relation to the spatial scale of the individuals and the population. If the species can move across the matrix, it needs to be determined whether individuals actually use matrix habitats, a likely event if the matrix contains some resources that the species can use. In this case, populations will be more or less continuous, but with variations in density in the different habitats, and their dynamics will be determined by patterns of habitat use (Boyce and McDonald 1999).

Some organisms may not be able to use matrix habitats, but may be able to cross them during dispersal. For these organisms, populations will be subdivided, but not disconnected. If rates of flow among fragments are high, the species will be present in most fragments across a landscape, although small fragments may be sink habitats. If rates of flow are low, then metapopulation models apply.

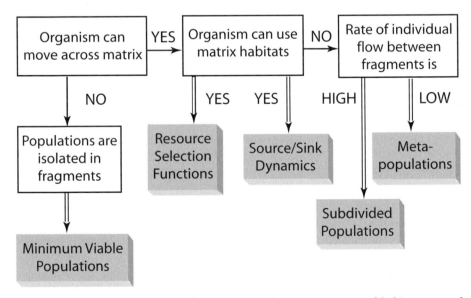

Fig. 10.4. Scheme showing how the interaction between patterns of habitat use and movement of a species may lead to different scenarios in a fragmented landscape, and the different conceptual frameworks (*highlighted boxes* and *double arrows*) that would apply to each situation

10.4 Framing the Evidence

Earlier attempts to organize the evidence on the effects of fragmentation into a scheme that would allow for generalizations have focused on listing general factors of vulnerability and mechanisms of extinction (Terborgh 1992; Turner 1996; Laurance et al. 1997). Turner (1996) proposed that fragmentation effects result from the action of six different classes of mechanisms: (1) the chance destruction of some habitats within the landscape; (2) restriction of population size; (3) prevention or reduction of immigration; (4) edge effects; (5) higher-order effects (i.e., interspecific interactions); and (6) immigration of exotics. Turner pointed out that the relative importance of each of these mechanisms remains uncertain. However, the kinetics of extinction in fragmented landscapes is bound to be complex when many different organisms are involved, and it cannot be expected that a single mechanism will prove to be of utmost importance. A simple strategy of accumulating studies and hoping for generalizations to emerge is unlikely to bear fruit. Instead, a more useful approach would be to test predictions derived from a conceptual framework.

The framework presented here provides a comprehensive view of fragments and their landscapes, and has the potential to explain cases in which results do not fit expectations derived from a fragment-based theory of fragmentation. We propose a model that considers the interaction between two domains of scale: *the spatial scale of fragmentation* – the scale at which fragmentation effects are operating and *the organismal scale* – the scale at which species respond to habitat changes. Each domain can be partitioned into two levels (Table 10.1). Fragmentation may have effects both at the regional level and at the fragment level, and may affect organisms both as individuals and as populations.

A wide variety of mechanisms of extinction can be framed in this scheme, by seeing them as different manifestations of the interaction between fragmentation scale and organismal scale, which has the end result of modifying the spatial structure of the population (Fig. 10.1). Mechanisms such as elimination of some habitats in the landscape (which translates into the regional extinction of species associated to those habitats) and habitat loss for large-bodied species (Turner 1996), as well as the disruption of regional migrations, operate at the regional level. In contrast, mechanisms such as isolation and exposure to edges (Turner 1996) operate at the fragment level. On the other hand, each type of mechanism may have a different effect on the species, depending on the characteristics of the individuals and the populations. Some mechanisms operate at one particular combination of scales, while others operate at two or more combinations (cells in Table 10.1). For example, changes in immigration rates operate at the regional level and have an effect at the population level. In contrast, changes in interspecific interactions may

Table 10.1. Interaction between the spatial and the organismal scales, each operating at two levels. The cells show examples of mechanisms of extinction that operate at different combinations of scales/levels

	Spatial scale Landscape	Spatial scale Fragment
Organismal scale Population	Low immigration rates Reduced population size Altered interspecific interactions	Reduced subpopulation size Isolation of populations
Organismal scale Individual	Restricted movement patterns and dispersal Disturbed regional migrations Disturbed interactions Altered resource availability	Edge avoidance or attraction Disturbed interactions Altered resource availability

occur at both levels of the spatial scale (regional and fragment) and have an effect both at the population and the individual levels.

Failure to consider factors that operate at the landscape level may produce what appear to be unexpected results in studies of forest fragmentation, because many animals are capable of interacting with the matrix or other landscape features outside forest fragments, or because they are functioning at a scale that is different from that at which fragmentation occurs. Therefore, their response to fragment size is not always straightforward. For example, shortly after fragmentation, visitation rates of euglossine bees to artificial baits in central Amazonia declined with fragment size (Powell and Powell 1987). However, a few years later, Becker et al. (1991) found greater bee abundance in 10-and 100-ha fragments than in continuous forest. The unexpected response of bees to fragmentation may be driven by their ability to move large distances and to use the secondary vegetation that eventually surrounded the fragments. In addition, the distribution of bees may be more responsive to the distribution of key resources, such as resins, than to habitat structure and patch size.

The spatial and temporal availability of resources is one aspect that has been overlooked as an a priori factor in studies of fragmentation. For example, the availability of breeding habitat, in particular aquatic habitats, determines the presence of many species of frogs in forest fragments (Kattan 1993; Tocher et al. 1997). In cloud forest fragments in the Colombian Andes, the persistence of some frog species depends on the presence of permanent streams, and the response of the individuals to these landscape features (Kattan 1993). Hummingbirds are another group of animals that are primarily resource-driven rather than habitat-driven. Not surprisingly, these nectar-feeding birds remain unaffected by fragmentation, provided that the landscape is sufficiently heterogeneous and provides adequate resources (Kattan et al. 1994; Stouffer and Bierregaard 1995).

The above examples show that fragment size and degree of isolation are not necessarily the best predictors of fragmentation effects. Here, we have synthesized different conceptual frameworks that apply to the study of forest fragmentation, and emphasized that the interaction between the scale of fragmentation and the scale of the organisms is key to understanding how fragmentation affects the spatial distribution of individuals and populations. To the extent that we can identify which factors operate at different spatial scales, and the responses of individuals and populations to these factors, we may be able to predict the effects of fragmentation. This, of course, entails collecting additional information on the basic ecology of the species, and taking into account many variables that have been considered marginal (such as body size, patterns of resource use, and ability to move through the matrix), and therefore, there are no simple solutions to the very complex problem of fragmentation.

References

Alvarez-López H, Kattan GH (1995) Notes on the conservation status of resident diurnal raptors of the middle Cauca Valley, Colombia. Bird Conserv Int 5:341–348

Andrade GI, Rubio H (1994) Sustainable use of the tropical rain forest: evidence from the avifauna in a shifting-cultivation habitat mosaic in the Colombian Amazon. Conserv Biol 8:545–554

Andrén H (1994) Effects of habitat fragmentation on birds and mammals in landscapes with different proportions of suitable habitat: a review. Oikos 71:355–366

Becker P, Moure JS, Peralta FJA (1991) More about euglossine bees in Amazonian forest fragments. Biotropica 23:586–591

Bierregaard RO, Stouffer PC (1997) Understory birds and dynamic habitat mosaics in Amazonian rainforests. In: Laurance WF, Bierregaard RO Jr (eds) Tropical forest remnants: ecology, management, and conservation of fragmented communities. University of Chicago Press, Chicago, pp 138–155

Block WM, Brennan LA (1993) The habitat concept in ornithology. Theory and applications. In: Power DM (ed) Current ornithology, vol 11. Plenum Press, New York, pp 35–91

Boyce MS, McDonald LL (1999) Relating populations to habitats using resource selection functions. Trends Ecol Evol 14:268–272

Brown KS, Hutchings RW (1997) Disturbance, fragmentation, and the dynamics of diversity in Amazonian forest butterflies. In: Laurance WF, Bierregaard RO Jr (eds) Tropical forest remnants: ecology, management, and conservation of fragmented communities. University of Chicago Press, Chicago, pp 91–110

Chiarello AG (1999) Effects of fragmentation of the Atlantic forest on mammal communities in south-eastern Brazil. Biol Conserv 89:71–82

Dale VH, Pearson SM, Offerman HL, O'Neill RV (1994) Relating patterns of land use change to faunal biodiversity in the central Amazon. Conserv Biol 8:1027–1036

Debisnki DM, Holt RD (2000) A survey and overview of habitat fragmentation experiments. Conserv Biol 14:342–355

Estrada A, Coates-Estrada R, Dadda AA, Cammarano P (1998) Dung and carrion beetles in tropical rain forest fragments and agricultural habitats at Los Tuxtlas, Mexico. J Trop Ecol 14:577–594

Feinsinger P (1994) Habitat "shredding". In: Meffe GK, Carroll CR (eds) Principles of conservation biology. Sinauer Associates, Sunderland, MA, pp 258–260

Goodwin BJ, Fahrig L (1998) Spatial scaling and animal population dynamics. In: Peterson DL, Parker VT (eds) Ecological scale: theory and applications. Columbia University Press, New York, pp 193–206

Hanski I, Gilpin ME (eds) (1997) Metapopulation biology: ecology, genetics, and evolution. Academic Press, San Diego

Hanski I, Simberloff D (1997) The metapopulation approach, its history, conceptual domain, and application to conservation. In: Hanski I, Gilpin ME (eds) Metapopulation biology: ecology, genetics, and evolution. Academic Press, San Diego, pp 5–26

Harrison S, Bruna E (1999) Habitat fragmentation and large-scale conservation: what do we know for sure? Ecography 22:225–232

Holt RD (1993) Ecology at the mesoscale: the influence of regional processes on local communities. In: Ricklefs RE, Schluter D (eds) Species diversity in ecological communities. University of Chicago Press, Chicago, pp 77–88

Jullien M, Thiollay JM (1996) Effects of rain forest disturbance and fragmentation: comparative changes of the raptor community along natural and human-made gradients in French Guiana. J Biogeogr 23:7–25

Kattan GH (1993) The effects of forest fragmentation on frogs and birds in the Andes of Colombia: implications for watershed management. In: Doyle JK, Schelhas J (eds) Forest remnants in the tropical landscape: benefits and policy implications. Smithsonian Migratory Bird Center, Washington, DC, pp 11–13

Kattan GH, Alvarez-López H (1996) Preservation and management of biodiversity in fragmented landscapes in the Colombian Andes. In: Schelhas J, Greenberg R (eds) Forest patches in tropical landscapes. Island Press, Washington, DC, pp 3–18

Kattan GH, Alvarez-López H, Giraldo M (1994) Forest fragmentation and bird extinctions: San Antonio eighty years later. Conserv Biol 8:138–146

Koenig WD, Van Buren D, Hooge PN (1996) Detectability, phylopatry, and the distribution of dispersal distances in vertebrates. Trends Ecol Evol 11:514–517

Kozakiewickz M (1995) Resource tracking in space and time. In: Hansson L, Fahrig L, Merriam G (eds) Mosaic landscapes and ecological processes. Chapman and Hall, London, pp 136–148

Laurance WF, Bierregaard RO Jr (eds) (1997) Tropical forest remnants: ecology, management, and conservation of fragmented communities. University of Chicago Press, Chicago

Laurance WF, Bierregaard RO, Gascon C et al. (1997) Tropical forest fragmentation: synthesis of a diverse and dynamic discipline. In: Laurance WF, Bierregaard RO Jr (eds) Tropical forest remnants: ecology, management, and conservation of fragmented communities. University of Chicago Press, Chicago, pp 502–514

Leck CF (1979) Avian extinctions in an isolated tropical wet-forest preserve, Ecuador. Auk 96:343–352

Leigh EG (1997) Book review: tropical forest remnants. Trends Ecol Evol 12:414

Levey DJ, Stiles FG (1992) Evolutionary precursors of long-distance migration: resource availability and movement patterns in Neotropical landbirds. Am Nat 140:447–476

Lovejoy TE, Bierregaard RO, Rylands AB, et al. (1986) Edge and other effects of isolation on Amazon forest fragments. In: Soulé ME (ed) Conservation biology: the science of scarcity and diversity. Sinauer Associates, Sunderland, MA, pp 257–285

MacNally R, Bennet AF (1997) Species-specific predictions of the impact of habitat fragmentation: local extinction of birds in the box-ironbark forests of central Victoria, Australia. Biol Conserv 82:147–155

Malcolm JR (1997) Biomass and diversity of small mammals in Amazonian forest fragments. In: Laurance WF, Bierregaard RO Jr (eds) Tropical forest remnants: ecology, management, and conservation of fragmented communities. University of Chicago Press, Chicago, pp 207–221

Matthysen E, Adriaensen F, Dhondt AA (1995) Dispersal distances of nuthatches, *Sitta europaea*, in a highly fragmented forest habitat. Oikos 72:375–381

McCullough DR (ed) (1996) Metapopulations and wildlife conservation. Island Press, Washington, DC

Medina CA, Escobar F, Kattan GH (2002) Diversity and habitat use of dung beetles in a restored Andean landscape. Biotropica 34:181–187

Murcia C (1995) Edge effects in fragmented forests: implications for conservation. Trends Ecol Evol 10:58–62

Newmark WD (1991) Tropical forest fragmentation and the local extinction of understory birds in the eastern Usambara Mountains, Tanzania. Conserv Biol 5:67–78

Newmark WD (1995) Extinction of mammal populations in western North American national parks. Conserv Biol 9:512–526

Newmark WD (1996) Insularization of Tanzanian parks and the local extinction of large mammals. Conserv Biol 10:1549–1556

Peacock MM, Smith AT (1997) The effect of habitat fragmentation on dispersal patterns, mating behavior, and genetic variation in a pika (*Ochotona princeps*) metapopulation. Oecologia 112:524–533

Pickett STA, Cadenasso ML (1995) Landscape ecology: spatial heterogeneity in ecological systems. Science 269:331–334

Powell AH, Powell GVN (1987) Population dynamics of male euglossine bees in Amazonian forest fragments. Biotropica 19:176–179

Powell GVN, Bjork R (1995) Implications of intratropical migration on reserve design: a case study using *Pharomachrus mocinno*. Conserv Biol 9:354–362

Power AG (1996) Arthropod diversity in forest patches and agroecosystems of tropical landscapes. In: Schelhas J, Greenberg R (eds) Forest patches in tropical landscapes. Island Press, Washington, DC, pp 91–110

Pulliam HR (1996) Sources and sinks: empirical evidence and population consequences. In: Rhodes OE, Chesser RK, Smith MH (eds) Population dynamics in ecological space and time. University of Chicago Press, Chicago, pp 45–69

Redford KH (1992) The empty forest. BioScience 42:412–422

Renjifo LM (1999) Composition changes in a sub-Andean avifauna after long-term forest fragmentation. Conserv Biol 13:1124–1139

Renjifo LM (2001) Effect of natural and anthropogenic landscape matrices on the abundance of sub-Andean bird species. Ecol Appl 11:14–31

Rosenzweig ML (1996) Species diversity in space and time. Cambridge University Press, Cambridge

Saunders DA, Hobbs RJ, Margules CR (1991) Biological consequences of ecosystem fragmentation: a review. Conserv Biol 5:18–32

Schelhas J, Greenberg R (eds) (1996) Forest patches in tropical landscapes. Island Press, Washington, DC

Simberloff D (1988) The contribution of population and community biology to conservation science. Annu Rev Ecol Syst 19:473–511

Simberloff D (1997) Biogeographic approaches and the new conservation biology. In: Pickett STA, Ostfeld RS, Shachak M, Likens GE (eds) The ecological basis of conservation. Chapman and Hall, New York, pp 274–284

Stern SJ (1998) Field studies of large mobile organisms: scale, movement and habitat utilization. In: Peterson DL, Parker VT (eds) Ecological scale: theory and applications. Columbia University Press, New York, pp 289–307

Stouffer PC, Bierregaard RO (1995) Effects of forest fragmentation on understory hummingbirds in Amazonian Brazil. Conserv Biol 9:1085–1094

Terborgh J (1974) Preservation of natural diversity: the problem of extinction prone species. BioScience 24:715–722

Terborgh J (1992) Maintenance of diversity in tropical forests. Biotropica 24:283–292

Terborgh J, Winter B (1980) Some causes of extinction. In: Soulé ME, Wilcox BA (eds) Conservation biology: an evolutionary-ecological perspective. Sinauer Associates, Sunderland, MA, pp 119–133

Thiollay JM (1989) Area requirements for the conservation of rain forest raptors and game birds in French Guiana. Conserv Biol 3:128–137

Tocher MD, Gascon C, Zimmerman BL (1997) Fragmentation effects on a central Amazonian frog community: a ten-year study. In: Laurance WF, Bierregaard RO Jr (eds) Tropical forest remnants: ecology, management, and conservation of fragmented communities. University of Chicago Press, Chicago, pp 124–137

Turner IM (1996) Species loss in fragments of tropical rain forest: a review of the evidence. J Appl Ecol 33:200–209

Viña A, Cavelier J (1999) Deforestation rates (1938–1988) of tropical lowland forests on the Andean foothills of Colombia. Biotropica 31:31–36

Whitmore TC (1997) Tropical forest disturbance, disappearance, and species loss. In: Laurance WF, Bierregaard RO Jr (eds) Tropical forest remnants: ecology, management, and conservation of fragmented communities. University of Chicago Press, Chicago, pp 3–12

Wiens JA (1976) Population responses to patchy environments. Annu Rev Ecol Syst 7:81–120

Wiens JA (1994) Habitat fragmentation: island v landscape perspectives on bird conservation. Ibis 137:S97–S104

Wiens JA (1995) Landscapes mosaics and ecological theory. In: Hansson L, Fahrig L, Merriam G (eds) Mosaic landscapes and ecological processes. Chapman and Hall, London, pp 1–26

Wiens JA (1996) Wildlife in patchy environments: metapopulations, mosaics, and management. In: McCullough DL (ed) Metapopulations and wildlife conservation. Island Press, Washington, DC, pp 53–84

Wiens JA (1997a) The emerging role of patchiness in conservation biology. In: Pickett STA, Ostfeld RS, Shachak M, Likens GE (eds) The ecological basis of conservation. Chapman and Hall, New York, pp 93–107

Wiens JA (1997b) Metapopulation dynamics and landscape ecology. In: Hanski IA, Gilpin ME (eds) Metapopulation biology: ecology, genetics, and evolution. Academic Press, San Diego, pp 43–62

Willis EO (1974) Populations and local extinctions of birds on Barro Colorado Island, Panama. Ecol Monogr 44:153–169

Willis EO (1979) The composition of avian communities in remanescent woodlots in southern Brazil. Pap Avulsos Zool 33:1–25

With KA, King AW (1997) The use and misuse of neutral landscape models in ecology. Oikos 79:219–229

With KA, Gardner RH, Turner MG (1997) Landscape connectivity and population distributions in heterogeneous environments. Oikos 78:151–169

11 Reflections on Landscape Experiments and Ecological Theory: Tools for the Study of Habitat Fragmentation

R.D. HOLT and D.M. DEBINSKI

11.1 Introduction

Habitat destruction and fragmentation are widely recognized as some of the most serious aspects of global change (Saunders et al. 1991; Fahrig and Merriam 1994). Dealing with the far-flung consequences of habitat fragmentation mandates the fusion of a wide range of scientific perspectives. In the study of habitat fragmentation, as with any scientific endeavor, there are three basic tools: (1) observation and correlation; (2) theory and modeling; (3) experimental manipulation. In recent decades, a huge amount of literature on habitat fragmentation has been generated (e.g., Andren 1994; Leach and Givnish 1996; Laurance and Bierregaard 1997). There are hundreds of descriptive studies of fragmentation (e.g., Blake 1991; Aizen and Feinsinger 1994; Hanski et al. 1995) and a large and growing body of relevant theory (e.g., Hess 1996; Wahlberg et al. 1996). Yet, to date, there are barely over a score of fragmentation experiments, past and present, across all biomes worldwide (Margules 1996; Debinski and Holt 2000). Many experiments are quite recent in their initiation, with publications just starting to appear. In contrast to other areas of ecology, such as interspecific interactions like predation and competition (e.g., Hairston 1989), it seems fair to say that our understanding of habitat fragmentation has developed largely apart from the standard scientific method of experimentation. Indeed, authoritative syntheses of ecological experimentation barely even mention habitat fragmentation (Scheiner and Gurevitch 1993; Underwood 1997).

Manipulations of entire landscapes are necessarily large in scale, laborious, and costly (Steinberg and Kareiva 1997). Yet such experiments do exist, and encouragingly, in increasing numbers. This paper is motivated by several issues. First, we believe there is much to be gained from carefully designed experimental approaches to habitat fragmentation, but given the expense and logistical difficulties it is essential to synthesize results across studies, and to

Ecological Studies, Vol. 162
G.A. Bradshaw and P.A. Marquet (Eds.)
How Landscapes Change
© Springer-Verlag Berlin Heidelberg 2003

explicitly relate them to ecological theory. Second, it is a pity that more investigators do not seize the opportunity to engage in these ongoing experiments. By writing this paper and a companion piece (Debinski and Holt 2000), we hope to make ecologists more aware of these experiments, and to stimulate a wider range of investigators to consider participating in these grand-scale experiments, which potentially provide arenas for addressing a wide range of issues in spatial ecology. Third, we feel it is essential to be conscious of the limitations of experimental approaches to fragmentation. A full understanding of the dynamics of fragmental landscapes requires an intellectual perspective merging the insights of observational studies, experiment, and theory.

Here, we do not attempt a complete review of the fragmentation literature, but rather present with broad brush strokes a more personal assessment of the state of the science. To place fragmentation experiments in a broader conceptual context, we first outline in summary fashion core ecological theories that seem pertinent to habitat fragmentation, and in particular those that have motivated experiments on fragmentation. Then we sketch some potential strengths of experimental approaches to fragmentation. Some are just those of any ecological experiment (Hairston 1989; Underwood 1997), whereas others have particular relevance to the study of fragmentation. We discuss the interaction between empirical studies and ecological theory in the context of feedback towards a better understanding of fragmentation effects. We present a selective summary of a recent survey of fragmentation experiments around the globe (Debinski and Holt 2000), and try to encapsulate which parts of ecological theory were explicitly utilized, either as general motivating factors, or to justify features of experimental design. As a case study, we briefly discuss a long-term study of an experimentally fragmented landscape in eastern Kansas. This study suggests that many important consequences of habitat fragmentation are not apparent in short-term experiments. This leads us to the important topic of articulating limitations in experimental approaches to the study of habitat fragmentation. We conclude by arguing that much value could arise from the deliberate fusion of observation, theory, and experimentation, but that this fusion has rarely, if ever, satisfactorily been carried out. Underwood (1997, p. 22) notes that "biological science in general and ecology in particular would be well-served if the underlying models (world-views, paradigms, biases, constraints, etc.) were explicit." We hope that the ideas we present here will help facilitate a more conscious linking of theory, observation, and experimentation in the study of habitat fragmentation.

11.2 Theoretical Context

It is useful to begin with a consideration of theory. Experimental and observational studies in ecology are (or should be!) motivated either implicitly or explicitly by theoretical constructs. In turn, ecological models should be "checked" by empirical studies. In ecology, there are many sorts of theory. Some theories (e.g., hierarchy theory) are largely verbal. Usually, however, when one refers to "theory", one has in mind mathematical constructs that deliberately simplify the world so as to highlight some of its essential features. In general, the three roles of theory are: (1) to provide a clear conceptual framework for carrying out empirical studies; (2) to suggest concrete hypotheses and experiments; and (3) to clarify the kinds of data needed to address particular questions. In the study of fragmentation, there is another clear role for theory, which is to connect among scales differing radically in magnitude (Levin 1992).

In principle, almost any area of theoretical ecology could be brought to bear, one way or another, on habitat fragmentation. We suggest there are four core areas of ecological theory that directly pertain to habitat fragmentation, and which have helped to motivate and guide the design of field experiments. These can be crudely, but usefully labeled by the specific environmental factors emphasized in the theory, as follows: (1) area effects; (2) dispersal effects; (3) heterogeneous landscape effects; and (4) interspecific interactions and food web effects. The first two have conceptual roots in island biogeography (MacArthur and Wilson 1967; Robinson and Quinn 1992) and metapopulation biology (Levins 1969; Hanski and Gilpin 1997). The third and fourth arise from recent developments in landscape and community ecology. The topics overlap, but also provide useful points of departure.

11.2.1 Area Effects

The most basic effect of fragmentation is to reduce the original area of a particular habitat type, leaving remnants varying greatly in size. There are two distinct ways in which fragment area can directly affect species composition in the remnant community (leaving aside for the moment effects on colonization rates, but see below).

1. Reduced area almost always leads to lower habitat diversity within the fragment, relative to the original landscape. This implies that on smaller fragments, some habitats may be vanishing to the point of becoming rare, or absent altogether (Williamson 1981). If the original community contains habitat specialists needing one of these absent habitats, the remnant community will lack these species. More subtly, some habitat generalists may be absent, if they are obligate habitat generalists. For instance, many species

pass through life stages with genetically hard-wired ontogenetic habitat shifts. Claude Gascon (pers. comm.; Tocher et al. 1997), for instance, suggests that many frog extinctions in small rainforest patches in the Manaus experiment reflect the absence of bodies of water, rather than small size per se. Many "terrestrial" frogs need water for laying eggs and larval development, and so frog populations cannot survive beyond a single generation without a range of habitats that includes aquatic habitats. These observations suggest that the relationship between area and habitat diversity, and therefore species richness, has a strong autecological base. There is some theory development along these lines (e.g., Holt 1997), but on balance surprisingly little, compared to the next topic.

2. If a species has a fixed density (e.g., because of rigid territory size requirements), reduced area implies lower absolute population size. Smaller population sizes face increased extinction risk, even in favorable environments where a species might be expected to persist. There is a huge amount of theoretical literature on extinction dynamics of small populations (e.g., Stacey and Taper 1992; Hanski and Gilpin 1997; Klok and de Roos 1998). These theoretical studies of extinction risk in small populations have helped focus attention on a prime variable in most landscape experiments, which is the size of patches created experimentally. Many experiments mentioned below and discussed in Debinski and Holt (2000) focus on patch size effects.

11.2.2 Dispersal Effects

Fragmentation usually implies altered dispersal patterns, within and among fragments, relative to the original landscape (Doak et al. 1992). There are two distinct ways the dynamics of dispersal can change due to fragmentation:

1. In remnant patches, colonization may replenish losses due to ongoing local population declines and extinctions (Brown and Kodric-Brown 1977; Fahrig and Paloheimo 1988). The more fragmented a landscape is, the greater the average distance among patches will be. All else being equal, this implies lower recolonization rates on freshly empty patches and lower densities on occupied patches, thus reduced occupancy of patches in potentially habitable areas and overall lower abundances in occupied patches. These effects of greatly reduced dispersal can lead to a high regional extinction risk (With and Crist 1995), although this risk may be offset by corridors.

    Dispersal may also be greater onto larger patches because they are bigger "targets", facilitating recolonization following extinction (MacArthur and Wilson 1967; Brown and Kodric-Brown 1977). Without very detailed study, it can be difficult to discriminate this effect of habitat area from the more widely studied effect of area upon extinction rates.

2. Most species disperse during their life cycle. Fragmentation disrupts whatever dispersal was ongoing in the original landscape, an effect with many important and distinct consequences. In particular, unfragmented habitats often consist of a mosaic of landscape patches, differing inter alia both qualitatively (e.g., presence/absence of predators) and quantitatively (e.g., availability of nest sites). Dispersal permits species to exploit spatiotemporal variability by "averaging" across local conditions (McPeek and Holt 1992; Holt 1993). Fragmentation can strongly disrupt spatial mechanisms essential to persistence (e.g., of fugitive species in a patchy environment; Tilman et al. 1994; Tilman and Lehman 1997).

A large amount of literature pertains to these issues, though there has been much more attention given to the first consequence of fragmentation for dispersal (simple reductions in colonization/movement rates) than to the second (endangering species which exploit spatial heterogeneity in their life histories). Wolff (1999) provides a useful conceptual model that synthesizes how evolutionary history, ecological specialization, and social system bear on how dispersal patterns respond to fragmentation. Many of these ideas provide pointers for future theoretical exploration. In experiments on habitat fragmentation a consideration of position effects or landscape context effects (e.g., Debinski et al. 2002), both of fragments relative to each other as well as to more distant source pools, should be a central design feature.

11.2.3 Heterogeneous Landscape Effects

Habitat fragments are not islands, but instead patches of one general habitat type, embedded in a (possibly complicated) array of alternative habitat types (Saunders et al. 1991; McIntyre and Barrett 1992). Depending upon the scale at which the landscape is perceived, patchiness can be more or less evident to any particular taxon (Wiens 1989; Dunning et al. 1992; Danielson and Anderson 1999). This implies that the design of fragmentation experiments tends to target subsets of focal species rather than entire communities. In addition, investigators often assume that the scale they have chosen is correct, when in reality the choice of correct scale requires trial and error. Finally, because organisms and materials disperse asymmetrically in heterogeneous landscapes (Polis et al. 1997), habitat fragments and the surrounding matrix are coupled. Flows between distinct habitats have two distinct consequences:
1. There can be landscape controls on local dynamics. For instance, in source-sink dynamics, abundance in a sink habitat reflects source productivity (Pulliam 1988; Holt 1993). Habitats high in productivity are likely to export nutrients, materials, and organisms to less productive habitats (Polis et al. 1997). Compared to specialists, habitat generalists may persist at a higher abundance in each habitat patch type they utilize, for several distinct rea-

sons. A habitat generalist can buffer localized temporal variation in resource or predator abundance in a given habitat. Moreover, habitat generalists may be less likely to lose dispersing individuals as they move through unfavorable habitats. Habitat generalists are less sensitive to fragmentation than are habitat specialists (Hinsley et al. 1996). Moreover, landscape structure influences the rate of dispersal between habitable patches (Peles et al. 1999; Debinski et al. 2002), an effect that has been convincingly documented for a wide variety of taxa in the Biological Dynamics of Forest Fragmentation Project in central Amazonia (Gascon et al. 1999; Mesquita et al. 1999).

2. Following fragmentation, there is increased opportunity for invasions of "exotics." For instance, Harrison (1997) has shown that plant assemblages on small patches of serpentine soil are enriched by "spill-over" from the surrounding community, while losing some distinctive species present on large expanses of serpentine. Some woodland bird species become endangered on small fragments because of an influx of brood parasites and generalist predators, whose numbers are sustained by the surrounding landscape (Fahrig and Merriam 1994).

An additional landscape effect, distinct from spatial flows, arises because "edges" often have distinct properties, reflecting physical boundaries between habitats (Murcia 1995). The "width" of edges may also have important influences on ecological dynamics within a patch. Investigators may assume that edges are sharply defined, but in fact there are often gradients in edge "width" or edge effects perceived by species within patches. The detailed physical structure of plant architecture strongly affects the degree to which edges permit penetration of fragments, versus buffering fragments from the surrounding matrix (e.g., Didham and Lawton 1999).

A limited amount of theory exists addressing these issues (e.g., Wiens 1995), but this area is still quite poorly developed in terms of explicit theory. Fragmentation experiments need to be viewed holistically, including description and analysis of the habitat surrounding experimentally created habitat fragments. Landscape context analyses may be the next, crucial step towards a better mechanistic understanding of how habitat fragmentation affects ecological communities.

11.2.4 Interspecific Interaction and Food Web Effects

The final area of ecological theory relevant to fragmentation consists of analyses of food web dynamics and multispecies interactions. It is a commonplace observation that all species exist embedded in a network of interacting species (Pimm 1982). This implies that any area, dispersal, or landscape-level effect experienced by a given species may indirectly influence other species that interact with the directly affected species. This is important even when the research focus is on a single species. Batzli et al. (1999) argue that the interplay of multiple limiting factors (food, direct density-dependence, predation) influences the response of rodent species to fragmentation. More broadly, recent theory (e.g., Holt 1993, 1997) and empirical studies (e.g., Kruess and Tscharntke 1994; Post et al. 2000) suggest that food chains may be constrained in length by habitat area. Spatial dynamics may be crucial to the persistence of strong predator-prey interactions (Wilson et al. 1998; Holyoak 2002). The disappearance of a top trophic level can unleash shifts in interspecific interactions throughout a food web. Ostfeld et al. (1999) show that species-specific impacts of voles and mice on tree recruitment can substantially influence succession in heterogeneous landscapes; changes in rodent mortality regimes due to fragmentation could thus have major cascading effects on plant community dynamics. Likewise, Terborgh et al. (1997) argue that in Neotropical rainforest, top predators have a major indirect impact upon forest dynamics, influencing strongly the abundance and behavior patterns of mid-sized mammalian seed predators (e.g., agoutis) and raiders of bird nests (e.g., coatimundi). Top predators disappear on islands or isolated habitat patches, particularly if there is direct mortality superimposed on them by hunting. Terborgh et al. (1997) suggest that the absence of top predators has led to a systematic increase in seed predation in many areas of rain forest, favoring recruitment of tree species with unpalatable or low profitability seeds. Thus, there could be major shifts in tree community structure emerging over the next century in forest fragments, indirectly driven by the direct effect of fragmentation upon large top predators.

The general message is that all multispecies theory in ecology pertains, at least in principle, to the study of habitat fragmentation. There is a vast amount of literature here, though rather little has been explicitly tied to habitat fragmentation. One serious hurdle is that empirically it may be difficult to assess many community-level effects, except at the crudest level (e.g., presence-absence of species). The likelihood of complex impacts of fragmentation percolating through webs of interacting species makes it essential that fragmentation experiments attempt to focus on more than a few taxa.

In addition to multispecies issues, the understanding in detail of the ecological mechanisms underlying fragmentation effects is a challenging frontier. Most current experiments on habitat fragmentation have been motivated by very general, qualitative ecological theory, largely focused on area effects and colonization dynamics. This means that the existing designs are not explicit relative to many potentially operating mechanisms.

11.3 What Is a Fragmentation Experiment?

For our purposes, we define a "fragmentation experiment" as a deliberately created spatial design of habitat patches in a landscape. In some (but not all) cases, the surrounding matrix is also created or otherwise controlled by the experimenter. A fragmentation experiment is by necessity a "whole system" experiment, where almost any system component can change, because all species present experience the landscape structure created by the experimenter. In practice, some experiments focus on one or a few species (e.g., Kareiva's 1987 exemplary study of the influence of patchiness on aphid-predatory beetle population dynamics). Single-species approaches are sensible if one is examining short-term effects of fragmentation, for example, on behavior or demography, but may mislead in long-term experiments because of the opportunity for feedback through numerous system components.

11.4 Why Do Experiments on Fragmentation?

When feasible, manipulative field experiments in ecology (whether of fragmentation, or anything else) have many advantages over purely descriptive, correlative studies (Hairston 1989; Underwood 1997). One major advantage should be to provide a feedback or "check" to theoretical models. Here, we briefly discuss some of these advantages with respect to fragmentation; after presenting the experiments, we then address some limitations of fragmentation experiments (these issues are discussed in more detail in Debinski and Holt 2000; Holt and Bowers 1999).

1. *Knowledge of initial conditions.* In observational studies, it may be difficult to know what a landscape looked like prior to creation of habitat fragments, or the original species composition of the fragments. Unlike descriptive studies, which start with fragments already in place, a fragmentation experiment creates an array of patches at a specific time, permitting pre-treatment surveys to determine initial conditions. This can be important, for instance, in designing stratified sampling regimes that take historical preconditions of a site into account.

2. *Controls.* Ideally, experiments have controls, against which one measures treatment effects. In descriptive studies of fragmentation, it is often difficult to identify appropriate controls. Moreover, empirical studies in practice have problems maintaining controls (e.g., Debinski and Holt 2000), because matrix habitats separating fragments have their own dynamics.

3. *Specified treatments.* Unlike descriptive studies, in an experiment, one can define landscape attributes, such as patch size, position or landscape context in a landscape.

4. *Replication.* One advantage of treatment specification is that one can ensure replication. By contrast, in descriptive studies, it may be difficult to find comparable patches (e.g., equal in area or similar in management history).

5. *Synchronicity in patch initiation.* In anthropogenic landscapes, different patches may have been created at different, and unknown, times in the past. In experimental studies, one establishes specific dates of fragmentation effects.

6. *Randomization across the landscape.* A key element of experimental design is the spatial interspersion of different treatments, and of treatments and controls. This problem may be particularly severe in the case of descriptive studies of habitat fragmentation, and so we dwell on it here. The basic issue is that humans utilize landscapes nonrandomly (Turner et al. 1996), so anthropogenic fragmentation is nonrandom (e.g., in patch size and isolation) relative to pre-existing environmental gradients, and often highly so. For instance, for economic and logistic reasons, settlers typically clear areas that are flat, with fertile soils, and near likely transportation routes (e.g., rivers), rather than areas that are hilly, with rocky soils, and isolated from such routes. The effect of nonrandom landscape usage by humans is that small, rather than large, fragments are likely to be those left in the parts of the landscape favored by humans.

As an example of the existence of nonrandom placement of patches across the landscape in a descriptive study of habitat fragmentation, consider the interesting study by Laurance (1990, 1995) of rainforest mammals on fragments in the Atherton Tableland of Australia. Laurence identified ten fragments of forest, ranging in size from 1.4–590 ha, separated by agricultural lands from large contiguous areas that served as controls. As Laurance himself notes, all the fragments are located along streams, in steep canyons, areas where it is presumably more difficult to clear-cut forest; by contrast, the controls were not usually along streams. If in the preexisting landscape, there were characteristic differences in mammal communities between stream and nonstream habitats, this might determine present-day patterns. Moreover, because of the history of human occupation in the Atherton Tableland, if one takes the map in Laurance (1990, 1995), draws a polygon around the forest fragments, and then draws a similar polygon around the controls, the controls collectively span a larger area. One basic fact about the earth's surface is that heterogeneity of all sorts (e.g., in soil types, community composition, etc.) increases with area (Williamson 1981). The controls, taken as a group, are likely to be more heterogeneous than the treatments (fragments), as a group. Given that many species are habitat specialists, it is plausible to expect greater total species richness for controls, than for fragments. A diminution in species richness in the fragments might thus reflect idiosyncrasies in how the fragments were created, rather than fundamental effects of fragmentation, per se.

In an experimentally fragmented landscape, one can minimize this problem by ensuring the random spatial interspersion of patches differing in variables such as size.

11.5 A Global Survey of Fragmentation Experiments

We recently attempted to identify all past and present fragmentation experiments and in Debinski and Holt (2000) describe the design, specific objectives, and major findings of these experiments. Here, we summarize major features of our survey.

We identified 21 studies. Figure 11.1 shows the global distribution of the fragmentation experiments (two studies were both based at essentially the same location in Ohio, and so lumped together for the purpose of mapping). Fragmentation experiments are not randomly distributed across the globe, but instead largely restricted to either North America, or western Europe. This is not surprising, given the current distribution of academic ecologists and their funding agencies. Of the 21 studies, 5 focused at the population level (1–2 species monitored), and 16 at the community level (multiple taxa or functional groups monitored). There is a distinct biome bias. Nine studies were carried out in forest: one in tropical rainforest (the famous Manaus project initiated by Tom Lovejoy), five in temperate forest, and three in boreal forest. The other studies are in grassland or old fields. These habitat biases reflect obvious logistical constraints. For instance, with the notable exception of the Manaus project, the forest projects are integrated with silviculture and forestry and, hence, linked to profit-making activities that facilitate patch creation and maintenance. In like manner, grasslands and old fields are relatively easy to modify by mowing, providing an inexpensive mechanism for manipulating landscape structure.

With respect to temporal scale, as of 1980, there was exactly one study underway (the Manaus project). In 1990, there were 6, and in 1996 17 studies were underway. Our sense of these experiments, generated from the literature, discussions with scientists involved in setting them up, and our own experience (see below), is that it takes some years for these experiments to begin generating interesting results, particularly for systems involving large spatial scales and multiple taxa. Assuming the current set of experiments continues, one might expect a rich fruit of research results to appear in coming decades.

There is a strong negative relationship between patch size and replication in these experiments (Debinski and Holt 2000). Presumably reflecting logistical, fiscal, and other constraints, our survey found considerably more replication at small patch sizes, than at large sizes. Plot sizes larger than 1 ha usually have very little replication. It is obviously much easier to create and maintain small patches, on the order of 0.01 ha in size.

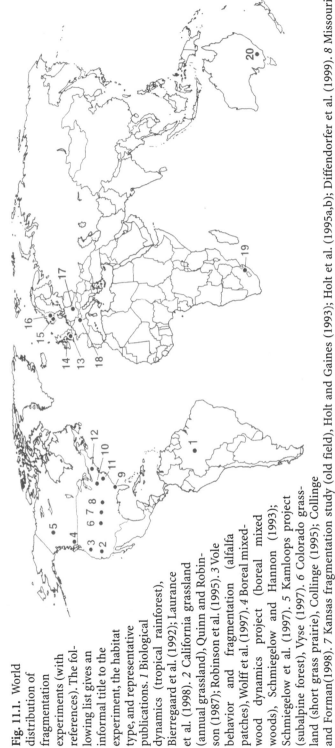

Fig. 11.1. World distribution of fragmentation experiments (with references). The following list gives an informal title to the experiment, the habitat type, and representative publications. *1* Biological dynamics (tropical rainforest), Bierregaard et al. (1992); Laurance et al. (1998). *2* California grassland (annual grassland), Quinn and Robinson (1987); Robinson et al. (1995). *3* Vole behavior and fragmentation (alfalfa patches), Wolff et al. (1997). *4* Boreal mixed-wood dynamics project (boreal mixed woods), Schmiegelow and Hannon (1993); Schmiegelow et al. (1997). *5* Kamloops project (subalpine forest), Vyse (1997). *6* Colorado grassland (short grass prairie), Collinge (1995); Collinge and Forman(1998). *7* Kansas fragmentation study (old field), Holt and Gaines (1993); Holt et al. (1995a,b); Diffendorfer et al. (1999). *8* Missouri Ozark forest ecosystem project (hardwood forest), Kurzejaski et al. (1993). *9* Savanna River site corridor project (forest clear-cuts), Haddad (1997); Haddad and Baum (1999). *10* Miami University fragmentation project (old field), Collins and Barrett (1997); Crist and Golden (pers. comm.). *11* Blandy farm fragmentation study (old field), Bowers and Dooley (1993). *12* Predator–prey interactions and fragmentation (goldenrod monoculture), Kareiva (1987). *13* Long Ashton (croplands), W. Powell (pers. comm.). *14* Moss ecosystem fragments (boulder field), Gonzalez et al. (1998). *15* Evensted research station (meadows), Ims et al. (1993). *16* Root vole sex ratio (meadows), Aars et al. (1995). *17* German fragmentation study (clover patches), Krues and Tscharntke (1994). *18* Swiss Jura Mountains (calcareous grassfield), Baur and Erhardt (1995). *19* Groenvaly experiment (grassland), Van Jaarsveld et al. (1998). *20* Wog Wog study (eucalypt forest), Margules (1992); Davies and Margules (1998)

How did ecological theory actually bear on the initiation or design of these experiments? Often, it is difficult to find explicit specifications of the theoretical underpinnings for a particular experimental design. However, our overall sense of the literature is that area effects have played a dominant role in guiding the development of fragmentation experiments. For instance, in the case of the Manaus project, the early publications show a strong influence of island biogeography theory and in particular emphasize effects of fragment area on the likely course of community decay or relaxation, following isolation (e.g., Lovejoy et al. 1984; Bierregaard et al. 1992). A similar grounding in island biogeography can be found in other studies as well (e.g., Margules 1992; Margules et al. 1994). Theoretical perspectives on how fragmentation alters the dispersal of demography of focal taxa have been the primary motive of a few studies (e.g., Haddad 1999; Dooley and Bowers 1998). There have been rather few attempts made to explicitly integrate ecological theory with fragmentation experiments. Rather than examine each study one by one, we focus briefly on this issue for a study with which we are personally familiar.

11.6 A Case Study: The Kansas Experimentally Fragmented Landscape

This study, initiated in 1984, is among the longer, continuously running fragmentation experiments. As described in Holt et al. (1995a,b), in contrast to many fragmentation experiments, the basic ecological questions motivating the study were the effect of patch size and position on the rate and pattern of secondary succession. In some respects, this experiment pertains more directly to succession theory and restoration ecology than to conservation biology. Elsewhere (e.g., Foster and Gaines 1991; Robinson et al. 1992; Holt et al. 1995a,b; Diffendorfer et al. 1995, 1996; Schweiger et al. 1999, 2000; Yao et al. 1999) we describe this study in some detail, so here we only sketch key design features and findings pertinent to the issue of the relationship of experimentation and theory in the study of habitat fragmentation.

Patches were created and maintained by intensive mowing in a plot of land formerly used for agriculture. Patches were allowed to undergo succession, based on the community present in the original seed bank and subsequent colonists. Given our interest in succession, our focal organisms naturally include vascular plants, but we have also monitored the small mammal community in detail. The choice of patch sizes and separation were governed by prior knowledge about home ranges, local abundance, and average dispersal distances. The surrounding landscape is heterogeneous, with woods south and west, and brome meadows and cultivated fields north and east. We thus expected gradients in vegetation establishment and so stratified placement of each patch size by distance to the woods. The smaller patches are arranged in clusters; sample stations in these clusters can be compared with similar grids of sample stations within the large patches.

Our research design was motivated by the ecological theories discussed above, as follows:

Area effects: during secondary succession following field abandonment, plant species absent from the original seed bank, but present nearby can colonize and eventually dominate. We hypothesized that colonization-extinction dynamics might be an important dimension of succession, and that succession might vary with patch size. At the time we initiated this study (1984) an explicit "island biogeographical" interpretation had not been applied to terrestrial plant succession. For small mammals with low densities, or large home range requirements, we expected our small patches to be unfavorable and predicted such species to be differentially missing in these patches.

Dispersal effects: distances among patches and to presumptive sources in the surrounding landscape were based on prior information as to what constituted significant barriers to dispersal for the plants and small mammals at our site. We attempted to choose distances that would hamper dispersal, but not so greatly as to decouple dynamics on different patches.

Interspecific interactions and food web effects: patch size could influence small mammal abundance and/or behavior. In turn, small mammals acting as herbivores and seed predators can severely impact plant numbers (Crawley 1996), so a direct effect of patch size on small mammal abundance could translate into an indirect effect on plant dynamics. Generalist consumers can eliminate preferred or slow-growing species while remaining sustained by less preferred or more productive prey (Holt and Lawton 1994). Depending on how consumer preference correlates with competitive ability, selective herbivory could either facilitate or slow down plant succession (Davidson 1993; Hulme 1996).

Landscape heterogeneity effects: to deal with heterogeneity in the surrounding landscape (viewing broad-scale heterogeneity as a nuisance variable in the experiment), we interspersed patch sizes in a stratified random design. However, our experimental design did not directly assess effects of landscape heterogeneity.

We now summarize core findings from this project, first from the early years of succession (1984–1990), and then more recently (1991–present).

11.6.1 Core Findings, 1985–1990

1. Contrary to our initial hypothesis there was little effect of patch size on overall successional rate or pattern (Robinson et al. 1992; Holt et al. 1995a).
2. Despite the absence of major effects of patch size on the overall trajectory of succession, there were population-level effects for particular plant species, and subtle effects on spatial structuring in the community. For instance, clonal plant species persisted better in our permanent sample quadrats on large, than on small, patches (Robinson et al. 1992; Heisler 1998).
3. There were strong, and at times surprising, effects of patch size on the small mammal community. As expected, some species (e.g., the large-bodied cotton rat *Sigmodon hispidus*) were largely restricted to large patches. However, other species (e.g., the smaller-bodied prairie vole, *Microtus ochrogaster* and deer mouse, *Peromyscus maniculatus*) were actually denser on smaller patches (Foster and Gaines 1991; Diffendorfer et al. 1995; Schweiger et al. 2000). Studies in other patch systems have likewise revealed enhanced densities of small mammals on smaller patches (Bowers and Matter 1997).

11.6.2 Core Findings, 1991–Present

1. There continue to be strong, species-specific effects of patch size on small mammal abundance; some species are more abundant on large patches, others on small patches. Several rodent species characteristic of woody vegetation have invaded, but are largely restricted to those large patches that are near contiguous woodland (Schweiger et al. 1999, 2000). The butterfly community on the site also shows a nested distributional pattern, with some species largely restricted to large patches (Holt et al. 1995b; D. Debinski, unpubl. data). Given the high mobility of butterflies, this likely reflects behavioral responses rather than population dynamics.
2. Woody plant invasion has accelerated and is occurring more rapidly into larger patches (Yao et al. 1999). In contrast to the first phase, there is now a substantial effect of patch size on the rate of succession. There is also a pronounced distance effect, with succession occurring more slowly in patches more distant from the forest (Yao et al. 1999).

As noted above, the original design of the experiment was motivated by theoretical ideas regarding area effects, dispersal effects, and indirect effects. Although ecological theory helped motivate the experimental design and to explain observed patterns, we must admit that to date the relationship between theory and experimentation has been rather loose. Explicit mathematical models are always radical simplifications of complex ecological systems, and the "whole-system" nature of fragmentation experiments makes it difficult to use them as "tests" of some particular mathematical model, except in the most general way. We suggest that many fragmentation experiments to date have had a similarly loose relationship between theory and experiment: theory has helped motivate experiments, but the experiments are not truly "tests" of theory. Indeed, it is difficult to see how experiments such as ours, which examine responses by an entire system over long time periods to fragmentation, could even, in principle, be used to "test" simple ecological models. By contrast, a tighter integration with theory is much more feasible when addressing particular mechanisms in short-term experiments focused on one to a few species, such as those of Bowers et al. (1996) and Wolff (1999) aimed at behavioral responses.

11.7 Limitations in Experimental Fragmentation Studies

There are many obvious limitations in experimental studies of fragmentation (Holt and Bowers 1999; W. Laurance, pers. comm.), particularly when one wishes to "scale up" to larger spatial arenas. For instance, there are logistical constraints in the design and execution of fragmentation experiments, including costs of set-up and sampling (which scale linearly with area), the "people power" available to conduct research at a site, constraints on land management, and pre-existing heterogeneities. More important than any of these, there are conceptual limitations that arise because different ecological processes operate at distinct spatial scales.

In the Kansas experiment, there was a particular farmer's field, owned by the University of Kansas Endowment Association, available to researchers to set up the experiment. In an ideal world, we would have exerted control over the surrounding landscape, but in practice for us the landscape was a "given". The predetermined shape and area of the available experimental field automatically set up interlinked constraints among the design desiderata in experimental design. For instance, for fixed patch sizes, increasing distance among patches automatically reduces the degree of feasible replication.

The intellectual scope of the Kansas project has been governed largely by availability of interested researchers and the constraints of spatial scale. For instance, we (RDH and DMD) are personally interested in birds and butterflies and have supervised students carrying out projects examining patch use by these taxa. However, the high mobility of these groups makes it likely that single individuals can use much of the experimental landscape, so these studies necessarily bear more on individual habitat selection and patch utilization, than on population or community-level effects of fragmentation. The spatial scale of our system would have been apt for examining insect population dynamics, but by chance our colleagues did not include specialists in this area, and instead included experts on rodent population dynamics. A number of publications thus examine effects of patch size on rodent populations, with an emphasis on how fragmentation disrupts dispersal dynamics. Only recently, Wilson (1998) has shown that variation in vertebrate predator activity may account for some of these patterns. The general point is that fragmentation experiments are "whole system" manipulations, yet person-power and fiscal limitations almost always restrict the range of taxa and processes investigated. These "real world" limitations constrain the range of processes, systems, and landscape features it is feasible to address with experimentation.

Our fragments are maintained by regular mowing in the interstitial habitat. Other possible devices for separating patches suggested at the outset ranged from fanciful (e.g., paving the entire area, except the patches, with Astroturf) to logistically difficult (e.g., surrounding patches with a coarse-meshed fence, then using sheep to maintain a low turf between patches, as in Quinn and Robinson 1987). Landscape manipulation can introduce artifacts. Mowing is a massive periodic disturbance that can have strong indirect effects on vegetation dynamics within patches. For instance, because of sheet erosion in the interstitial areas, runoff onto patch edges has disproportionately more impact on smaller patches.

In the Kansas study, as in all experimental studies of habitat fragmentation (Debinski and Holt 2000), the spatial scale is limited. This raises the crucial problem of how to scale up from these model systems to "real" landscapes of interest in conservation. Diffendorfer et al. (1996) showed that with small mammals one could successfully extrapolate from patterns of abundance by patch size in our system, to larger patches outside our site. Bowers and Matter (1997) however, show that in broad comparisons, the relationship of small mammal abundance to patch size depends upon the range of patch areas considered. There are thus difficulties in extrapolation, even for a single set of taxa. Furthermore, it is an open question whether or not one can use insights gleaned from the study of one set of taxa, to interpret patterns observed in other, disparate taxa. For instance, mobile organisms may experience a fragmentation experiment largely as minor patchiness. Fragmentation experiments conducted on a "fine" scale may fail to address issues of community or population dynamics adequately, simply because the patches do not support more than a few individuals of select species.

The temporal scale of fragmentation experiments is limited. Most experiments reviewed in Debinski and Holt (2000) have been extant only a few years, and some (e.g., Kruess and Tscharntke 1994; Bowers et al. 1996) are deliberately short-term. It is likely that effects of fragmentation loom larger at longer time scales (as in community "relaxation"). Successional dynamics in the Kansas study illustrate this effect. As noted above, in the first 6 years of the study there was no evident effect of patch size on plant succession, but in the last few years such effects on woody plant colonization have become pronounced. Moreover, the matrix may have successional dynamics that influence processes in the fragments. A perpetual isolation of patches from a surrounding matrix may not be a good model for "real" systems. For instance, clear-cuts grow back, so in fragmented forest environments the level of fragmentation and the heterogeneity of landscape units will have a temporal dynamic. A challenging task for future experimental work is to devise sensible designs that deliberately incorporate the spatial and temporal heterogeneity of the matrix landscape. Experiments typically create "sharp" edges; in real landscapes, edges are often fuzzy ecotones, and the sharpness of edges is likely to vary over time. Experiments have also focused on simple geometries (e.g., rectangles, linear corridors); in real landscapes, geometries are much more complex (stringy, fractal, etc.; see Adler and Nuernberger 1994 for pertinent theory). No doubt there are other limitations that have not occurred to us. There are many important features of real-world habitat fragmentation that are difficult to simulate in a clean experimental design.

11.8 Conclusions

As in many other areas of ecological science, the theory of fragmentation is much more advanced than are rigorous empirical tests of theory. Over the past few decades, we have witnessed an evolution in the conceptual focus of empirical fragmentation studies from island biogeography, emphasizing area effects, to metapopulation dynamics, which stresses colonization effects, to a landscape ecology perspective, where landscape context, heterogeneity, and interactions with the matrix seem increasingly important. There is also an increasing appreciation of synergistic effects, often mediated by complex food web interactions. No longer are we simply interested in counting the number of species present on a patch; we are now delving into understanding processes that explain such patterns. The landscape context of the patch, the history of the patch, and the behavior of the organisms in the patches are all clearly key determinants of fragmentation effects.

Despite the current urgent need to understand better how fragmentation operates as a driver of species extinction, there are still just a handful of experimentally designed fragmentation studies. The lack of experimental studies is primarily due to logistical constraints, but it may also be explained by sociology, namely, the tendency for most ecologists to work alone or in small groups. Because the costs of establishing and maintaining a fragmentation experiment are high, and the benefits accruing potentially large, it is urgent that ecologists join forces and engage in collaborative research projects, using the habitat fragmentation experiments now scattered around the globe (Fig. 11.1), or designing new experiments that improve on the old ones. These experiments in turn could benefit greatly, we believe, from more explicit ties to ecological theory, and the results from such experiments can in turn point the direction towards new areas of theoretical development. Experimental landscapes are almost always caricatures of real landscapes, which hardly ever are comprised of habitable patches with geometrically simple shapes embedded in a completely uninhabitable matrix, but instead are complex mosaics of many habitat types with complex shapes and fuzzy edges. However, such caricatures are useful, particularly if one wishes to have a bridge between the abstract "perfect crystals" (May 1973) of ecological theory and the messy reality addressed in purely descriptive studies.

Acknowledgement. We in particular thank all the participants in the Kansas project for our long-term collaborative efforts. We thank the National Science Foundation for its continued support, and the organizers of the IAI-Amigo workshop, Pablo Marquet and Gay Bradshaw, for the opportunity to present this paper. We also thank Mike Bowers for an exceptionally thoughtful review of the manuscript. This is journal paper no. J-19059 of the Iowa Agriculture and Home Economics Experiment Station, Ames, Iowa, project 3377, and supported by Hatch and State of Iowa funds.

References

Aars J, Andreassen HP, Ims RA (1995) Root voles: litter sex ratio variation in fragmented habitat. J Anim Ecol 64:459–472

Adler FR, Nuernberger B (1994) Persistence in patchy irregular landscapes. Theor Popul Biol 45:41–75

Aizen MA, Feinsinger P (1994) Forest fragmentation, pollination, and plant reproduction in a Chaco dry forest, Argentina. Ecology 75:330–351

Andren H (1994) Effects of habitat fragmentation on birds and mammals in landscapes with different proportions of suitable habitat: a review. Oikos 71:355–366

Batzli GO, Harper SJ, Lin Y-TK, Desy EA (1999) Experimental analyses of population dynamics: scaling up to the landscape. In: Barrett GW, Peles JD (eds) Landscape ecology of small mammals. Springer, Berlin Heidelberg New York, pp 107–127

Baur B, Erhardt A (1995) Habitat fragmentation and habitat alterations: principal threats to most animal and plant species. GAIA 4:221–226

Bierregaard RO Jr, Lovejoy TE, Kapos V, dos Santos AA, Hutchings RW (1992) The biological dynamics of tropical rainforest fragments: a prospective comparison of fragments and continuous forest. BioScience 42: 859–866

Blake JG (1991) Nested subsets and the distribution of birds on isolated woodlots. Conserv Biol 5:58–66

Bowers MA, Dooley JL (1993) Predation hazard and seed removal by small mammals: microhabitat versus patch scale effects. Oecologia 94:247–254

Bowers MA, Matter SF (1997) Landscape ecology of mammals: relationships between density and patch size. J Mammal 78: 999–1013

Bowers MA, Matter SF, Dooley JL, Dauten JL, Simkins JA (1996) Controlled experiments of habitat fragmentation: a simple computer simulation and a test using small mammals. Oecologia 108:182–191

Brown JH, Kodric-Brown A (1977) Turnover rates in insular biogeography: effect of immigration on extinction. Ecology 58:445–449

Collinge SK (1995) Spatial arrangement of patches and corridors in the landscape: consequences for biological diversity and implications for landscape architecture. PhD Diss, Harvard University, Cambridge, MA

Collinge SK, Forman RTT (1998) A conceptual model of land conversion processes: predictions and evidence from a microlandscape experiment with grassland insects. Oikos 82:66–84

Collins RJ, Barrett GW (1997) Effects of habitat fragmentation on meadow vole (*Microtus pennsylvanicus*) population dynamics in experimental landscape patches. Landscape Ecol 12:63–76

Crawley MJ (1996) Plant ecology, 2nd edn. Blackwell, Oxford

Danielson BJ, Anderson GS (1999) Habitat selection in geographically complex landscapes. In: Barrett GW, Peles JD (eds) Landscape ecology of small mammals. Springer, Berlin Heidelberg New York, pp 89–103

Davidson DW (1993) The effect of herbivory and granivory on terrestrial plant succession. Oikos 68:23–35

Davies KF, Margules CR (1998) Effects of habitat fragmentation on carabid beetles: experimental evidence. J Anim Ecol 67:460–471

Debinski DM, Holt RD (2000) A survey and overview of habitat fragmentation experiments. Conserv Biol 14:1–13

Debinski DM, Ray C, Saveraid EH (2002) Species diversity and the scale of the landscape mosaic: do scales of movement and patch size affect diversity? Biol Conserv 98:179–190

Didham RM, Lawton TH (1999) Edge structure determines the magnitude of changes in microclimate and vegetation structure in tropical forest fragments. Biotropica 31:17–30

Diffendorfer JE, Gaines MS, Holt RD (1995) The effects of habitat fragmentation on movements of three small mammal species. Ecology 76:1610–1624

Diffendorfer JE, Holt RD, Slade NA, Gaines MS (1996) Small mammal community patterns in old fields: distinguishing site-specific from regional processes. In: Cody ML, Smallwood JA (eds) Long-term studies of vertebrate communities. Academic Press, New York, pp 421–466

Diffendorfer JE, Gaines MS, Holt RD (1999) Patterns and impacts of movements at different scales in small mammals. In: Barrett GW, Peles JD (eds) Landscape ecology of small mammals. Springer, Berlin Heidelberg New York, pp 63–88

Doak DF, Marino PC, Kareiva PM (1992) Spatial scale mediates the influence of habitat fragmentation on dispersal success: implications for conservation. Theor Popul Biol 41:315–336

Dooley JL Jr, Bowers MA (1998) Demographic responses to habitat fragmentation: experimental tests at the landscape and patch scale. Ecology 79:969–980

Dunning JB, Danielson BJ, Pulliam HR (1992) Ecological processes that affect populations in complex landscapes. Oikos 65:169–175

Fahrig L, Merriam G (1994) Conservation of fragmented populations. Conserv Biol 8:50–59

Fahrig L, Paloheimo J (1988) Determinants of local population size in patchy habitats. Theor Popul Biol 34:194–213

Foster J, Gaines MS (1991) The effects of a successional habitat mosaic on a small mammal community. Ecology 72:1358–1373

Gascon C, Lovejoy TE, Bierregaard RO Jr et al. (1999) Matrix habitat and species richness in tropical forest remnants. Biol Conserv 91:223–229

Gonzalez A, Lawton JH, Gilbert FS, Blackburn TM, Evans-Freke I (1998) Metapopulation dynamics in a microecosystem. Science 281:2045–2047

Haddad NM (1997) Do corridors influence butterfly dispersal and density? A landscape experiment. PhD dissertation. University of Georgia, Athens, GA

220   R.D. Holt and D.M. Debinski

Haddad NM (1999) Corridor and distance effects on interpatch movements: a landscape experiment with butterflies. Ecol Appl 9:612–622
Haddad NM, Baum KA (1999) An experimental test of corridor effects on butterfly densities. Ecol Appl 9:623–633
Hairston NG Sr (1989) Ecological experiments: purpose, design, and execution. Cambridge University Press, Cambridge
Hanski IA, Gilpin ME (eds) (1997) Metapopulation biology: ecology, genetics, and evolution. Academic Press, San Diego
Hanski I, Pakkala T, Kuussaari M, Lei G (1995) Metapopulation persistence of an endangered butterfly in a fragmented landscape. Oikos 72:21–28
Harrison S (1997) How natural habitat patchiness affects the distribution of diversity in California serpentine chaparral. Ecology 78:1898–1906
Heisler DA (1998) Population dynamics of clonal plants in fragmented landscapes: empirical analyses and simulation models. PhD Diss, University of Kansas, Lawrence, KS
Hess GR (1996) Linking extinction to connectivity and habitat destruction in metapopulation models. Am Nat 148:226–236
Hinsley SA, Bellamy PE, Newton I, Sparks TH (1996) Influences of population size and woodland area on bird species distributions in small areas. Oecologia 105:100–106
Holt RD (1993) Ecology at the mesoscale: the influence of regional processes on local communities. In: Ricklefs R, Schluter D (eds) Species diversity in ecological communities. University of Chicago Press, Chicago, IL, pp 77–88
Holt RD (1997) From metapopulation dynamics to community structure: some consequences of spatial heterogeneity. In: Hanski IA, Gilpin ME (eds) Metapopulation biology: ecology, genetics, and evolution. Academic Press, San Diego, pp 149–165
Holt RD, Bowers MA (1999) Experimental design at the landscape scale. In: Barrett GW, Peles JD (eds) Landscape ecology of small mammals. Springer, Berlin Heidelberg New York, pp 263–286
Holt RD, Gaines MS (1993) The influence of regional processes on local communities: Examples from an experimentally fragmented landscape. In: Powell T, Levin S, Steele J (eds) Patch dynamics in marine and terrestrial environments. Springer, Berlin Heidelberg New York
Holt RD, Lawton JH (1994) The ecological consequences of shared natural enemies. Annu Rev Ecol Syst 25:495–520
Holt RD, Debinski D, Diffendorfer J et al. (1995a) Perspectives from an experimental study of habitat fragmentation in an agroecosystem. In: Glen D, Greaves M, Anderson H (eds) Ecology and integrated farming systems. Wiley, New York, pp 147–175
Holt RD, Robinson GR, Gaines MS (1995b) Vegetation dynamics in an experimentally fragmented landscape. Ecology 76:1610–1624
Holyoak M (2000) Habitat fragmentation causes changes in food web structure. Ecol Lett 3:509
Hulme PE (1996) Herbivory, plant regeneration, and species coexistence. J Ecol 84:609–615
Ims RA, Rolstad J, Wegge P (1993) Predicting space use responses to habitat fragmentation: can voles *Microtus oeconomus* serve as an experimental model system (EMS) for capercaille grouse *Tetrao urogalius* in boreal forest? Biol Conserv 63:261–268
Kareiva P (1987) Habitat fragmentation and the stability of predator-prey interactions. Nature 326:388–390
Klok C, de Roos AM (1998) Effects of habitat size and quality on equilibrium density and extinction time of *Sorex araneus* populations. J Anim Ecol 67:195–209
Kruess A, Tscharntke T (1994) Habitat fragmentation, species loss, and biological control. Science 264:1581–1584

Kurzejaski EW, Clawson RL, Renken RB et al. (1993) Experimental evaluation of forest management: the Missouri Ozark Forest Ecosystem Project. Trans North Am Wildl Nat Resour Conf 58:599–609

Laurance WF (1990) Comparative responses of five arboreal marsupials to tropical forest fragmentation. J Mammal 71:641–653

Laurance WF (1995) Extinction and survival of rainforest mammals in a fragmented tropical landscape. In: Lidicker WZ Jr (ed) Landscape approaches in mammalian ecology and conservation. University of Minnesota Press, Minneapolis, pp 46–63

Laurance WF, Bierregaard RO Jr (eds) 1997. Tropical forest remnants: ecology, management, and conservation of fragmented communities. University of Chicago Press, Chicago

Laurance WF, Ferreira LV, Rankin-de Merona JM et al. (1998) Effects of forest fragmentation on recruitment patterns in Amazonian tree communities. Conserv Biol 12:460–464

Leach MK, Givnish TJ (1996) Ecological determinants of species loss in remnant prairies. Science 273:1555–1561

Levin SA (1992) The problem of pattern and scale in ecology. Ecology 73:1943–1967

Levins R (1969) Some demographic and genetic consequences of environmental heterogeneity for biological control. Bull Entomol Soc Am 15:237–240

Lovejoy TE, Rankin JM, Bierregaard RO Jr et al. (1984) Ecosystem decay of Amazon forest remnants. In: Nitecki MH (ed) Extinctions. University of Chicago Press, Chicago, pp 295–326

MacArthur RH, Wilson EO (1967) The theory of island biogeography. Princeton University Press, Princeton, NJ

Margules CR (1992) The Wog Wog habitat fragmentation experiment. Environ Conserv 19:316–325

Margules CR (1996) Experimental fragmentation. In: Settele J, Margules CR, Poschlod P, Henle K (eds) Species survival in fragmented landscapes. Kluwer, Dordrecht, pp 128–137

Margules CR, Milkovits GA, Smith GT (1994) Contrasting effects of habitat fragmentation on the scorpion *Cercophonius squama* and an amphipod. Ecology 75:2033–2042

Margules CR, Austin MP, Davies KF, Meyers JA, Nicholls AO (1998) The design of programs to monitor forest biodiversity: lessons from the Wog Wog habitat fragmentation experiment. In: Dallmeier F, Comiskey JA (eds) Forest biodiversity research, monitoring, and modeling. Parthenon Publishing, Paris, pp 183–196

May RM (1973) Stability and Complexity in Model Ecosystems. Princeton University Press, Princeton, NJ

McIntyre S, Barrett GW (1992) Habitat variegation: an alternative to fragmentation. Conserv Biol 6:146–147

McPeek MA, Holt RD (1992) The evolution of dispersal in spatially and temporally varying environments. Am Nat 6:1010–1027

Mesquita RCG, Delamonica P, Laurance WF (1999) Effect of surrounding vegetation in edge-related tree mortality in Amazonian forest fragments. Biol Conserv 91:129–134

Murcia C (1995) Edge effects in fragmented forests: implications for conservation. Trends Ecol Evol 10:58–62

Ostfeld RS, Manson RH, Canham CD (1999) Interactions between meadow voles and white-footed mice at forest-oldfield edges: competition and net effects on tree invasion of oldfields. In: Barrett GW, Peles JD (eds) Landscape ecology of small mammals. Springer, Berlin Heidelberg New York, pp 229–247

Peles JD, Bowne DR, Barrett GW (1999) Influence of landscape structure on movement patterns of small mammals. In: Barrett GW, Peles JD (eds) Landscape ecology of small mammals. Springer, Berlin Heidelberg New York, pp 41–62

Pimm SL (1982) Food webs. Chapman and Hall, London

Polis GA, Anderson WB, Holt RD (1997) Towards an integration of landscape ecology and food web ecology: the dynamics of spatially subsidized food webs. Annu Rev Ecol Syst 28:289–316

Post DM, Pace ML, Hairston NG Jr (2000) Ecosystem size determines food-chain length in lakes. Nature 405:1047–1049

Pulliam HR (1988) Sources, sinks, and population regulation. Am Nat 132:652–661

Quinn JF, Robinson GR (1987) The effects of experimental subdivision on flowering plant diversity in a California annual grassland. J Ecol 75:837–856

Robinson GR, Quinn JF (1992) Habitat fragmentation, species diversity, extinction, and design of nature reserves. In: Jain SK, Botsford LW (eds) Applied population biology. Kluwer, Dordrecht, pp 223–248

Robinson GR, Holt RD, Gaines MS et al. (1992) Diverse and contrasting effects of habitat fragmentation. Science 257:524–526

Robinson GR, Quinn JF, Stanton ML (1995) Invasibility of experimental habitat islands in a California winter annual grassland. Ecology 76–786–794

Saunders DA, Hobbs RJ, Margules CR (1991) Biological consequences of ecosystem fragmentation: a review. Conserv Biol 2:340–347

Scheiner SM, Gurevitch J (eds) (1993) Design and analysis of ecological experiments. Chapman and Hall, London

Schmiegelow FKA, Hannon SJ (1993) Adaptive management, adaptive science and the effects of forest fragmentation on boreal birds in northern Alberta. Trans North Am Wildl Nat Resour Conf 58:584–598

Schmiegelow FKA, Machtans CS, Hannon SJ (1997) Are boreal birds resilient to forest fragmentation: an experimental study of short-term community responses. Ecology 78–1914–1932.

Schweiger EW, Diffendorfer J, Pierotti R, Holt RD (1999) The relative importance of small-scale and landscape-level heterogeneity in structuring small mammal distributions. In: Barrett GW, Peles JD (eds) Landscape ecology of small mammals. Springer, Berlin Heidelberg New York, pp 175–210

Schweiger EW, Diffendorfer J, Holt RD, Pierotti R (2000) The interaction of habitat fragmentation, plant, and small mammal succession in an old field. Ecol Monogr 70:383–400

Stacey P, Taper ML (1992) Environmental variation and the persistence of small populations. Ecol Appl 2:18–29

Steinberg EK, Kareiva P (1997) Challenges and opportunities for empirical evaluations of "spatial theory". In: Tilman D, Nareiva P (eds) Spatial Ecology. Princeton University Press, Princeton, NJ, pp 318–332

Terborgh J, Lopez L, Tello J, Yu D, Bruni AR (1997) Transitory states in relaxing ecosystems of land bridge islands. In: Laurance WF, Bierregaard RO Jr (eds) Tropical forest remnants: ecology, management, and conservation of fragmented communities. University of Chicago Press, Chicago, pp 256–274

Tilman D, Lehman CL (1997) Habitat destruction and species extinctions. In: Tilmans D, Nareiva P (eds) Spatial ecology. Princeton University Press, Princeton, NJ, pp 233–249

Tilman D, May RM, Lehman CL, Nowak MA (1994) Habitat destruction and the extinction debt. Nature 370:66–68

Tocher MD, Gascon C, Zimmerman BL (1997) Fragmentation effects on a central Amazonian frog community: a ten-year study. In: Laurance WF, Bierregaard RO Jr (eds) Tropical forest remnants: ecology, management, and conservation of fragmented communities. University of Chicago Press, Chicago, pp 124–137

Turner MG, Wear DN, Flamm RO (1996) Land ownership and land-cover change in the Southern Appalachian highlands and the Olympic Peninsula. Ecol Applic 6:1150--1172

Underwood AJ (1997) Experiments in ecology. Cambridge University Press, Cambridge

Van Jaarsveld AS, Ferguson JWH, Bredenkamp GJ (1998) The Groenvaly grassland fragmentation experiment: design and initiation. Agric Ecosyst Environ 68:139–150

Vyse A (1997) The Sicamous Ck. Silvicultural Systems Project: how the project came to be and what it aims to accomplish. In: Hollstedt C, Vyse A (eds) Workshop proceedings. Research publication 24 C. British Columbia Ministry of Forests, Victoria, Canada, pp 4–14

Wahlberg N, Moilanem A, Hanski I (1996) Predicting the occurrence of endangered species in fragmented landscapes. Science 273:1536–1538

Wiens JA (1989) Spatial scaling in ecology. Funct Ecol 3:385–397

Wiens JA (1995) Landscape mosaics and ecological theory. In: Hansson L, Fahrig L, Merriam G (eds) Mosaic landscapes and ecological processes. Chapman and Hall, London, pp 1–26

Williamson MH (1981) Island populations. Oxford University Press, Oxford

Wilson CP (1998) Predator utilization of a fragmented study site in northeast Kansas. Ms Thesis. University of Kansas, Lawrence, Kansas

Wilson HB, Hassell MP, Holt RD (1998) Persistence and area effects in a stochastic tritrophic model. Am Nat 151:587–596

With KA, Crist TO (1995) Critical thresholds in species' responses to landscape structure. Ecology 76:2446–2459

Wolff JO (1999) Behavioral model systems. In: Barrett GW, Peles JD (eds) Landscape ecology of small mammals. Springer, Berlin Heidelberg New York, pp 11–40

Wolff JO, Schauber EM, Edge ED (1997) Effects of habitat loss and fragmentation on the behavior and demography of gray-tailed voles. Conserv Biol 11:945–956

Yao J, Holt RD, Rich PM, Marshall WS (1999) Woody plant colonization in an experimentally fragmented landscape. Ecography 22:715–728

12 Spatial Autocorrelation, Dispersal and the Maintenance of Source-Sink Populations

T.H. KEITT

12.1 Introduction

It is often assumed that population growth is limited by an upper bound or carrying capacity of the environment, below which a population increases and above which the population decreases (Murdoch 1994). Pulliam (1988) recognized that population growth may be regulated by an alternative mechanism: dispersal among habitats of varying quality. He referred to those areas where birth rates exceed death rates as demographic "source" populations and those areas where death rates exceed birth rates as "sinks". Coupled together via dispersal, population sources and sinks present a number of interesting scenarios. For example, sink populations may persist despite high mortality rates because of immigration from nearby source habitats. Furthermore, source-sink population structure can lead to situations in which the majority of individuals in a population are poorly adapted to the habitats in which they occur (Dias 1996).

There are a number of factors that can influence the survival of source-sink populations and these relate both to the extent and spatial arrangement of source and sink habitats, and the life-history traits of the species of concern. Primary, of course, are the demographic parameters of a given species: all things being equal, a species with a high rate of reproduction and low mortality has a lower risk of extinction than a species with high mortality and low reproduction. However, persistence can also be influenced by behavioral attributes, such as whether a species typically disperses long distances to find new habitat areas or tends to remain close to home. Species that disperse long distances may be more successful in finding unoccupied territories or, in the case of plants, a forest gap. On the other hand, a species that disperses short distances may avoid becoming lost or ending up in a large area of inhospitable habitat. Another attribute of dispersal is directed vs. passive dispersal. Directed dispersal occurs when an organism actively searches for high quality habitats, whereas with passive dispersal, propagules land in a random pat-

Ecological Studies, Vol. 162
G.A. Bradshaw and P.A. Marquet (Eds.)
How Landscapes Change
© Springer-Verlag Berlin Heidelberg 2003

tern some distance from where they started. A classic example of passive dispersal is the "seed shadow" surrounding trees whose seeds are dispersed primarily by wind and gravity.

12.2 Spatial Autocorrelation

The outcome of dispersal events depends not only on dispersal behavior, but also on the spatial arrangement of source and sink habitats. In landscapes where source habitats often occur in close proximity to sink habitats, source and sink habitats will be strongly coupled, even for organisms that disperse relatively short distances. The proximity of source and sink habitat can be quantified in terms of a "two-point" autocorrelation function. Consider pairs of points chosen from the landscape at random, but constrained to be distance k apart. The (sample) autocovariance function is

$$g_k = \sum_{x=k+1}^{n} (y_x - \bar{y})(y_{x-k} - \bar{y}) / n \qquad (12.1)$$

where y_x is a point located at x and $\bar{y}$ is the mean value of y across the landscape (Diggle 1990). For a two-dimensional landscape, x should of course be a vector pair of coordinates, but I will use the simpler one-dimensional notation above. The autocorrelation function is then

$$r_k = \frac{g_k}{g_0} \qquad (12.2)$$

where g_0 is the variance of the y's. A large autocorrelation at a particular scale (distance) in a landscape indicates that moving that distance will result in only a small change in habitat quality. On the other hand, a small autocorrelation means that the habitat quality changes rapidly. Generally, as the degree of spatial autocorrelation increases, landscapes become less fragmented, having fewer, but larger patches. Because spatial correlation implies larger patches, individuals dispersing away from source habitats in a correlated landscape will tend to encounter higher quality habitats and be less likely to disperse into a demographic sink.

The fractal dimension (Mandelbrot 1982) D of a landscape is directly related to the autocorrelation function. In fact, saying that a landscape is fractal requires the autocorrelation function to take a particular form. Let the mean value of the landscape $\bar{y} = 0$, then

$$v_k = g_0(1 - r_k) \tag{12.3}$$

where v_k is the variance of point a distance k apart (Diggle 1990). In the geostatistics literature, v_k is known as the "variogram" (Cressie 1993). For a fractal landscape with dimension D, then

$$v_k \propto k^{2H} \tag{12.4}$$

where $H = 3 - D$ (because, in this case, we have two spatial dimensions and one dimension for habitat quality). The symbol H is known as the "Hurst exponent". Thus, the autocorrelation function can be related directly to the fractal dimension by

$$r_k \propto 1 - \frac{k^{2H}}{g_0} \tag{12.5}$$

where, again, $H = 3 - D$.

The important thing to keep in mind is that the four factors mentioned, local population growth rate, dispersal range, active vs. passive dispersal, and landscape structure, all interact because they affect the density of individuals that reside in population sinks vs. population sources, and thus the overall viability of the species in a given landscape. In this paper, I evaluate the impact of these four factors on population viability in a spatially explicit model of source-sink dynamics. The results are presented in the form of an "impact table," a device for communicating the effects of landscape alteration on population viability. For readers interested in the mathematical details of the source-sink model, a more detailed analysis is presented in Appendices A and B (Sects. 12.6, 12.7).

12.3 Models and Methods

12.3.1 Population Processes

Models are often constructed because we gain insight from building and analyzing the model, even if the model is not an exact replica of nature. I begin with the simplest of spatial population models, the so-called "BIDE" model in which

Local population growth = **B+I–D–E** $\tag{12.6}$

where **B** is the local birth rate, **D** is the death rate, **I** is the immigration rate, and **E** is the emigration rate. Thus, a demographic sink is a patch in which **B<D**; for a source, **B>D**. An interesting property of the model is that a demographic source can decline to extinction if **E–I>B–D**, i.e., excess emigration overcomes local population growth. [In Pulliam's (1988) original model, sources could never go extinct, because emigration only occurred after the source population reached its carrying capacity.]

The BIDE model can be extended to a network of habitat patches, each with its own rates of birth, death, immigration, and emigration. In the current paper, I will only consider a situation in which each patch is assigned to one of two habitats, a source habitat (**B>D**) or a sink habitat (**D>B**). The rates of immigration to and emigration from each patch depend on the spatial arrangement of the source and sink patches across the landscape as described below.

The problem then is to determine the population growth rate of the entire network of patches. It is easy to show (see Sect. 12.6, Appendix A) that the long-term growth rate of the coupled populations depends only on the local growth rates in source and sink patches and the fraction of population occurring in each patch. Letting λ be the finite growth rate of the sum of all source and sink populations, then

$$\lambda = \sum_{i}^{M} \alpha_i v_i \tag{12.7}$$

where M is the number of patches, α_i is the finite rate of increase ($=B_i-D_i$) in patch i and V_i is the fraction of the entire population that resides in patch i at any given moment. For the purposes of this paper, note that if $\lambda>1$, the aggregate population is considered viable; if $\lambda<1$, all populations will become extinct.

In all modeling scenarios, the finite growth rate of source patches was equal to 1.2 (20 % increase per year). Sink patches had a growth rate equal to 0.2.

12.3.2 Landscape Model

In order to evaluate the effect of landscape pattern on population processes, it is necessary to define a model describing spatial pattern. Here, I use a model based on fractal geometry (Mandelbrot 1982) that incorporates both habitat density and habitat fragmentation. The technical details of the model can be found in Appendix A (Sect. 12.6). The model has two parameters: the first, p, controls the total amount of source habitat on the landscape; the second, H, controls the number of patches among which the source habitat is distributed. The two parameters can be varied independently so it is possible to emulate a wide variety of landscape scenarios (Fig. 12.1). Simulated landscape can be

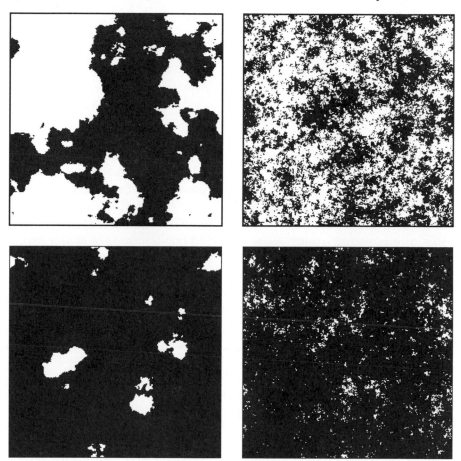

Fig. 12.1. Fragmented landscapes with 50 % (*top panels*) and 5 % (*bottom panels*) remaining habitat. Remaining habitat is either grouped into several large patches (*left*) or many small patches (*right*)

strongly autocorrelated (H→1.0) or uncorrelated (H→0.0) and the density of source habitat p may be varied continuously between 0.0 and 1.0.

An important property of the landscape models is that, within the constraints imposed by the tuning parameters, the landscapes are random with respect to the spatial arrangement of patches. Thus, each successive landscape is *statistically independent* from the previous landscape. If I had chosen some ad hoc, nonrandom algorithm, it would be impossible to know whether the results were biased by an undescribed and uncontrolled aspect of the spatial model.

In all simulations presented here, landscapes were 32 × 32 arrays of habitat patches. Each habitat patch was designated either as a source patch or a sink patch according to the fractal landscape model. Periodic boundary conditions were used such that in individual leaving one edge of the landscape would appear on the opposite edge.

12.3.3 Dispersal Model

Most organisms exhibit a leptokurtic dispersal pattern, i.e., most individuals disperse a short distance, whereas a few disperse very long distances (Neubert et al. 1995). The shape of the dispersal curve, particularly in the long-distance tail, is extremely important in determining the rate of spread of an organism (Kot et al. 1996). Most data suggest either a power-law or negative exponential function. Dispersal models based on Fickian diffusion (i.e., a random walk) generally underestimate the frequency of long-distance dispersal events, often with profound consequences on model predictions (Kot et al. 1996). A more fruitful approach is to model the dispersal curve directly.

I modeled dispersal using a negative exponential dispersal function. The dispersal function determines the probability that an individual will disperse a given distance across the landscape (Fig. 12.2). Two scenarios were examined: short-distance dispersal corresponding to a mean dispersal distance equal to the width of a single habitat patch or 1/32 the width of the entire landscape; for long-distance dispersal, the mean dispersal distance was 32 patch widths or the entire width of the landscape.

A second aspect of the dispersal models was active vs. passive dispersal. In passive dispersal, individuals disperse in a completely random direction, independent of any landscape features. Passive dispersal models systems such as seed dispersal in plants (assuming wind and other factors do not bias the dispersal direction). In active dispersal, individuals bias their dispersal such that they land in source patches more often then sink patches. The extent of the bias towards or away from a given patch was proportional to its "quality,"

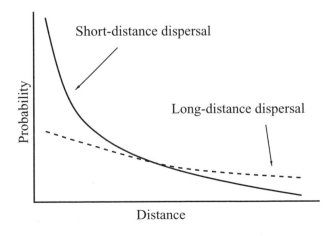

Fig. 12.2. Probability functions for two dispersal scenarios – short- and long-distance dispersal strategies – used in the model

Table 12.1. Summary of model parameters

Parameter	Treatments
Source habitat density	Medium density (50 % source habitat) Low density (5 % source habitat)
Landscape autocorrelation	Autocorrelated (source habitat in several large patches) Uncorrelated (source habitat in many small patches)
Dispersal distance	Short distance (mean distance equals one patch width) Long distance (mean distance equals landscape width)
Dispersal behavior	Passive (random direction) Active (dispersal biased towards source habitats)

defined here as the local population growth rate. Active dispersal emulates a situation in which an organism searches for high quality habitat.

12.3.4 Modeling Scenarios

The combination of the landscape and dispersal models resulted in four model parameters: source habitat density, habitat autocorrelation, mean dispersal distance, and active vs. passive dispersal behavior. I evaluated the impact of each of these parameters on population viability using a full factorial experimental design. For each of the parameters, two values were chosen (see Table 12.1). All factorial combinations of the parameter values resulted in 16 different scenarios. Ten replicate landscapes were evaluated for each scenario.

12.4 Results and Discussion

Results of the analysis are presented in the form of impact tables (Tables 12.2 and 12.3). Each column of the impact table lists a landscape attribute (the columns correspond to the landscapes shown in Fig. 12.1). The rows are species attributes; here, dispersal distance. Two tables are shown, one for passive dispersal and one for active dispersal. The entries in the table are qualitative assessments of the impact of landscape alteration. Imagine we begin with a landscape completely filled with source habitat and then remove 50 % of the landscape leaving behind many small source habitat patches. The relative impacts under this scenario are given in column two of the tables.

Several general patterns emerged from the analysis. As expected, population viability declined as the amount of source habitat was reduced. Land-

Table 12.2. Impact table for passive dispersal

	50 % Habitat Remaining, several large patches	50 % Habitat Remaining, many small patches	5 % Habitat Remaining, several large patches	5 % Habitat Remaining, several large patches
Short dispersal distance	Low	Moderate	Moderate	High
Long dispersal distance	Moderate	Moderate	Very High	Very High

Table 12.3. Impact table for active dispersal

	50 % Habitat Remaining, several large patches	50 % Habitat Remaining, many small patches	5 % Habitat Remaining, several large patches	5 % Habitat Remaining, many small patches
Short dispersal distance	Low	Low	Low	Moderate
Long dispersal distance	Low	Low	High	High

scape configuration also affected viability: when the source habitat was clumped into larger patches (increased autocorrelation), the aggregate population growth rate was higher. Somewhat surprisingly, long-distance dispersal resulted in lower population viability. There was no mortality penalty for long-distance dispersal; rather, individuals that disperse long distances away from source habitat more often end up in sink habitats, thus lowering the overall growth rate of the populations. Generally, as one would expect, passive dispersal resulted in lower viability for much the same reason as long-distance dispersal: more individuals landed in the sink.

The relative magnitude of effects from different parameters and parameter interactions is given in Table 12.4. The parameters that had the greatest effect on population viability were source habitat density, followed by dispersal distance, and passive vs. active dispersal. Several interaction effects followed the three most influential factors. Decreasing source habitat density inflated the effect of long-distance dispersal on viability. In short, when there was little habitat available on the landscape, individuals were better off staying close to home. There was a three way interaction between search strategy, habitat density, and dispersal distance, indicating that both dispersal behavior and habitat density need to be considered in viability studies.

Table 12.4. Most influential parameters and interactions

Rank	Parameter or interaction
1	Source habitat density
2	Dispersal distance
3	Search (passive vs. active)
4	Density × distance
5	Search × density × distance
6	Hurst exponent
7	Hurst × distance

The influence of spatial autocorrelation on viability was the sixth largest effect, followed by an interaction between autocorrelation and dispersal distance. The interaction between the Hurst exponent and dispersal distance occurred because more correlated landscapes favor short distance dispersal. It is interesting that the impact of spatial autocorrelation occurred so low in the ranking, below all other single parameters and several interactions. It is possible that the importance of landscape configuration was underestimated in the current source-sink model. The model does not consider the possibility of high intensity environmental disturbances that can cause local extinctions, i.e., classical metapopulation dynamics (Levins 1969; Hanski and Gilpin 1991). In the presence of local extinction and colonization, the importance of long-distance dispersal may be much greater than represented here.

12.5 Management Implications

Given that the total area of source habitat is the most important factor affecting population viability, should we care about landscape geometry? It depends somewhat on the situation. Clearly, one needs to consider the life-history traits, such as mean dispersal distance and dispersal mode, of the species of concern before making a blanket statement regarding the importance of landscape configuration. These results do suggest that in situations where we are able to preserve large areas of high quality, source habitats, we should do so.

It is important, however, to realize that virtually all habitat management decisions involve tradeoffs between competing and sometimes conflicting goals. In an ideal world, we would set aside all remaining habitats supporting endangered species as reserves. However, in reality, we must pick and choose. If we are constrained, owing to budgetary, social, or political factors, in the amount of habitat we can preserve, landscape configuration can be an important consideration in maximizing the effectiveness of conservation efforts. For example, given a target of preserving 50 % of the available habitat for a species, the analysis presented here suggests that the species will have greater viability if the habitat is clumped into several large patches.

The interesting problems arise, of course, when there are tradeoffs to be made between total area and landscape connectivity. Should one maximize habitat area at the expense of connectivity? There is no simple answer, because both area and connectivity are important. The answer will generally depend on how a species uses the landscape. However, even simple analysis of landscape pattern can be useful in prioritizing conservation decisions, and in general, landscape connectivity can be enhanced with little or no loss in the total habitat area preserved.

12.6 Appendix A: Mathematical Models

12.6.1 Fractal Landscapes

Landscapes were modeled as segmented fractional-Brownian surfaces (sfBs; Keitt and Johnson 1995). SfBs were constructed by first creating a fraction-Brownian surface and then slicing the surface at a particular elevation. All points above the slice were assigned to one class and those below to another class (see Fig. 12.1). Binary sfBs were indexed by two parameters, p, which determined the area assigned to one of the classes, and H, which set the fractal dimension of the surface (Mandelbrot 1982; Feder 1988; Peitgen and Saupe 1988).

A fractional-Brownian surface is most easily defined in terms of its Fourier transform (Hastings and Sugihara 1993). For fractal patterns, the power-spectrum (square of the Fourier coefficients) scales as a power-law of the frequency:

$$S(f) = kf^{\beta} \tag{12.8}$$

where $S(f)$ is the power at frequency f, k is a normalization constant, and β is a scaling exponent related to the fractal dimension of the surface. For a two-dimensional surface $\beta=2H+1$, where H is known as the "Hurst exponent" (Mandelbrot 1982). The Hurst exponent also determines the fractal dimension $D=3-H$.

Fractional-Brownian surfaces are easily created by generating random Fourier coefficients whose variance decays as a power-law function of frequency. An inverse Fourier transform is then applied to produce the fractal landscape.

12.6.2 Stochastic Landscape Networks

A stochastic landscape network describes the probability of an individual dispersing from one habitat patch to any other habitat patch in a landscape. The network can be normalized in terms of a matrix T whose elements t_{ij} are transition probabilities from patch i to patch j. We require that each row of T sum to one, because (ignoring for the moment reproduction and mortality) the sum of all individuals leaving a patch must equal the sum of individuals entering other patches. The matrix T thus defines a Markov chain.

For the simple lattices used here, filling the elements of T was simply a matter of computing the distance between two patches and then modeling the probability of dispersal as a function of distance. Here, dispersal was modeled by a negative exponential

$$p(d) = \theta e^{-\theta d} \tag{12.9}$$

where $p(d)$ is the probability of dispersing a distance d and θ is the dispersal coefficient. The mean dispersal distance was equal to $1/\theta$. For the artificial landscapes used here, I simply assigned $t_{ij}=p(d_{ij})/\sum_j p(d_{ij})$, where d_{ij} is the distance between patch i and patch j. This approximation was justified by the fact that each of the 32×32 cells on the landscape were considered an individual patch. Thus, all patches were the same size and had a compact shape. In cases where patches are defined with different sizes and sinuous or oblong shapes, corrections to t_{ij} need to be made to account for the irregular patch geometries.

One further modification needs to be made to T in order to incorporate active dispersal. As defined above, T models passive dispersal; the patch transition probabilities only depend on the distance between patches. However, if organisms search for high quality habitats, transitions to better habitat patches should be higher than to poor habitats. I introduce the parameter s to represent the degree to which dispersal is biased towards high quality habitats. I then define

$$t_{ij} = \frac{(1-\sigma+\sigma\alpha_j)p(d_{ij})}{(1-\sigma)\sum_j p(d_{ij}) + \sigma\sum_j \alpha_j p(d_{ij})} \tag{12.10}$$

where α_j is the growth rate in patch j. This function simply biases the transition probabilities towards patches with higher growth rates. When $\sigma=0$, then $t_{ij}=p(d_{ij})/\sum_j p(d_{ij})$ as before. When $\sigma=1$, t_{ij} is biased in proportion to the quality of patch j. The degree of bias in t_{ij} is a linear function of σ.

Given T we can construct the full BIDE population model and determine its overall growth rate. For a single patch, the BIDE model can be written as

$$n'_j = n_j \sum_i^M t_{ij}\alpha_i \qquad (12.11)$$

where n_j is the local population size in patch j, n'_j is the population size in the following generation, and M is the number of patches in the system. The entire model can be written in matrix form

$$\vec{n}' = AT\vec{n} \qquad (12.12)$$

where A is a matrix whose diagonal elements are the local population growth rates in each patch, and $\vec{n}$ is a vector containing the local population sizes. Thus, the overall growth rate of the set of populations is the largest eigenvalue λ of the matrix AT.

It is possible to derive λ as a function of the distribution of individuals among patches and the growth rate in each patch as follows. As long as T is positive, there exists an eigenvector $\vec{v}$ associated with λ such that

$$AT\vec{v} = \lambda\vec{v} \qquad (12.13)$$

We may choose any scaling of $\vec{v}$, so let chose $\Sigma_i v_i = 1$. Summing across the rows of Eq. (12.13), we have

$$\sum_j \sum_i t_{ij}\alpha_i v_i = \lambda \sum_j v_j. \qquad (12.14)$$

Noting that $\Sigma_j v_j = 1$ and that $\Sigma_i t_{ij} = 1$, and rearranging a bit gives the result in Eq. (12.7).

12.7 Appendix B: Statistical Analysis and Results

Mean growth rates on which the impact tables were based are shown in Table 12.5. The precision estimate is one standard error.

The results of an ANOVA on the model output is shown in Table 12.6. The given p-values are not particularly meaningful, however, the F-value gives a ranking of the effect of each parameter and parameter combinations on population viability.

Table 12.5. Mean value of λ for factorial treatments

		$p=0.50$		$p=0.05$	
		$H=1$	$H=0$	$H=1$	$H=0$
$\sigma=0$	$\theta=32$	1.0695±0.0351	0.9548±0.0455	0.7261±0.0511	0.5073±0.06097
	$\theta=1$	0.7064±0.0027	0.7021±0.0016	0.2521±0.0011	0.2507±0.0007
$\sigma=1$	$\theta=32$	1.1659±0.0065	1.1322±0.0062	1.0599±0.0247	0.9662±0.0535
	$\theta=1$	1.0617±0.0020	1.0587±0.0010	0.4694±0.0079	0.4468±0.0046

Table 12.6. ANOVA results comparing model treatments

	Df	Sum of Sq	Mean Sq	F Value	Pr(F)
Search	1	3.002692	3.002692	3616.080	0.0000000
Density	1	6.291388	6.291388	7576.589	0.0000000
Hurst	1	0.151507	0.151507	182.457	0.0000000
Distance	1	4.335539	4.335539	5221.200	0.0000000
Search × density	1	0.030375	0.030375	36.580	0.0000000
Search × Hurst	1	0.021660	0.021660	26.084	0.0000010
Density × Hurst	1	0.020432	0.020432	24.605	0.0000020
Search × distance	1	0.002159	0.002159	2.600	0.1090616
Density × distance	1	0.685271	0.685271	825.257	0.0000000
Hurst × distance	1	0.115287	0.115287	138.838	0.0000000
Search × density × Hurst	1	0.000297	0.000297	0.358	0.5505592
Search × density × distance	1	0.417326	0.417326	502.577	0.0000000
Search × Hurst × distance	1	0.031927	0.031927	38.449	0.0000000
Density × Hurst × distance	1	0.013578	0.013578	16.352	0.0000854
Search × density × Hurst × distance	1	0.002770	0.002770	3.335	0.0698753
Residuals	144	0.119574	0.000830		

Acknowledgement. The support of the Santa Fe Institute is gratefully acknowledged. This paper resulted from an IAI Sponsored workshop: "Habitat Fragmentation and Degradation," December 9–13, 1996, Marbella, Chile. Discussions with G. Bradshaw, P. Marquet, B. Macready, and E. van Nimwegen were influential in many aspects of this paper.

References

Cressie NAC (1993) Statistics for spatial data, revised edn. Wiley, New York

Dias PC (1996) Sources and sinks in population biology. Trends Ecol Evol 11:326–330

Diggle PJ (1990) Time series: a biostatistical introduction. Oxford statistical science series, vol 5. Oxford Science Publications, Oxford

Feder J (1988) Fractals. Plenum Press, New York

Hanski I, Gilpin M (1991) Metapopulation dynamics: brief history and conceptual domain. Biol J Linn Soc 42:3–16

Hastings HM, Sugihara G (1993) Fractals: a user's guide for the natural sciences. Oxford University Press, Oxford

Keitt TH, Johnson AR (1995) Spatial heterogeneity and anomalous kinetics: emergent patterns in diffusion limited predator-prey interactions. J Theor Biol 172:127–139

Kot M, Lewis MA, Vandendreissche P (1996) Dispersal data and the spread of invading organisms. Ecology 77:2027–2042

Levins R (1969) Some demographic and genetic consequences of environmental heterogeneity for biological control. Bull Entomol Soc Am 15:237–240

Mandelbrot B (1982) The fractal geometry of nature. Freeman, New York

Murdoch WW (1994) Population regulation in theory and practice. Ecology 75:271–287

Neubert MG, Kot M, Lewis MA (1995) Dispersal and pattern formation in a discrete-time predator-prey model. Theor Popul Biol 48:1–73

Peitgen H, Saupe D (1988) The science of fractal images. Springer, Berlin Heidelberg New York

Pulliam HR (1988) Sources, sinks, and population regulation. Am Nat 132:652–661

13 Patch Dynamics, Habitat Degradation, and Space in Metapopulations

P.A. MARQUET, J.X. VELASCO-HERNÁNDEZ, J.E. KEYMER

13.1 Introduction

Metapopulation theory is the leading theoretical framework to analyze the consequences of habitat fragmentation. The process of habitat fragmentation entails the creation of discrete habitat patches inhabited by local populations that interact with each other through the exchange of dispersing individuals behaving as a metapopulation system (Hanski and Simberloff 1997). Levins (1969, 1970) proposed the first metapopulation model that assumes a set of identical habitat patches with local subpopulations becoming extinct and the empty patches being recolonized from the currently occupied ones. This patch-occupancy metapopulation model provided a simple and fruitful way to understand the basic dynamical properties of metapopulations. This success is reflected in its many subsequent modifications and application to describe single-species (e.g., Hanski 1985, 1991; Gotelli 1991), two-species (e.g., Horn and MacArthur 1972; Slatkin 1974; Hanski 1983; Nee and May 1992; Nee et al. 1997), and multi-species interactions (e.g., Tilman et al. 1994; Holt 1997). However, there are three key features associated with the fragmentation of habitats that need to be incorporated into this type of metapopulation model if we are interested in gaining further insights into this process. These features are patch dynamics, habitat degradation, and the spatial dimension.

The first process underscores the fact that patches have dynamics of their own, being created and destroyed at different rates and these dynamics can affect the persistence of a metapopulation. Habitat degradation, on the other hand, makes reference to the fact that after creation, patches change in quality, usually becoming less suitable for species populations as different biotic and abiotic processes start to operate upon the patches (e.g., edge effects; Murcia 1995), meaning that patches are of different quality through time. Finally, the process of fragmentation, or more generally habitat loss, is essentially spatial. As several recent studies have shown (e.g., Dytham 1995; With

Ecological Studies, Vol. 162
G.A. Bradshaw and P.A. Marquet (Eds.)
How Landscapes Change
© Springer-Verlag Berlin Heidelberg 2003

and Crist 1995; Bascompte and Solé 1996; Keitt et al. 1997; Tilman et al. 1997), the spatial pattern of habitat loss does affect metapopulation dynamics and persistence.

In this chapter we analyze the dynamic consequences of incorporating patch dynamics and habitat degradation in a simple patch occupancy metapopulation model of the Levins type. We explicitly consider the dynamics of the species (how individuals occupy patches), and the dynamics of the patches (how patches of different types are created, occupied, and become extinct), coupling patch and species dynamics. We also analyze some threshold parameters that determine the successful colonization of a metapopulation and show that, for simple models of the patch occupancy type, these parameters also determine the existence of an equilibrium level of occupied patches. Finally, we show the importance of incorporating space into metapopulation models by analyzing the dynamics of a cellular automata model derived from our patch occupancy model based on ODEs (ordinary differential equations).

13.2 Levins' Original Model

The pioneer of metapopulation models is the one proposed by Levins (1969). This model assumes that N, the total number of available patches, is a constant. Let U and O denote the number of unoccupied and occupied patches, respectively. Levins' model assumes that the organisms achieve their carrying capacity instantly upon colonization of an empty patch, thus reaching their demographic equilibrium within each patch. Assume that at this equilibrium, each individual in the patch produces a total of β propagules per unit time. Therefore, βO represents the total number of propagules produced by all the individuals in the occupied patches. These propagules find unoccupied patches at a rate proportional to their frequency U/N; thus unoccupied patches are "lost" to colonization at a rate of $-\beta OU/N$ per unit time, and occupied patches increase by the same number per unit time. If we assume that occupied patches become extinct at a rate e then eO is the number of occupied patches that become extinct per unit time. Furthermore, this model assumes that occupied patches become unoccupied and immediately available for colonization at the same rate at which they become extinct, implying a closed system without an independent patch dynamics. The equations that govern this system are:

$$\frac{d}{dt}U = -\beta O\frac{U}{N} + eO$$

$$\frac{d}{dt}O = \beta O\frac{U}{N} - eO \tag{13.1}$$

Dividing both equations by N and defining $O/N=p$, we note that $U/N=1-p$, and the equations reduce to the Levins metapopulation model:

$$\frac{dp}{dt} = \beta p(1-p) - ep \qquad (13.2)$$

Since all patches and colonizing individuals are identical, one can dynamically follow the proportion of occupied patches instead of their actual number, and characterize the whole dynamic with two parameters: β and e. This model makes no distinction between the dynamics of patches and that of the organisms that occupy them.

Levins' model predicts that colonization of empty patches is successful whenever $\beta/e>1$. This condition also determines the existence of a nontrivial equilibrium point $p^*=1-e/\beta$ that is globally asymptotically stable.

13.3 Incorporating Patch Dynamics and Habitat Degradation (Model 2)

To incorporate patch dynamics and habitat degradation into Levins' model we consider a pristine landscape with two different types of patches that are occupied by a single type of organism (Fig. 13.1; see Marquet and Velasco-Hernández 1997 for further details). These patches are of different quality, which in a fragmented landscape could be associated with fragment size. The first type is a *source* patch, or one where organisms have, on average, a higher propagule production rate and a minimal natural extinction rate. The second patch type is a *sink* patch where, on average, organisms have low propagule

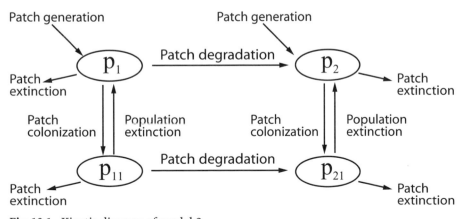

Fig. 13.1. Kinetic diagram of model 2

production rate and an extinction rate higher than in source patches (see Holt 1985; Pulliam, 1988).

The numbers of empty source and sink patches are denoted in this model by p_1 and p_2, respectively. We follow the temporal dynamics of these empty patches assuming that the total number of patches is not constant, but varies with time. Empty patches of both types can be colonized, generating two additional patch types denoted by p_{11} and p_{21} (number of occupied source and sink patches, respectively).

As mentioned earlier, we do not consider the total number of patches, defined in this model as $p=p_1+p_2+p_{11}+p_{21}$, to be constant. Habitat fragmentation entails the creation of patches, which in our model is introduced through a generation process of empty patches of both types. Patch dynamics enters our model by considering that each patch can be generated and can become extinct with or without a local population on it. Similarly, a species can colonize empty patches of both types and can become extinct for two reasons: either the patch that it was inhabiting went extinct, or it was not able to sustain a viable population in the patch, which implies that the patch becomes empty. Thus, patch dynamics can affect the time to extinction of the metapopulation. Habitat degradation is incorporated by allowing the conversion, at a rate k, of source into sink patches.

Let $p=p_1+p_2+p_{11}+p_{21}$ the total patch population. We have:

$$\frac{dp_1}{dt} = q\Lambda - (e+k)p_1 + h_1 p_{11} - \beta\frac{p_1}{p}(\sigma p_{21} + p_{11})$$

$$\frac{dp_2}{dt} = (1-q)\Lambda + kp_1 + h_2 p_{21} + ep_2 - \beta\frac{p_2}{p}(p_{11} + \sigma p_{21})$$

$$\frac{dp_{11}}{dt} = -(e+h_1)p_{11} + \beta\frac{p_1}{p}(\sigma p_{21} + p_{11}) - k_1 p_{11}$$

$$\frac{dp_{21}}{dt} = -(e+h_2)p_{21} + \beta\frac{p_2}{p}(\sigma p_{21} + p_{11}) + k_1 p_{11} \qquad (13.3)$$

In our model patch dynamics is driven by several parameters. Λ is the production rate of uncolonized patches, of which a fraction q are source p_1 patches. The rate e and k are the extinction and degradation rates. We refer to the extinction rate e as the background extinction rate since it is associated with the empty patches. A degraded source patch becomes a sink patch at a rate k. Thus, the number of degraded uncolonized source patches per unit time is kp_1. The dynamics of subpopulations occupying these patches is driven in turn by the following parameters. The constants h_1 and h_2 are the extinction rates of the subpopulations in either patch type. Therefore, occupied patches are recovered as empty source or sink patches at the rates h_1 and h_2. Thus, the number of colonized source patches that disappear per unit time is $(e+h_1)$ p_{11} The corresponding rate for colonized sink patches is $(e+h_2)$ p_{21} Degradation from occupied source to occupied sink patches occurs at a rate k_1 per unit time.

This model assumes that the colonization of empty source (or sink) patches is a frequency-dependent process proportional to the relative frequency of empty patches of both types, namely, p_1/p and p_2/p. The colonization or propagule production rate for the organism is higher in the source p_1 than in the sink p_2 patches, that is $\beta p_{11} > \sigma \beta p_{21}$ (the propagule production rate of organisms living on source patches is always greater than the propagule production rate of organisms on sink patches). The coefficient σ measures the reduction in the propagule production or colonization rate for individuals in sink patches $(0 \leq \sigma \leq 1)$. It is therefore, a measure of "sinkiness."

13.4 The Invasion Threshold

Threshold parameters are valuable theoretical tools for the qualitative evaluation of key metapopulation processes, and provide a useful and simple way to compare patch occupancy metapopulation models (Marquet and Velasco-Hernández 1997; Hernández-Suarez et al. 1999). In particular, the invasion threshold we show below provides information on the likelihood of invasion and colonization of a pristine metapopulation composed of empty patches. This threshold parameter corresponds to the basic reproductive number used in epidemiology, where it measures the number of secondary infections that a single infectious individual produces when introduced into a completely susceptible population (Diekman et al. 1990). If this number is above one, the disease spreads in the host population. Otherwise, no epidemic outbreak ensues and the disease dies out. The basic reproductive number, usually denoted by the symbol R_0, is an invasion criterion: it determines if a pathogen will be able to survive in a host population once it is introduced. It is important to point out that R_0 is computed assuming that all individuals in the population are susceptible since it measures the ability of the pathogen to spread initially. In general, R_0 does not provide information on the long-term persistence of the disease, although in simple cases it can.

In a metapopulation context, R_0 may be interpreted as the number of newly colonized patches arising from a single colonization event in an otherwise empty habitat or set of patches (Gyllenberg et al. 1997; Marquet and Velasco-Hernández 1997). As in the case of epidemics, one has to assume that at the beginning of the invasion the number of empty patches is large and that extinction and colonization have a negligible impact on the total number of empty patches.

In many situations, including Levins' model and the one analyzed here, this threshold parameter gives information on both of these processes: the likelihood of successful invasion and the existence and stability properties of equilibrium points where occupied patches are always present. In this latter case, they provide information on the persistence of occupied patches and the robustness and resilience of this state under perturbations.

13.5 The Threshold Parameter in Levins' Metapopulation Model

In the Levins metapopulation model (1969, Eq. 13.2), successful invasion of empty patches takes place only if the threshold parameter β/e is greater than one, where β and e are the propagule production and extinction rates, respectively. We interpret this threshold condition as saying that for a successful invasion of an *empty habitat* to occur, the number of propagules produced by *one* average occupied patch during its lifetime must be enough to allow for the colonization of *more* than one empty patch initially (i.e., on average each newly colonized patch gives rise to more than one additional colonized patch). Note that a successful invasion means only that, in the beginning of the process, there is an increase in the number of newly occupied patches. After more time has elapsed, this initial increase may lead to the persistence of occupied patches, or may lead to their extinction. In general, threshold parameters do not give information on this long term dynamic. However, in Levins' model, the threshold parameter does.

Note that in Levins' model the parameter β/e is associated with the eigenvalue of the corresponding linearized system at the equilibrium point when the proportion of empty patches is one. Also, the steady state with occupied patches is given by

$$p_0^* = 1 - e/\beta \tag{13.4}$$

Thus, for values of $\beta/e < 1$, only the steady state $p_e^* = 0$ exists and is stable. When $\beta/e > 1$, there is a bifurcation of the previous equilibrium point. The steady state $p_e^* = 0$ is now unstable and a new equilibrium $p_o^* > 0$ is asymptotically stable.

13.6 Threshold Parameters for Model 2

In the case of model 2, we find a threshold invasion criterion analogous to the one found for Levins' model. From now on, the threshold parameter for model 2 is denoted by the symbol R_0 and is given by the expression:

$$R_0 = \frac{\beta eq}{(e+h_1+k_1)(e+k)}\left(1-\sigma\frac{e+h_1}{e+h_2}\right)+\frac{\beta\sigma}{e+h_2} \tag{13.5}$$

This R_0 is the equivalent of Levins' threshold parameter. In our case, the existence of two types of patches that differ in both propagule production and extinction properties makes this an average of two numbers. Note that if there is no patch degradation ($k=0$), if only one type of patch is produced ($q=1$), if propagule production rates are equal in both patches ($\sigma=1$), and if the extinction rates of sink and source patches are equal ($e+h_1=e+h_2$), then R_0 becomes the threshold parameter of Levins' model.

The quantity $1/(e+k)$ can be interpreted as the average lifetime of a type 1 patch (source) before degrading to the other type (sink). Analogously, the quantities $1/(e+h_1+k_1)$ and $1/(e+h_2)$ are the average lifetime of type 1 and 2 occupied patches before extinction, respectively. Therefore, $\beta/(e+k)(e+h_1+k_1)$ and $\beta\sigma/(e+h_2)$ represent the propagule production rate of a p_{11} and p_{21} pair during its life span before extinction (when invading an empty habitat) respectively. R_0 is computed by averaging two parameters describing each type of patch. Thus, R_0 is the average number of successful colonization attempts of empty patches produced by an *average* occupied patch during its average lifetime when invading an empty habitat: if $R_0>1$, initially empty patches are invaded successfully.

In Levins' model, the nontrivial equilibrium exists only if invasion is successful ($R_0>1$). In our model, the same property holds. Thus, R_0 is able to describe not only invasion success, but also the existence of an equilibrium point where all patch types are present (it represents a steady state where the metapopulation shows a mixture of both types of empty patches, and both types of occupied patches). In Fig. 13.2, we show the level curve $R_0=1$ using β and e as parameters; all other parameter values are fixed. The values of β and

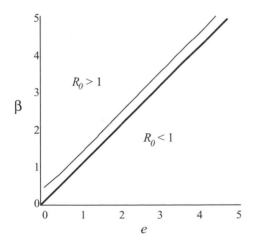

Fig. 13.2. Contour plot of R_0 as a function of β and e. The *line* $\beta=e$ represents Levins' threshold parameter

e that give $\beta/e=1$ in the Levins' model are on the line $\beta=e$. Note that the level curve of $R_0=1$ for our model is always to the left of the line $\beta=e$ of Levins' model, and therefore for a given e, the magnitude of β required to put the threshold parameter above one must be always higher in our model than the one required by Levins' model. Therefore, an increase in patch heterogeneity increases the propagule production rate needed to achieve $R_0>1$ (successful colonization). In this regard, the presence of sink patches has diluted the quality of the landscape rendering it more difficult to invade and persist.

However, while in Levins' model R_0 decreases monotonically as e increases, in our model R_0, seen as a function of e, is a peaked right-skewed function with a maximum (Fig. 13.3). This is a consequence of patch degradation. The propagule production rate of an occupied source patch is βq. An occupied source patch may be lost because the patch itself is lost (e), because it loses its subpopulation (h_1), or because it degrades and becomes a sink-occupied patch (k_1). Therefore, the unadjusted lifetime propagule production of an occupied source patch is $\beta q/(e+h_1+k_1)$. We have to correct this estimate because once a source patch is degraded, its propagule production rate is no longer βq. A source patch has an average lifespan of $1/(e+k)$, of which only a fraction $e/(e+k)$ is spent as an occupied, undegraded source patch. Therefore, the discounted propagule production of an occupied source patch is given by

$$\frac{\beta q}{e+h_1+k_1} \times \frac{e}{e+k} \tag{13.6}$$

Thus, there is an optimal rate at which patches can be lost from the system that improves invasion and persistence.

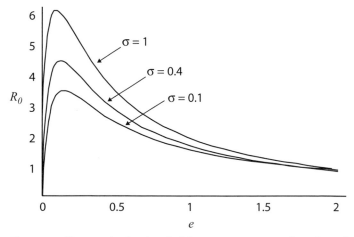

Fig. 13.3. Changes in the threshold parameter R_0 as a function of e. Different functions correspond to different values of the parameter σ

(a)

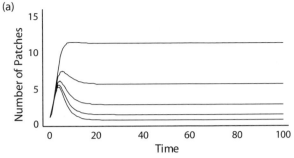

(b)

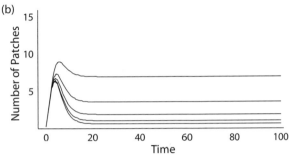

Fig. 13.4. Time series of the number of source (**a**) and sink (**b**) patches for different values of Λ (from *top* to bottom $\Lambda=10.5, 2.5, 1.25, 0.6$)

The production rate of uncolonized patches Λ does not affect the invasion rate or persistence of the metapopulation. However, it does have an effect upon the equilibrium proportion of occupied patches of both types (Fig. 13.4). All other things being equal, and for a given value of Λ, source patches are always found in higher numbers at equilibrium.

13.7 A Spatially Explicit Metapopulation Model

The explicit consideration of space, and the development of theoretical approaches to study its effects upon the structure and functioning of ecological systems, is currently recognized as one of the great advances, as well as a theoretical and empirical challenge, in contemporary ecology (Levin 1992; Durret and Levin 1994; Kareiva 1994; Tilman and Kareiva 1997).

One of the major consequences of considering the spatial dimension, and in particular local spatial interactions, has been the realization that space by itself is capable of giving rise to emergent self-organized spatial patterns with unexpected dynamical consequences, as seen in interactive particle systems (Durrett 1988), coupled map lattices (e.g., Kaneko 1989), and cellular automata models (Wolfram 1983, 1984). All three of these spatially explicit modeling approaches have been applied to biological systems, but probably the most widely used correspond to cellular automata models (CA; see Ermentrout and

Eldestein-Keshet 1993 for a review). In ecology, CA models have been applied to population dynamics (Caswell and Etter 1993; Molofsky 1994; Darwen and Green 1996; Keymer et al. 1998), community ecology (Czárán and Bartha 1992), predator prey dynamics (Van der Laan et al. 1995), forest dynamics (Auger 1995), competition (Green 1989), and biogeography (Carey 1996).

In this chapter, we incorporate space into the above formulation (Model 2) using the stochastic cellular automata formalism described in a previous work (Keymer et al. 2000). Basically, we consider a two-dimensional lattice with periodic boundary conditions (a 2-D torus) where each cell, or spatial position within this lattice, can be found in one of the following states: matrix (nonhabitat), empty source, empty sink, occupied sink, and occupied source state. We call these *state 0*, *state 1*, *state 2*, *state 3*, and *state 4*, respectively. Patch dynamics are introduced by considering that at any particular time step, cells in a matrix habitat state can change to either empty source or sink habitat states (a transition from state 0 to state 1 or 2, respectively, a process analogous to patch creation in model 2). These locations can in turn undergo extinction, becoming a matrix habitat (a transition from state 1 or 2 to 0, a process analogous to patch extinction in model 2). Once a source habitat cell is created it can be colonized, changing to an occupied source state (a transition from state 1 to state 4). Patches in this state can degrade becoming an occupied sink location (a transition from state 4 to state 3). All transitions at any particular time step and spatial location are shown in Fig. 13.5a. In this figure $p_{i,j}$ represents the probability of each path (event). These probabilities are directly derived from model 2, and are defined by the matrix P:

$$P = \begin{pmatrix} 1-\Lambda & \Lambda q & \Lambda(1-q) & 0 & 0 \\ e & (1-e)\left[1-\left(k(1-c)+c(1-k)+ck\right)\right] & (1-e)k(1-c) & (1-e)ck & (1-e)c(1-k) \\ e & 0 & (1-e)(1-c) & (1-e)c & 0 \\ e & 0 & (1-e)h_2 & (1-e)(1-h_2) & 0 \\ e & (1-e)h_1(1-k_1) & (1-e)k_1h_1 & (1-e)k_1(1-h_1) & (1-e)\left[1-\left(h_1(1-k_1)+k_1(1-h_1)+k_1h_1\right)\right] \end{pmatrix}$$

$$(13.7)$$

where c is the colonization probability at each particular spatial location and at one particular time step. This c is not a parameter, but a function of each particular neighborhood configuration surrounding a focal cell (in this particular case we used the so-called Moore neighborhood, which is composed of the eight nearest neighbors). This functional relationship is given by the following equation:

$$c = c\left(\Omega(r),t\right) = \beta P_{source}\left(\Omega(r),t\right) + \beta \sigma P_{sink}\left(\Omega(r),t\right) \tag{13.8}$$

where $P_{source}(\Omega(r),t)$ and $P_{sink}(\Omega(r),t)$ are the proportion of cells in a source and sink state, respectively, at time t in the neighborhood $\Omega(r)$ surrounding the spatial location r on the lattice.

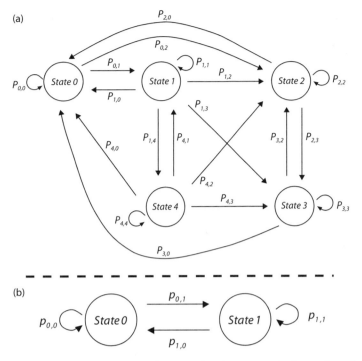

Fig. 13.5. State transition diagram for the CA equivalent of model 2 (a), and for the reduced model (b) when $q=1, \sigma=1, k=0, k_1=0, h_1=h_2=1$, and $\beta=0$

13.8 Spatial Habitat Dynamics

Habitat dynamics are driven by the conversion of matrix habitat into source and sink patches and by the extinction of these back to matrix habitat. These transitions do not depend on particular spatial configurations, and are independent of spatial location and neighborhood configurations. Thus, habitat dynamics can be analyzed analytically as the result of a random process across space. To illustrate this we consider a particular case of the model. Suppose there is only one type of suitable habitat and no population dynamics, that is we set $q=1, \sigma=1, k=0, k_1=0, h_1=h_2=1, \beta=0$. The previous graph now collapses to a new graph describing habitat formation and extinction (into matrix habitat) at each particular spatial location (Fig. 13.5b). The transition matrix associated with this graph is:

$$\Phi = \begin{pmatrix} 1-\Lambda & \Lambda \\ e & 1-e \end{pmatrix} \qquad (13.9)$$

If we now represent the probability of each state (habitat or matrix) at time t by a vector $V(t)$, we can show by applying Markov chain theory that $V(t)=\Psi^t V_0$, where Ψ is the transpose matrix of Φ (Feller 1966). The time-dependent probability of habitat at each spatial location is given by

$$P_{habitat}(t) = \frac{1}{\Lambda+e}\left(e+\left(p_0\Lambda-p_1e\right)\left(1-e-\Lambda\right)^t\right)$$

(13.10)

where p_0 and p_1 are the initial probabilities of matrix and habitat locations respectively. Taking the asymptotic limit when t goes to infinity (this limit exists for all Λ, e except for $\Lambda=1$, $e=1$), we get the expected probability of having habitat at a particular spatial location:

$$P_{habitat}^{*} = \frac{\Lambda}{\Lambda+e}$$

(13.11)

The proportion of habitat in the landscape is a function of Λ and e only (Fig. 13.6). Note that this probability is independent of space. Hence, we can treat the habitat proportion in the whole lattice as given by a binomial process with parameters n (number of spatial locations) and μ (equal to $P_{habitat}^{*}$). Therefore, the habitat proportion in the whole lattice behaves in time as a binomial noise with expected value $\dfrac{\Lambda}{\Lambda-e}$ and variance $\dfrac{\Lambda e}{\left(\Lambda+e\right)^2}$.

For two-dimensional lattices with a Moore neighborhood, it has been shown that a threshold percolation value of site occupancy probability (p_c) exists and equals 0.4 (Stauffer 1985; Plotnik and Gardner 1993). In our model, at this threshold the habitat percolates through the lattice, forming a large and connected cluster of habitat. Below the threshold the habitat is fragmented in many different unconnected patches. Thus, we can change the spatial pattern in our lattice by choosing Λ and e, in such a way as to get a value of $P_{habitat}^{*}$ above or below the critical threshold.

In our model patches are being created and go extinct in time. This implies that the spatial pattern dictated by $P_{habitat}^{*}$ will change in time, but the topological properties of each landscape component (patches) will be the same (since Λ and e do not change). These landscape dynamics are determined by the mean lifetime of a habitat patch at each spatial location $(1/e)$. A small $1/e$ will imply that the landscape changes faster. Thus, in our model habitat dynamics have a spatial and a temporal component. The spatial component will be given by the Λ and e, which affect the dominant spatial pattern. On the other hand, the quantity $1/e$ provides a temporal scale for landscape change,

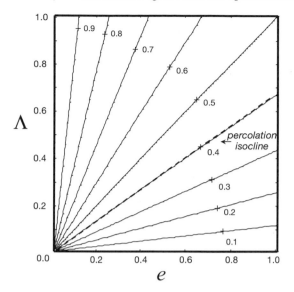

Fig. 13.6. The proportion of habitat ($P_{habitat}$*), as a function of Λ and e

or how fast the spatial pattern changes while retaining its essential properties (i.e., connected or fragmented habitat distribution). Different combinations of spatial pattern and rate of landscape change provide different habitat templates to model metapopulation dynamics in heterogeneous environments.

13.9 The Interaction Between Metapopulation Dynamics and Habitat Dynamics

Studies of population dynamics and fragmentation on static landscapes have shown that habitat connectivity is very important to the long-term persistence of a metapopulation (Dytham 1995; Bascompte and Solé 1996). While our model agrees with this result it also provides some insights on the effect of landscape dynamics upon species metapopulation persistence. In this regard our most important result is that long-term metapopulation persistence depends on the relationship between the scale of metapopulation dynamics (given by β, h_1, and h_2) and landscape dynamics (given by Λ and e).

To illustrate this, consider a particular species which colonize neighboring patches at a rate $\beta=0.9$, and undergoes extinction at rate $h_1=0.1$. These parameters determine the time scale at which metapopulation processes take place. As shown in Fig. 13.7, the persistence of this species is dependent upon Λ and e, which affect the expected proportion of available habitat $\dfrac{\Lambda}{\Lambda+e}$, and the temporal scale for landscape change ($1/e$). For static or slowly changing

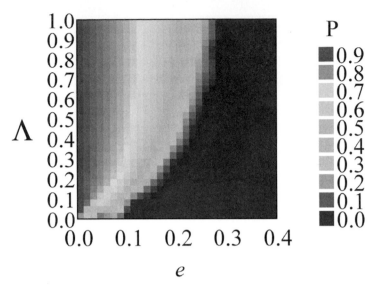

Fig. 13.7. Proportion of occupied habitat patches for a species with β=0.9, h_1=0.1 for different landscapes dynamics as a function of Λ and e

landscapes (low e) the proportion of occupied sites increases with Λ as a consequence of an increase in habitat availability (see Fig. 13.6); however, for fast-changing landscapes this dependency breaks down and persistence is no longer dependent on habitat availability, but on landscape dynamics. Figure 13.7 underscores the fact that species with different life-histories (in terms of dispersal ability and extinction proneness) are restricted to a particular subset of landscapes suitable for their persistence, which are defined by landscape dynamics and habitat suitability (Keymer et al. 2000). Thus, in addition to landscape pattern (With and Crist 1995; Keitt et al. 1997), the persistence of species is highly dependent on landscape dynamics. This suggests the need for the establishment of landscape monitoring programs as a basic step towards preventing the extinction of species.

Recently, Wiens (1997) has argued the need to merge metapopulation dynamics and landscape dynamics. Our model provides a simple approach to achieve this goal. However, while we incorporate the effects of landscape spatial patterning and dynamics into the model, we still assume a featureless matrix surrounding patches. Further modification of this model should focus on how the structure of the matrix affects movement of individuals across the landscape, and the dynamics of subpopulations within the matrix.

Acknowledgement. This work was supported through grant FONDECYT 1951036, 1990144 and FONDAP-FONDECYT 1501-0001.

References

Auger P (1995) Automates cellulaires et dynamique spatiale. Modélisation de la dynamique forestière. Rev Ecol (Terre et Vie) 50:261–271

Bascompte J, Solé RV (1996) Habitat fragmentation and extinction thresholds in spatially explicit models. J Anim Ecol 65:465–473

Carey PD (1996) Disperse: a cellular automaton for predicting the distribution of species in a changed climate. Global Ecol Biogeogr Lett 5:217–226

Caswell H, Etter RJ (1993) Ecological interactions in patchy environments: from patch-occupancy model to cellular automata. Lect Notes Biomath 96:93–109

Czárán T, Bartha S (1992) Spatiotemporal dynamics of plant populations and communities. Trends Ecol Evol 7:38–42

Darwen PJ, Green DG (1996) Viability of populations in a landscape. Ecol Modeling 85:165–171

Diekman O, Hesterbeek JAP, Metz JAJ (1990) On the definition and computation of the basic reproductive ratio R_0 in models for infectious diseases in heterogeneous populations. J Math Biol 28:365–382

Durrett R (1988) Crabgrass, measles, and gypsy moths: an introduction to interacting particle systems. Math Intelligencer 10:37–47

Durrett R, Levin S (1994) The importance of being discrete (and spatial). Theor Popul Biol 46:363–394

Dytham C (1995) The effect of habitat destruction pattern on species persistence: a cellular model. Oikos 74:340–344

Ermentrout GB, Edelstein-Keshet L (1993) Cellular automata approaches to biological modelling. J Theor Biol 160:97–133

Feller W (1966) An introduction to probability theory and its applications, vol 1, 3rd edn. Wiley, New York

Gotelli NJ (1991) Metapopulation models: the rescue effect, propagule rain, and the core-satellite hypothesis. Am Nat 138:768–776

Green DG (1989) Simulated effects of fire, dispersal and spatial pattern on competition within forest mosaics. Vegetatio 82:139–153

Gyllenberg M, Hanski I, Hastings A (1997) Structured metapopulation models. In: Hanski I, Gilpin ME (eds) Metapopulation biology. Academic Press, San Diego, pp 93–122

Hanski I (1983) Coexistence of competitors in a patchy environment. Ecology 64:493–500

Hanski I (1985) Single species spatial dynamics may contribute to long-term rarity and commonness. Ecology 66:335–343

Hanski I (1991) Single-species metapopulation dynamics: concepts, models and observation. Biol J Linn Soc 42:17–38

Hanski I, Simberloff D (1997) The metapopulation approach, its history, conceptual domain, and application to conservation biology. In: Hanski I, Gilpin ME (eds) Metapopulation biology. Academic Press, San Diego, pp 5–26

Hernández-Suarez CM, Marquet PA, Velasco-Hernández JX (1999) On threshold parameters and metapopulation persistence. Bull Math Biol 60:1–14

Holt RD (1985) Populations dynamics in two-patch environments: some anomalous consequences of an optimal habitat distribution. Theor Popul Biol 28:181–208

Holt RD (1997) From metapopulation dynamics to community structure. Some consequences of spatial heterogeneity. In: Hanski I, Gilpin ME (eds) Metapopulation biology. Academic Press, San Diego, pp 149–164

Horn HS, MacArthur RH (1972) Competition among fugitive species in an harlequin environment. Ecology 53:749–752

Kaneko K (1989) Spatiotemporal chaos in one- and two-dimensional coupled map lattices. Physica D 37:60–82

Kareiva P (1994) Space: the final frontier for ecological theory. Ecology 75:1

Keitt TH, Urban DL, Milne BT (1997) Detecting critical scales in fragmented landscapes. Conserv Ecol [on line] 1(1):4 Available from the Internet. URL: http://www.consecol.org/vol1/iss1/art4

Keymer JE, Marquet PA, Velasco-Hernández JX, Levin SA (2000) Extinction thresholds and metapopulation persistence in dynamic landscapes. Am Nat 156:478–494

Keymer JE, Marquet PA, Johnson AR (1998) Pattern formation in a patch occupancy metapopulation model: a cellular automata approach. J Theor Biol 194:79–90

Levin SA (1992) The problem of pattern and scale in ecology. Ecology 73:1943–1967

Levins R (1969) Some demographic and genetic consequences of environmental heterogeneity for biological control. Bull Entomol Soc Am 15:237–240

Levins R (1970) Extinction. In: Gesternhaber M (ed) Some mathematical problems in biology. American Mathematical Society, Providence, Rhode Island, pp 77–107

Marquet PA, Velasco-Hernández JX (1997) A source-sink patch occupancy metapopulation model. Rev Chil Hist Nat 70:371–380

Molofsky J (1994) Population dynamics and pattern formation in theoretical populations. Ecology 75:30–39

Murcia C (1995) Edge effects in fragmented forests: implications for conservation. Trends Ecol Evol 10:58–62

Nee S, May RM (1992) Dynamics of metapopulations: habitat destruction and competitive coexistence. J Anim Ecol 61:37–40

Nee S, May RM, Hassell MP (1997) Two-species metapopulation models. In: Hanski I, Gilpin ME (eds) Metapopulation biology. Academic Press, San Diego, pp 123–147

Plotnik RE, Gardner RH (1993) Lattices and landscapes. In: Gardner RH (ed) Lectures on mathematics in the life sciences: predicting spatial effects in ecological systems. Volume 23. American Mathematical Society, Providence, Rhode Island, pp 129–157

Pulliam HR (1988) Sources, sinks and population regulation. Am Nat 132:267–274

Slatkin M (1974) Competition and regional coexistence. Ecology 55:128–134

Stauffer D (1985) Introduction to percolation theory. Taylor and Francis, London

Tilman D (1994) Competition and biodiversity in spatially structured habitats. Ecology 75:2–16

Tilman D, Kareiva P (eds) (1997) Spatial ecology. The role of space in population dynamics and interspecific interactions. Princeton University Press, Princeton, NJ

Tilman D, May RM, Lehman CL, Nowak MA (1994) Habitat destruction and the extinction debt. Nature (Lond) 371:65–66

Tilman D, Lehman CL, Yin C (1997) Habitat destruction, dispersal, and deterministic extinction in competitive communities. Am Nat 149:407–435

Van Der Laan JD, Lhotka L, Hogeweg P (1995) Sequential predation: a multi-model study. J Theor Biol 174:149–167

Wiens JA (1997) Metapopulation dynamics and landscape ecology. In: Hanski I, Gilpin ME (eds) Metapopulation biology. Academic Press, San Diego, pp 43–62

With KA, Crist TO (1995) Critical thresholds in species' responses to landscape structure. Ecology 76:2446–2459

Wolfram S (1983) Statistical mechanics of cellular automata. Rev Mod Phys 55:601–644

Wolfram S (1984) Universality and complexity in cellular automata. Physica D 10:1–35

14 How Much Functional Redundancy Is Out There, or, Are We Willing to Do Away with Potential Backup Species?

F.M. Jaksic

14.1 The Issue

One of the major consequences of habitat fragmentation is the loss of species biodiversity. Inasmuch as species have a degree of functional redundancy, such biodiversity losses may be tolerable to the extent that they do not seriously compromise ecosystem function (Walker 1992). By definition, guilds of resource users are composed of species with a high degree of functional redundancy (Jaksic et al. 1996).

Root (1967) coined the term guild as "a group of species that exploit the same class of environmental resources in a similar way." He also pointed out that "this term groups together species, without regard to taxonomic positions, that overlap significantly in their niche requirements." Without realizing it, Root opened the debate on species redundancy in ecosystems (Jaksic 1981; Walker 1992). That is, if any guild member (a given focal species) declines or even disappears, its guild-mate (a presumably redundant species) may respond numerically by increasing its abundance. In this case, there would be a smooth takeover of functional roles by guild-mates, or a shifting balance of consumer abundances depending on how exactly resources fluctuate (e.g., Wiens 1990a,b).

14.2 Soft Evidence for Redundancy

The idea that some vertebrate species may be functionally redundant has been lurking in the literature for quite a while. My chief expertise is on predator ecology, and thus I will make my points with reference to the literature with which I am most familiar. Errington (1932, 1933) showed that the diets of several raptors were almost identical in intensively studied areas of southern

Ecological Studies, Vol. 162
G.A. Bradshaw and P.A. Marquet (Eds.)
How Landscapes Change
© Springer-Verlag Berlin Heidelberg 2003

Wisconsin. Craighead and Craighead (1956) noted the same in southern Michigan. Orians and Kuhlman (1956) observed that in Wisconsin, red-tailed hawks (*Buteo jamaicensis*) and great horned owls (*Bubo virginianus*) used habitat features and prey resources in an almost indistinguishable manner, except for the hawk being diurnal and the owl being nocturnal.

Jaksic (1982) computed food-niche overlaps among entire sets of sympatric raptors – both diurnal and nocturnal – and found that in southern Michigan, southern Wisconsin, northern Utah, southern Spain, and central Chile, the overlap values among synchronous (diurnal and nocturnal) raptors were statistically indistinguishable from those among asynchronous raptors (diurnal versus nocturnal). That is, some owls overlapped more with some hawks than with their respective synchronous mates. Carothers and Jaksic (1984) and Jaksic (1988) proposed that interference competition kept these two sets of predators temporally apart. Marti et al. (1993) updated and discussed this issue at length. My conclusion is that some raptor species pairs, trios or even larger groupings appear to be redundant at least in their use of food resources, regardless of where and when they hunt for prey. It is interesting that the same conclusion was reached by Huey and Pianka (1983), in a study of food-niche overlaps among insectivorous lizards.

14.3 Somewhat Harder Evidence for Redundancy

Perhaps the most compelling evidence for functional redundancy in ecological communities comes from worldwide and long-term studies of predatory vertebrate assemblages. These are assemblages composed of all the major groups of predatory vertebrates: snakes (Serpentes), hawks (Falconiformes), owls (Strigiformes), and carnivores (Carnivora). They are optimal study systems with regard to food-niche patterns, because their diets can be examined to extremely detailed levels (i.e., species, see Greene and Jaksic 1983), and thus the problem of "aliasing" is easily avoided. Aliasing is shorthand for pooling disparate species into broader taxonomic categories, thus overestimating the similarity of consumers that use particular resources (MacNally 1995). Using fine-tuned – prey species-level recognition – computations of food-niche overlap among predators in central Chile, Jaksic et al. (1981) detected a number of guilds with very high similarity in prey use. Jaksic and Delibes (1987) expanded this analysis to a southern Spanish predatory assemblage, standardized guild nomenclature across continents, and detected guilds as large as seven species in three orders. The most diverse guild – composed of up to four orders – was detected in central California (Jaksic and Medel 1990).

As for long-term studies of guild dynamics, the best data on the issue of functional redundancy come from two simultaneous studies conducted in north-central Chile: Aucó and Fray Jorge. Habitat changes in these localities are driven by El Niño Southern Oscillation (ENSO) episodes, wherein ENSO years with high precipitation and high productivity are preceded and followed by prolonged droughts and much-reduced primary and secondary productivity. At these two sites, Jaksic and coworkers have gathered extremely detailed data on the temporal changes in the diets of a number of vertebrate predators, chiefly owls, hawks, and foxes. These predators live in a fluctuating environment, prey-wise (Wiens 1993).

At the Fray Jorge study site, as precipitation levels increase and decrease, so does mammal prey abundance and the population densities of predators at the site (Jaksic et al. 1997). Although all predator species are resident at the site, their respective abundances vary in parallel with prey abundance. This is because each predator species tracks mammal prey abundance by increasing or decreasing its respective numbers roughly in proportion to mammalian densities (Silva et al. 1995). By analyzing diet composition through time and applying cluster analysis, Jaksic et al. (1997) were able to detect the presence of two trophic guilds, a carnivorous one (made up of the barn owl *Tyto alba* and great horned owl *Bubo virginianus*) and an omnivorous one (made up of several species). The carnivorous owls were highly redundant in diet, preying on essentially the same prey species in about the same proportions. It would not be too unrealistic to predict that the disappearance of either of these two owl species would only result in density compensation on the part of the other guild-mate, without much – if any – consequence for ecosystem function.

Essentially the same points may be raised from data obtained in the Aucó study site. Here, mammalian prey abundance fluctuates with precipitation levels, and though not measured as accurately in this case, predator species abundance appears to follow the same patterns (Jaksic et al. 1993, 1996). The interesting aspect of this study is that the species richness of predators also changes. As prey populations declined from the beginning of the study, different predator species abandoned the site, with only a hard core of guild-mates remaining there. As prey populations recovered, exiled predator species returned to the site. Once again, the temporary disappearance of some predator species did not seem to have had any impact in ecosystem function, at least in a 7-year time horizon.

By analyzing diet composition through time in the Aucó study site and applying cluster analysis, Jaksic et al. (1996) were able to detect the presence of two trophic guilds, a carnivorous one (made up once again by barn owl and great horned owl) and an omnivorous one (composed of several species). The carnivorous owls were highly redundant in diet, preying on essentially the same prey species in about the same proportions. Figure 14.1 merits analysis because it demonstrates the insights and shortcomings of the analysis of redundancy. The omnivorous guild may be deemed an artifact of aliasing: the American kestrel (*Falco sparverius*), austral pygmy owl (*Glaucidium nanum*), burrowing owl (*Athene cunicularia*), and foxes (*Pseudalopex culpaeus* and *P. griseus*) cluster closely chiefly because they prey extensively on insects, which

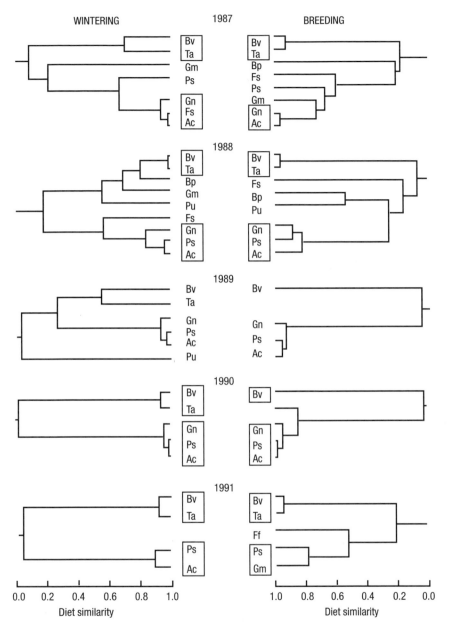

Fig. 14.1. Trophic guild structure of predators throughout 14 biological seasons (wintering and breeding) at Aucó (north-central Chile). Diet similarities close to 1 indicate high overlaps in diet; those close to 0 indicate distinct diets. Trophic guilds recognized are demarcated by *rectangles*, using the criterion of >50% diet similarity (note that similarity usually exceeds 75%). Species abbreviations are: Bv, *Bubo virginianus*; Ta, *Tyto alba*; Gm, *Geranoaetus melanoleucus*; Ps, *Pseudalopex culpaeus* and *griseus*; Gn, *Glaucidium nanum*; Fs, *Falco sparverius*; Ac, *Athene cunicularia*; Bp, *Buteo polyosoma*; Pu, *Parabuteo unicinctus*; Ff, *Falco femoralis*; and El, *Elanus leucurus*

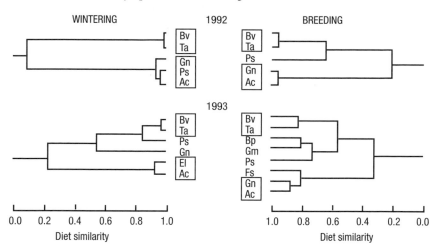

were categorized from specific up to ordinal levels. High overlap among these predators may only reflect a failure to identify exactly the prey species being used. This is not the case for the carnivorous guild, because barn and great horned owls preyed almost exclusively on vertebrates, mainly mammals, especially a single rodent species (*Phyllotis darwini*; see Jaksic et al. 1993, 1996). Therefore, there is no aliasing problem here, and the two owls may be said to be virtually redundant. The great horned owl was always present at the study site, but the barn owl left the area during the breeding seasons of 1989 and 1990, returning later to reconstitute the carnivorous guild with the great horned owl (Fig. 14.1).

14.4 How Will We Know What Is Redundant?

The old paradigm that no two species may use the same resources without one driving the other to competitive exclusion – and maybe extinction – has not exactly been demolished. Nevertheless, Schoener (1974) delivered a heavy blow to that belief when recognizing that similarities in one niche dimension may be compensated by differences in another: the hypothesis of complementarity among niche dimensions. A crippling blow came from Wiens (1977), when he convincingly argued that competition may be a sporadic phenomenon and that resource abundance may be the usual condition, at least in some ecosystems. In retrospect, Root's (1967) coining of the term guild recognized that two or more species may indeed be so similar to each other as to play essentially the same ecological role in communities. Although the connection of guild structure with ecosystem function is still fuzzy (e.g., Fuentes et al. 1995), it should not come as a surprise if such a connection is demonstrated. If indeed guilds are collections of species that play essentially the same functional role, then they may be considered as redundant ecological systems (Walker 1992).

At least with reference to the two long-term studies discussed above (Aucó and Fray Jorge), the apparent absence of changes in ecosystem function suggests that these systems might continue to function with fewer redundant predatory species. Nevertheless, what is the relationship between dietary opportunism (well demonstrated above, I hope) and functional redundancy? That is, if predator species respond opportunistically to prey availability, then their diets overlap and thus they form recognizable guilds. In that case, those species are bound to be gauged as functionally redundant. However, this may be more an illusion than reality, as the same species may diverge during conditions of prey scarcity, and redundancy then disappears (Wiens, pers. comm.). Indeed, this was an important point in Jaksic et al.'s paper (1996): the inability to classify species as redundant or nonredundant due to temporal fluctuations (Naeem, pers. comm.).

How shall we recognize redundancy when we see it? What is meant by ecosystem function? Of course, the first answer depends on what we come up with for the second. The major problem with Walker's (1992) redundancy hypothesis lies in defining in just what ecosystem functions the species are redundant. This issue is complicated, not only by the multiple functional roles of species in ecosystems – for example, a species may belong to several guilds defined by different criteria, but also by thresholds in biodiversity loss (Naeem et al. 1994). A certain level of redundancy may be necessary for system resilience, and only beyond that is apparent redundancy truly redundant (Wiens, pers. comm.). I will pursue this argument below.

14.5 What If Backup Species Are Necessary for Ecosystem Persistence?

Another problem with the minimalist approach of species redundancy is that if we accept that premise, therefore accepting a reduction in the number of redundant species, we may be increasing the risk of serious disruptions of ecosystem function. Simply stated, the question is whether we can afford to do away with redundant species, not knowing when we may need a "spare" in case of ecosystem malfunction in the near or distant future. As cogently stated by Naeem (1998), redundant species may act as backups or guarantors of reliable ecosystem function. Perhaps redundancy is a critical feature of ecosystems that must be preserved for them to function reliably and deliver the goods and services required for their own persistence in the face of a changing environment.

Empirical demonstrations of the value of species redundancy in ecosystem persistence, community structure, and the value of species conservation are few. The work of Kurzava and Morin (1998) is a good example of the complexities of determining what is a redundant species. They showed that two generalist predatory newts (*Notophthalmus viridiscens* and *Enneacanthus obesus*) found in different habitats had similar effects on prey community-structure patterns. When placed together in ponds with six species of larval anurans, they had similar impacts on composite community attributes such as prey species diversity and total prey biomass, but the two newts differed in their effect on prey species composition, each one determining the existence of a different assemblage of larval anurans. This case illustrates two important points: (1) "ecosystem or community function" may remain unchanged, even if some species are eliminated by others; (2) though still functioning, a net biodiversity loss may have occurred in an ecosystem.

Are we really willing to do away with redundant, potential backup species? I am no longer sure.

Acknowledgement. I happily acknowledge my intellectual debt to Peter Feinsinger and John Wiens in developing my views on ecological redundancy – this is not to be taken to mean that they agree with my views. Indeed, in paraphrasing Wiens' concerns about the redundancy hypothesis, I pay tribute to his foresight. I extend my gratitude to the conveners of this symposium, Gay Bradshaw and Pablo Marquet, for providing the freedom to develop my ideas on the topic in discussion. I developed this chapter using free time extended by a Presidential Chair in Science. The research itself was funded by grants from Fondecyt (196–0319) and The Mellon Foundation. I thank Ted Case and Shahid Naeem for a stimulating exchange of views on redundancy.

References

Carothers JH, Jaksic FM (1984) Time as a niche difference: the role of interference competition. Oikos 42:403–406

Craighead JJ, Craighead FC (1956) Hawks, owls, and wildlife. Stackpole Company, Harrisburg, Pennsylvania

Errington PL (1932) Food habits of southern Wisconsin raptors. Part I. Owls. Condor 34:176–186

Errington PL (1933) Food habits of southern Wisconsin raptors. Part II. Hawks. Condor 35:19–29

Fuentes ER, Montenegro G, Rundel PW et al. (1995) Functional approaches to biodiversity in the Mediterranean-type ecosystems of central Chile. In: Davis GW, Richardson DM (eds) Mediterranean-type ecosystems: the function of biodiversity. Ecological studies 109. Springer, Berlin Heidelberg New York, pp 185–232

Greene HW, Jaksic FM (1983) Food-niche relationships among sympatric predators: effects of level of prey identification. Oikos 40:151–154

Huey RB, Pianka ER (1983) Temporal separation of activity and interspecific dietary overlap. In: Huey RB, Pianka ER, Schoener TW (eds) Lizard ecology: studies of a model organism. Harvard University Press, Cambridge, MA, . pp 281–290

Jaksic FM (1981) Abuse and misuse of the term "guild" in ecological studies. Oikos 37: 397–400

Jaksic FM (1982) Inadequacy of activity time as a niche difference: the case of diurnal and nocturnal raptors. Oecologia 52:171–175

Jaksic FM (1988) Trophic structure of some Nearctic, Neotropical and Palearctic owl assemblages: potential roles of diet opportunism, interspecific interference and resource depression. J Raptor Res 22:44–52

Jaksic FM, Delibes M (1987) A comparative analysis of food-niche relationships and trophic guild structure in two assemblages of vertebrate predators differing in species richness: causes, correlations, and consequences. Oecologia 71:461–472

Jaksic FM, Medel RG (1990) Objective recognition of guilds: testing for statistically significant species clusters. Oecologia 82:87–92

Jaksic FM, Greene HW, Yáñez JL (1981) The guild structure of a community of predatory vertebrates in central Chile. Oecologia 49:21–28

Jaksic FM, Feinsinger P, Jiménez JE (1993) A long-term study on the dynamics of guild structure among predatory vertebrates at a semi-arid neotropical site. Oikos 67:87–96

Jaksic FM, Feinsinger P, Jiménez JE (1996) Ecological redundancy and long-term dynamics of vertebrate predators in semiarid Chile. Conserv Biol 10:252–262

Jaksic FM, Silva SI, Meserve PL, Gutiérrez JR (1997) A long-term study of vertebrate predator responses to an El Niño (ENSO) disturbance in western South America. Oikos 78:341–354

Kurzava LM, Morin PJ (1998) Tests of functional equivalence: complementary roles of salamanders and fish in community organization. Ecology 79:477–489

MacNally RC (1995) Ecological versatility and community ecology. Cambridge University Press, Cambridge

Marti CD, Korpimäki E, Jaksic FM (1993) Trophic structure of raptor communities: a three-continent comparison and synthesis. In: Power DM (ed) Current ornithology, volume 10. Plenum, New York, pp 47–137

Naeem S (1998) Species redundancy and ecosystem reliability. Conserv Biol 12:39–45

Naeem S, Thompson LJ, Lawler SP, Lawton JH, Woodfin RM (1994) Declining biodiversity can alter the performance of ecosystems. Nature 368:734–737

Orians G, Kuhlman F (1956) Red-tailed hawk and horned owl populations in Wisconsin. Condor 58:371–385

Root RB (1967) The niche exploitation pattern of the blue-gray gnatcatcher. Ecol Monogr 37:317–350

Schoener TW (1974) Resource partitioning in ecological communities. Science 185:27–39

Silva SI, Lazo I, Silva-Aránguiz E et al. (1995) Numerical and functional response of burrowing owls to long-term mammal fluctuations in Chile. J Raptor Res 29:250–255

Walker BH (1992) Biodiversity and ecological redundancy. Conserv Biol 6:18–23

Wiens JA (1977) On competition and variable environments. Am Sci 65:590–597

Wiens JA (1990a) The ecology of bird communities. Volume 1: foundations and patterns. Cambridge University Press, Cambridge

Wiens JA (1990b) The ecology of bird communities. Volume 2: processes and variations. Cambridge University Press, Cambridge

Wiens JA (1993) Fat times, lean times and competition among predators. Trends Ecol Evol 8:348–349

15 Predicting Distributions of South American Migrant Birds in Fragmented Environments: a Possible Approach Based on Climate

L. JOSEPH

15.1 Introduction

Migratory species pose unique challenges to conservation. They have three distributions: their breeding distribution, an often widely disjunct nonbreeding distribution to which they migrate after breeding (often termed the wintering distribution), and the intervening areas through which they move and stop over in passage. Their conservation, therefore, requires knowledge of distribution patterns in spatial and temporal terms. In the context of habitat fragmentation, migratory birds that breed in southern South America offer further challenges. Many migrate after breeding to the vastness of Amazonia, where rapid field surveys or studies of community ecology are not always feasible. Furthermore, Tuomisto et al. (1995) showed that Amazonian environments are more heterogeneous or, in a sense, more naturally fragmented than previously suspected. Potentially, this indicates even more difficulties in adequately describing the seasonal distributions of migrant birds there. Some method of predicting or modeling distributions of migrants in space and time in areas such as Amazonia would therefore be useful. For example, a modeled migratory pathway could be a layer in a geographic information system (GIS) along with layers showing extent of habitat fragmentation on local or regional scales. In that way, often-scarce resources might be more effectively channeled into fieldwork to obtain still badly needed data on the natural history and ecology of migrant birds. It is my general aim in this chapter, then, to explore a possible method of such prediction.

I will focus on migrant birds that breed in southern South America, so a brief review of their migration patterns is appropriate (see also Joseph 1996). The prevailing view of their migration patterns has combined elements of climate and latitude and has been accompanied by widespread use of the term "austral migration" (e.g., see introductory sections of Keast and Morton 1980; Hagan and Johnston 1992; De Graaf and Rappole 1995). Chesser (1994), in a

Ecological Studies, Vol. 162
G.A. Bradshaw and P.A. Marquet (Eds.)
How Landscapes Change
© Springer-Verlag Berlin Heidelberg 2003

pioneering and stimulating review of the subject, applied this combined latitudinal and climatic approach when he defined austral migrants as "species
that breed in temperate areas of South America (climatic and latitudinal elements) and migrate north, towards or into, Amazonia (the latitudinal element), for the southern winter."

Some authors, however, have decoupled the climatic and latitudinal elements when discussing these same birds' migration patterns. In solely climatic terms, two broad migration patterns have long been recognized. Wetmore (1926), Zimmer (1938), Gore and Gepp (1978), Narosky and Yzurieta
(1993), Chesser (1994) and Hayes et al. (1994) recognized these patterns,
which might be considered as two extremes of a climatically defined spectrum of long- and short-distance migration patterns among these species.
One pattern is characterized by species with nonbreeding distributions in
warm, humid lowland tropical zones. Examples are Swainson's flycatcher
Myiarchus swainsoni, tropical kingbird *Tyrannus melancholicus* and grey-
breasted martin *Progne chalybea* (for details see Ridgely and Tudor 1989,
1994; Stotz et al. 1996). The other extreme pattern involves birds with nonbreeding distributions in climatically temperate regions, some of which may
extend well into tropical latitudes; these regions variously include the high
Andes (e.g., all migratory ground tyrants *Muscisaxicola* spp.) and the northern parts of the so-called southern cone of South America (e.g., rufous-
chested dotterel *Charadrius modestus*, chocolate-vented tyrant *Neoxolmis
rufiventris*, black-crowned monjita *Xolmis coronata*). Conversely, Hayes
(1995) adopted an extreme latitudinal approach, defining "austral migration"
as the migration of any migratory birds breeding anywhere in the southern
hemisphere, not just in South America, and then undergoing a northward
migration of any extent.

My specific aim in this chapter is to explore the potential of a purely climatic approach in the description and prediction of seasonal changes in distribution of South American austral migrants (hereafter for simplicity used
sensu Chesser 1994). I address the aim through a preliminary test of the null
hypothesis that the climatic characteristics of a migrant's breeding range during the months when the birds are there will be the same as that of the nonbreeding and passage distributions during the months when the birds are in
those areas. The hypothesis is further explored by Joseph and Stockwell
(2000). Unless indicated otherwise, further use of the term "nonbreeding distribution" provisionally includes the passage distribution both for simplicity
and because migration routes of many austral migrants are not well known,
and overlapping breeding and nonbreeding distributions are common among
austral migrants (Chesser 1994). I recognize that this could be artificial and
misleading in some cases, and so in the case of two migrants, *Elaenia strepera*
and *E. albiceps chilensis*, for which appropriate data are available, I also
explore the effects of separating passage distribution from core nonbreeding
and breeding ranges. Where the null hypothesis is not rejected, and noting

that for many austral migrants breeding distributions are better known than nonbreeding distributions (e.g., Marantz and Remsen 1991), one might then try to predict the nonbreeding distribution more completely by asking where else and at what times of the year does the climatic envelope of the breeding distribution occur.

I will estimate the climatic envelopes, albeit crudely, with the daily mean temperature (DMT, °C) in January for the breeding distribution and the DMT in July for the nonbreeding distribution. Temperature was chosen as a representative climatic variable because it has been found to be a primary climatic correlate of the breeding distributions of migrant birds. Thus, Root (1988) noted that in North America range limits of wintering birds correlated with mean January (winter) temperatures and Chesser (1998) found that the ratio of migratory to total breeding tyrant-flycatchers in southern South America was strongly associated with mean temperature of the coldest month (among other variables). The present approach differs from other studies that have linked migratory birds' distributions to environmental parameters in that it asks whether climatic correlates of the breeding distribution can *predict* the physical location of the nonbreeding distribution (and vice versa). It also describes patterns generated by evolutionary processes that underlie how a migratory species contributes to the latitudinally based diversity gradients that were Newton and Dale's (1996a,b) focus of study of migrant birds in Europe and North America. Finally, in this regard, I note that photoperiod is an environmental parameter of bird migration that has long been studied for its control of the phenomenon (Berthold 1994). It helps explain *when* a bird migrates; my approach asks *where the bird goes* on migration.

15.2 Methods

I chose for study only those austral migrants for which I could locate published reviews of seasonal distribution (Table 15.1). For the royal tern *Sterna maxima*, Swainson's flycatchers *Myiarchus s. swainsoni* and *M. s. ferocior*, slaty elaenia *Elaenia strepera* and Patagonian tyrant *Colorhamphus parvirostris* the reviews were based on all specimen and reliable sight records known to the authors; textual data on localities and dates of collection/sighting were included, but varied in detail. For the lesser elaenia *E. chiriquensis albivertex*, white-crested elaenia *Elaenia albiceps chilensis*, gray-crowned tyrannulet *Serpophaga griseiceps* and lined seedeater *Sporophila lineola* the reviews gave reasonably detailed maps of specimen localities, but gave few or no textual data on localities or dates of collection and/or sightings. I did not include the Suiriri flycatcher *Suiriri suiriri* and the small-billed elaenia *E. parvirostris,* both of which were considered austral migrants by Chesser (1994) and reviewed taxonomically by Traylor (1982), because data on sea-

Table 15.1. Literature sources from which descriptions of breeding and nonbreeding distributions of nine austral migrants have been taken

Migrant	Primary source(s)	Secondary source(s)
Sterna maxima	Escalante (1968, 1985)	Narosky and di Giacomo (1993)
Colorhamphus parvirostris	Chesser and Marin (1994)	
Elaenia strepera	Marantz and Remsen (1991)	
Elaenia chiriquensis albivertex	Marini and Cavalcanti (1990)	Scherer-Neto and Straub (1995)
Elaenia albiceps chilensis	Marini and Cavalcanti (1990)	Traylor (1982); Araya and Millie (1991); Belton (1994); Ridgely and Tudor (1994)
Myiarchus s. swainsoni	Lanyon (1978)	
Myiarchus s. ferocior	Lanyon (1978)	Romanella (1993)
Serpophaga griseiceps	Straneck (1993)	
Sporophila lineola	da Silva (1995)	

sonal changes in distribution were not available or were simply too limited to be of use.

DMTs (Table 15.2) were taken from Schwerdtfeger's (1976) and Prohaska's (1976) tables of temperature data from recording stations within the circumscribed breeding and nonbreeding distributions of the study species in the source papers. Localities of climate stations (Fig. 15.1) were matched for the altitudinal range of each migrant as given in the source papers and in Stotz et al. (1996). The use of means rather than variances for each locality was dictated by availability of temperature data from most stations as DMTs already calculated for each month. For some localities, the values were given as mean monthly maxima and minima, so in these cases I took the DMT to be the average of the two extremes, following Griffiths (1976). I compared the mean January and July DMTs in the breeding and nonbreeding distributions, respectively, using nonparametric t-tests performed with Statistica for Windows (Version 5.0, StatSoft 1994). These months were chosen because January is when austral migrants are expected to be well into their breeding cycle, either with young in the nest or feeding fledged young, and July is when they are expected to have reached their winter destinations. For further comparisons, maps of January and July DMT isotherms from Hoffman (1975) and Martyn (1992) were compared with the breeding and nonbreeding ranges, respectively, by visual inspection.

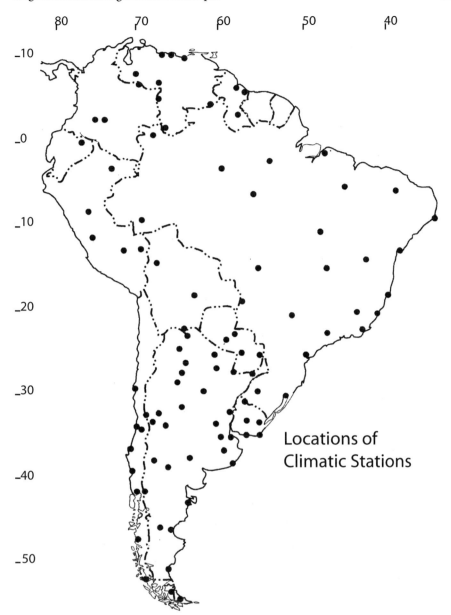

Fig. 15.1. Map of South America showing locations of climatic recording stations in Schwerdtfeger (1976) from which daily mean temperature data have been taken. For details of the localities see Table 15.2

Table 15.2. Locations and altitudes of climatic recording stations, and January and July daily mean temperatures (°C) at those stations and altitude in meters above sea level (m a.s.l.). In each row and according to the codes indicated below, the final column shows which localities were used for which species and whether in the breeding season (January) or nonbreeding season (July)

Climate Station	Latitude	Longitude	Altitude (m a.s.l.)	January DMT (°C)	July DMT (°C)	Migrants[a]
Venezuela						
Caracas	1030N	6656	1035	19.2	21.1	4B; 6B
Barcelona	1007N	6441	7	25.2	26.0	4B; 6B; 7A
San Fernando	0753N	6726	73	26.7	25.5	4B
Pto Ayacucho	0541N	6738	99	28.5	25.2	4B; 7B
Sta Elena	0436N	6107	907	21.6	20.9	4B; 7B
San Carlos	0154N	6703	95	26.3	25.4	4B; 7B
Guyana						
St Ignatius	0321N	5448	99	27.7	27.1	4B
New Amsterdam	0615N	5731	2	26.8	27.2	4B
Georgetown	0648N	5808	2	26.3	26.7	4B
Colombia						
Arauca	0704N	7040	122	27.1	25.7	4B; 6B; 7A; 7B
Barrancabermeja	0704N	7352	107	29.3	29.1	4B; 6B; 7A
Villavicencio	0409N	7336	423	26.6	25.2	4B; 5B; 6B; 7A; 8B; 9B
Espinal	0409N	7453	322	28.0	28.1	4B; 5B; 6B; 7A; 8B; 9B
Ecuador						
Puyo	0135	7754	950	21.5	20.5	5B; 6B; 7B; 8B; 9B
Perú						
Iquitos	0346	7320	104	27.0	25.0	5B; 6B; 7A; 9B
Cuzco	1333	7159	3312	13.0	10.0	8B
Tingo María	0908	75557	665	24.5	24.5	5B; 6B; 7A; 8A; 8B; 9B
Huancayo	1207	7520	3380	12.5	9.5	8A; 8B

Climate Station	Latitude	Longitude	Altitude (m a.s.l.)	January DMT (°C)	July DMT (°C)	Migrants[a]
Pto Maldonado	1238	6912	256	26.0	23.0	5B; 6B; 7A; 9B
Cajamarca	0708	7828	2621	15.0	12.5	8B
Bolivia						
Apolo	1443	6830	1382	21.0	18.0	8A; 8B
Santa Cruz	1747	6311	437	26.0	20.0	3A; 5A; 5B; 7A; 8A
Brazil						
Uapés	0008N	6705	85	25.5	24.3	4B; 7A; 7B; 9B
Sena Madurerra	0985	6840	135	25.4	22.9	7A; 7B; 9B
Manaus	0385	6001	48	26.2	26.8	4B; 7A; 7B; 9B
Belém	0128	4827	24	25.2	25.8	4B; 7A; 7B; 8B; 9B
Santarém	0225	5442	20	25.7	25.1	4B; 7A; 7B; 8B; 9B
Barra do Corda	0530	4516	81	25.5	24.2	4B; 7A; 7B
Quixeramobim	0512	3918	198	28.6	26.4	4B; 7A; 7B
Olinda	0801	3451	57	27.1	24.3	7A; 7B; 8B
Alto Tapajós	0720	5730	140	24.7	24.0	7A; 7B
Porto Nacional	1031	4843	237	25.2	24.3	7A; 7B
Formosa	1532	4718	0912	21.7	18.7	7A; 8A; 8B
Remanso	0941	4204	411	27.2	25.7	7A; 7B
Caetite	1403	4237	878	22.6	18.9	7A; 7B
Salvador	1255	3841	45	26.1	13.0	7A
Vitoria	2019	4020	31	25.3	20.5	7A; 9A
Cuiabá	1535	5606	171	26.4	22.5	7A; 8A; 9A
Corumbá	1900	5739	138	26.7	21.0	7A; 9A
Tres Lagoas	2047	5142	312	25.4	19.2	7A; 9A
Belo Horizonte	1956	4356	915	22.5	17.2	7A; 8A; 9A
Campinas	2253	4705	663	22.4	16.2	7A; 9A
Rio de Janeiro	2254	4310	30.5	25.1	30.2	1B; 7A; 8A; 8B; 9A

Table 15.2. (*Continued*)

Climate Station	Latitude	Longitude	Altitude (m a.s.l.)	January DMT (°C)	July DMT (°C)	Migrants[a]
Alegrete	2946	5547	204	24.7	12.9	4A; 8A
Porto Alegre	3002	5113	10	24.7	14.3	1B; 4A; 8B
Paraguay						
Mcal Estigarribia	2201	6036	181	29.7	19.8	4B; 9A
Pto Casado	2217	5752	87	29.3	20.3	4B; 9A
Asunción	2516	5738	64	29.3	18.3	9A
Pto Pdte Franco	2536	5434	1255	26.8	16.7	4A; 8A; 9A
Argentina						
La Quiaca	2206	6536	3459	12.4	4.0	8A
Las Lomitas	2442	6035	130	27.1	15.7	3A; 5A; 9A
Salta	2451	6529	1226	21.4	10.6	3A; 6A; 8A; 9A
Tucumán	2648	6512	481	24.5	12.6	3A; 5A; 6A; 8A; 9A
Pte Roque S. Peña	2649	6027	92	27.3	15.2	3B; 5A; 9A
Posadas	2725	5556	136	26.2	15.6	4A; 7A; 8A; 8B; 9A
Corrientes	2728	5849	60	27.4	15.7	3B; 4A; 7A; 8A; 8B; 9A
S. del Estero	2746	6418	199	27.3	12.9	3A; 5A; 8A; 9A
Catamarca	2826	6546	5547	27.7	11.4	3A; 5A; 8A; 8B; 9A
Ceres	2953	6157	88	26.1	12.3	3B; 5A; 9A
Concordia	3123	5802	38	25.7	12.5	3B; 9A
Córdoba	3124	6411	425	24.2	10.6	3A; 3B; 5A; 8A; 8B; 9A
San Juan	3136	6833	630	26.0	8.0	3A; 8A;
Mendoza	3253	6850	769	23.6	7.6	3A; 8A
Rosario	3255	6047	27	23.8	10.2	5A
San Luis	3316	6621	716	24.0	8.8	3B; 5A; 8B
Buenos Aires	3435	5829	25	23.7	10.5	5A; 7A
Junín	3435	6056	81	23.4	9.8	3B; 5A; 8B

Climate Station	Latitude	Longitude	Altitude (m a.s.l.)	January DMT (°C)	July DMT (°C)	Migrants[a]
Las Flores	3602	5906	34	22.8	9.1	5A
Macachín	3708	6341	142	23.7	7.4	3A; 8A
Mar del Plata	3808	5733	14	19.0	7.7	1A
Cipolletti	3857	6759	265	21.6	5.8	8A
Bariloche	4106	7110	836	14.5	2.3	8A
Trelew	4314	6518	34	20.6	6.0	1A
Pto Santa Cruz	5001	6832	12	14.3	2.2	5A
Chile						
La Serena	2954	7115	35	18.2	11.7	2B
Valparaíso	3301	7138	41	18.0	11.8	2B
Santiago	3327	7042	520	20.0	8.1	2B
Concepción	3640	7303	155	18.0	9.1	2A; 2B
Valdivia	3948	7314	9	17.0	7.7	2A; 2B; 8A
Pto Montt	4128	7257	13	15.2	7.6	2A; 2B; 8A
San Pedro	4743	7455	22	11.2	5.7	2A
Evangelistas	5224	7506	555	8.7	4.4	2A; 8A
Punta Arenas	5310	7054	8	11.7	2.5	2A; 8A
Uruguay						
Artigas	3024	5628	117	26.6	13.6	4A; 8A; 8B
Paso de los Toros	3249	5631	79	24.9	11.5	4A; 8B
Treinta y Tres	3311	5421	31	24.4	11.7	4A
Montevideo	3452	5612	22	22.5	10.5	1B; 4A; 8B
Punta del Este	3458	5457	16	21.6	11.4	1B; 4A; 8B

[a] Codes: A, breeding; B, nonbreeding; 1, *Sterna maxima*; 2, *Colorhamphus parvirostris*; 3, *Serpophaga griseiceps*; 4, *Myiarchus s. swainsoni*; 5, *M. s. ferocior*; 6, *Elaenia strepera*; 7, *Elaenia chiriquensis albivertex*; 8, *Sporophila lineola*; 9, *Elaenia albiceps chilensis*.

15.3 Results

I briefly summarize the published accounts of the birds' breeding and non-breeding distributions (for more details, see sources in Table 15.1) and relate them to differences in DMTs within those areas in January and July, respectively (see Tables 15.2 and 15.3; Figs. 15.2–15.4). The sequence in which the species accounts below are presented follows that in Fig. 15.2.

15.3.1 *Sterna maxima*

The southern South American population breeds in the southern Argentinean province of Chubut (e.g., Punta Leon) from October to March, and there is one old nest record from Buenos Aires province (see Escalante 1985). It migrates northwards to winter in Uruguay and far southern Brazil occa-

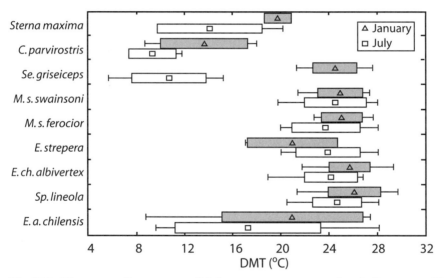

Fig. 15.2. Diagrammatic summary of daily mean temperature values in January and July for the nine austral migrants studied here. Means, ranges and one standard deviation are shown. Note that *E. strepera* has been recorded in the breeding season in northern Argentina to 2600 m a.s.l. and for that altitude within this part of Argentina, the January DMT had to be estimated as follows. DMT data were available from Salta at 1226 m a.s.l. (January DMT=21.4 °C) and La Quiaca at 3459 m a.s.l. (January DMT=12.4 °C). Tables 15.2 and 15.3 and this figure were compiled using an estimated January DMT at 2600 m a.s.l. of 17.0 °C, which is the mid-point between the values at Salta and La Quiaca. This estimate appears reasonable because isotherm maps of January and July DMTs in Hoffman (1975) show complex and sharp gradients in temperature and altitude between 1500 and 3000 m a.s.l. in this part of northern Argentina with some areas close to 3000 m a.s.l. having January DMTs close to 20 °C. The other values used were for Salta and Tucumán

Table 15.3. Summary statistics for the comparisons of January and July daily mean temperatures for each of the nine austral migrants studied here. *t* Student's *t* statistic; *df* degrees of freedom; NS not significant

Migrant	Number of climatic stations		Mean July DMT (°C)	Mean January DMT (°C)	*t*	*df*	*P*
	July	January					
Sterna maxima	4	2	14.0	19.8	1.72	4	NS
Colorhamphus parvirostris	6	6	9.3	13.6	2.54	10	0.029[a]
Serpophaga griseiceps	7	9	10.7	24.6	11.40	14	0.000[b]
Myiarchus s. swainsoni	21	10	24.6	25.0	0.43	29	NS
Myiarchus s. ferocior	7	14	23.8	25.1	1.37	19	NS
Elaenia strepera	10	3	23.9	21.0	1.54	11	NS
E. chiriquensis albivertex	17	33	24.1	25.7	2.73	48	0.008[b]
without Caetite 14°03′S, 878 m a.s.l.	16	33	24.5	25.7	2.29	47	0.03[a]
E. albiceps chilensis	24	28	16.9	21.8	2.86	50	0.006[b]
Sporophila lineola	11	22	24.7	26.1	1.81	31	NS

[a] 0.05>*P*>0.01.
[b] *P*<0.01.

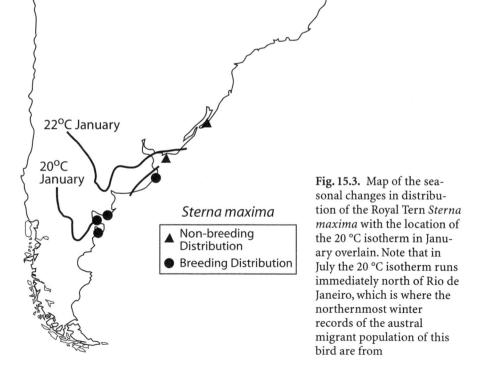

Fig. 15.3. Map of the seasonal changes in distribution of the Royal Tern *Sterna maxima* with the location of the 20 °C isotherm in January overlain. Note that in July the 20 °C isotherm runs immediately north of Rio de Janeiro, which is where the northernmost winter records of the austral migrant population of this bird are from

sionally reaching Rio de Janeiro. Some individuals remain in Uruguay and Brazil all year, but breeding has not been recorded in these countries. The 5.8 °C difference between the mean January DMT of the breeding distribution and the mean July DMT of the nonbreeding distribution (Fig. 15.2, Table 15.3) is not significant (but see Discussion). Figure 15.3 is a range map for the population and includes the location of the 20 °C isotherm in January. Note that the locality of the old breeding record from Buenos Aires province, the northernmost one shown on the map, is from an area with January DMTs just below 20 °C. In July, the 20 °C isotherm meets the coast at Rio de Janeiro, the extreme northern limit recorded for the population in the nonbreeding season (Hoffman 1975). Therefore, migrant individuals appear to have an upper limit to the January and July DMTs of both their breeding and nonbreeding distributions, respectively, of ca. 20 °C, whereas nonmigrant individuals spend the austral summer in areas with January DMTs greater than 20 °C.

15.3.2 *Colorhamphus parvirostris*

The breeding range extends from southern Chilean Tierra del Fuego to Concepción and the overlapping nonbreeding range extends approximately between Concepción and La Serena. Chesser and Marín (1994) showed that the altitudinal range is from sea level to 2000 m above sea level (a.s.l.) (*pace* Stotz et al. 1996). The 4.3 °C difference between the mean January DMT in the breeding distribution and the mean July DMT in the nonbreeding distribution is significant ($P=0.03$).

15.3.3 *Serpophaga griseiceps*

Straneck (1993) reviewed the taxonomic status of the *S. munda-subcristata* complex and concluded that the taxon *griseiceps*, which had been synonymized with *S. munda*, deserved species-level recognition, primarily on vocal criteria. As so construed, it ranges from Cochabamba, Bolivia south through Argentina to the center of La Pampa, the west of Córdoba and Chaco. Straneck indicated that it is migratory, at least partially, with records all year from Córdoba and San Luis, and nonbreeding season records in the provinces of Buenos Aires, Corrientes and Santa Fe. Tentatively accepting Straneck's (1993) taxonomic conclusions, we see that this species inhabits very different temperature regimes in its breeding and nonbreeding distributions (Fig. 15.2, Table 15.3). The January DMT in its breeding range averages 24.6 °C, whereas the July DMTs in its nonbreeding range average 10.7 °C, the difference being highly significant ($P=0.000$).

15.3.4 *Myiarchus swainsoni swainsoni* and *Myiarchus swainsoni ferocior*

The breeding range of *M. s. swainsoni* is from Sao Paulo, southeastern Brazil, to Uruguay and west to eastern Paraguay (Lanyon 1978). Most specimens from its nonbreeding range are from Venezuela with some records from Colombia, along the Amazon in Brazil and a few, presumably in passage, from western Paraguay and southern Bolivia. The breeding distribution of *M. s. ferocior* comprises most of northern Argentina from the provinces of Buenos Aires and La Pampa, and the Chaco of western Paraguay and south-eastern Bolivia; its recorded winter range is essentially western Amazonia in warm, humid lowlands from southern Bolivia through Peru to Limoncocha, Ecuador (Lanyon 1978). The altitudinal range of both subspecies does not exceed 1000 ma.s.l. (Stotz et al. 1996). There is no significant difference between the mean of the January DMT of the breeding distribution and that of the July DMT of the nonbreeding distribution for either subspecies; these means are all between 23.8 and 25.1 °C (Table 15.3). For further analyses, see Joseph and Stockwell (2000).

15.3.5 *Elaenia strepera*

The breeding range of *E. strepera* is in the Andes of northwestern Argentina in Jujuy and Tucumán and southern Bolivia in Tarija, and possibly also Santa Cruz. Marantz and Remsen (1991) showed that the core of the winter distribution is a relatively small part of northern Venezuela in Bolívar, Carabobo and Sucre with all other nonbreeding season specimens from Peru and Colombia and possibly central Bolivia likely being passage migrants. The altitudinal range of the breeding distribution is 800–2600 m a.s.l., whereas that of its core nonbreeding range in Venezuela is <1000 m a.s.l. (see Paynter 1982; Marantz and Remsen 1991).

The January DMTs in the breeding distribution and the July DMTs in the nonbreeding distribution are not significantly different, whether the latter includes only Venezuelan records or areas where the bird occurs in passage. In general, most values are greater than or equal to 20 °C, but higher altitude breeding localities extend the January range to below 20 °C. Figure 15.2 presents further details.

15.3.6 *Elaenia chiriquensis albivertex*

Marini and Cavalcanti (1990) suggested that this subspecies has a migrant population that breeds in central southern Brazil between 15° and 24°S and migrates to Amazonia and the Center-West of Brazil in the nonbreeding season. They found no records south of Brasilia between June and August. Presumably some birds are resident in Amazonia, as records there span the whole year. The 1.6 °C difference between the mean January DMT in the breeding range (25.7 °C) and the mean July DMT in the nonbreeding range (24.1 °C) is significant ($P=0.008$) and remains so when the relatively low July value of 18.9 °C from Caetite (878 m a.s.l.), which is questionably within the nonbreeding range, is excluded from the analysis ($P=0.03$).

15.3.7 *Sporophila lineola*

The main breeding range is southeastern Brazil, Paraguay and Argentina. In the nonbreeding season, records fall in two areas. One includes Minas Gerais, Goiás, Mato Grosso, Mato Grosso do Sul, Tocantins and Bolivia (see da Silva 1995), where there may be some sites at which the species is present all year and at which it breeds; the other is central and western Amazonia from which most records are between February and November, although the species is present all year. The 1.4 °C difference between the mean January DMT in the breeding distribution and the mean July DMT in the nonbreeding distribution is not significant (Fig. 15.2, Table 15.3).

15.3.8 *Elaenia albiceps chilensis*

Analyses for this migrant are based on data from recording stations within Marini and Cavalcanti's (1990) plot of records from the periods April to August and September to March, the latter including both the breeding season and times when migrants are in passage. Both the breeding and nonbreeding distributions of *E. a. chilensis* when so construed span localities with January and July DMTs, respectively, ranging from approximately 15 to 25 °C or higher, but the difference between mean January and July values is barely significant ($P=0.047$).

Despite our limited knowledge of austral migration, it is worth noting that *E. a. chilensis* is remarkable among austral migrants in having a breeding distribution from sea-level to 3300 m a.s.l. and a nonbreeding distribution from the high Andes to lowland Amazonia (for comparisons with other austral migrants see Ridgely and Tudor 1989, 1994). Figure 15.4 attempts to explore the data for *E. a. chilensis* in more detail based on the following considerations. First, Marini and Cavalcanti (1990) suggested that *E. a. chilensis* has two distinct migration routes. One is along the eastern slopes of the Andes to as far as central Colombia. The other is more diffuse, but runs counterclockwise with a northward passage from Argentina up and along the Atlantic seaboard of Brazil and into Amazonia with a return passage through Central Brazil and the basins of the Rio Paraguay and Rio Paraná. Marini and Cavalcanti (1990) noted that records in southern central Bolivia and central-northern Argentina span September–March, and Traylor (1982) gave the altitudinal range of breeding in this region as between 1500 and 3500 m a.s.l. Migrants in passage have been recorded in Peru between March and October, in southeastern Brazil between February and May and along the Rio Paraguay and in central Brazil mainly in October–December.

Second, the migratory and breeding status of *E. a. chilensis* in southern Bolivia is unclear. Traylor (1982) remarked that southern Bolivian specimens lack the distinctively modified wing tip of migrant populations to the south and so inferred that they were sedentary. Ridgely and Tudor (1994) noted the uncertainty surrounding southern Bolivian breeding records and the possibility that northern Argentinean records may be of birds in passage. Zimmer (1941) remarked that the migration of *E. a. chilensis* northward through Peru occurs only along, or over, the chain of the Andes. Accordingly, I excluded January DMTs from southern Bolivian climatic stations between 1500 and 3500 m a.s.l. on the grounds that I wished to maximize the likelihood of focusing on migrant populations. I tentatively "corrected" for the spatial and temporal dynamics of migration by dissecting the nonbreeding distribution to match localities with the times of the year when the birds were suggested by Marini and Cavalcanti (1990) as being present in them. For the Brazilian route, I examined April DMTs from Rio de Janeiro and Campinas, July DMTs from Belém, Olinda and Santarém, and October DMTs from Belo Horizonte, Formosa, Corrientes and Posadas. For the Andean route, I included April, July and October DMTs from Huancayo, Cuzco, Cajamarca and Apolo.

Figure 15.4 shows the results of this analysis. Along the Brazilian route, the range of DMTs now varies around ca. 22 °C, whereas along the Andean route the means and ranges for April, July and October are not surprisingly shifted towards lower values and mostly do not exceed 20 °C. Month by month differ-

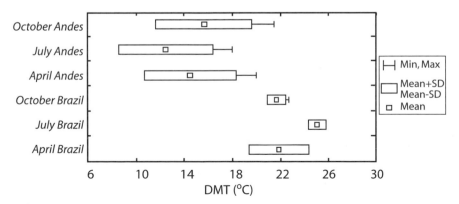

Fig. 15.4. Daily mean temperatures along the Andean and Brazilian migration routes of *Elaenia albiceps chilensis* tentatively "corrected" for the spatial and temporal dynamics of migration

ences between the Brazilian and Andean routes then become pronounced, but statistical comparisons are limited because of the small number of climate stations available for each pair-wise comparison (e.g., April: Andean vs. April: Brazilian, etc.). It would be valuable to repeat this analysis when further data on the seasonal changes of this bird's distribution are available.

15.4 Discussion

The central aim of this chapter was to explore whether a climatic approach to describing distribution patterns might provide the basis for predicting seasonal changes in distributions of austral migrants generally, but especially those in fragmented environments. The approach was to test the null hypothesis that there is no difference in the climatic characteristics of an austral migrant's breeding and nonbreeding distributions (for simplicity including passage distributions) in the times of the year when the birds are in those distributions. As a first and admittedly crude approximation, climatic characteristics have been estimated with January and July DMTs in the breeding and nonbreeding distributions, respectively. It appears that seasonal changes in distribution can indeed be usefully described on a case-by-case basis for each migrant. Three main broader findings emerged.

First, 20 °C marks either an approximate upper or lower limit to the July DMTs of the nonbreeding distributions of each migrant with the exception of *E. a. chilensis* when all its populations are considered together. Second, for the study species or their migrant populations that are temperate-tropical migrants, i.e., with nonbreeding distributions in the warm, humid lowlands and again with the exception of *E. a. chilensis*, the mean January and July DMTs of the breeding and nonbreeding distributions have closely similar values that are generally greater than ca. 20 °C. The null hypothesis is either not rejected (*E. strepera, M. s. swainsoni, M. s. ferocior, Sporophila lineola*) or can be rejected on the basis of only a small shift in DMT well above 20 °C (*E. c. albivertex*). Third, for those migrants with nonbreeding distributions in the cooler parts of the continent, the January and July DMTs of the breeding and nonbreeding distributions are relatively more dissimilar than are those of the previous group of migrants; July DMTs in their nonbreeding distributions are nonetheless generally less than ca. 20 °C. In these cases, the null hypothesis is either strongly rejected (*C. parvirostris, Serpophaga griseiceps*) or, in the case of *Sterna maxima*, not rejected, possibly due to few breeding sites restricting the sample size of climatic stations from which data can be used.

E. a. chilensis appears as a clear exception in that July DMTs in its nonbreeding distribution include values well above and below 20 °C. Similarly, the January DMTs in its breeding distribution show a comparable range. Even given our limited knowledge of austral migrants, its breeding distribution's great altitudinal range, its nonbreeding distribution spanning Andean and Amazonian areas and its apparent dual migration routes are all extraordinary among austral migrants. These factors prompt a more detailed analysis of the bird's migration. Interestingly, therefore, when one attempts to correct for the spatial and temporal dynamics of the bird's migration, a pattern emerges that is more consistent with the findings in the other species. Thus, the birds migrating along the Brazilian route fall into the category of migrants breeding and migrating generally above ca. 20 °C. In contrast, those migrating along the Andean route fall in the category of birds showing a more marked shift in the DMTs of their breeding and nonbreeding distributions, with nonbreeding areas generally having DMTs less than 20 °C in the months when the birds are there.

It is worth relating this last finding to other exceptional points concerning *E. a. chilensis* noted by earlier authors. Traylor (1982) had noted that in its Argentinean breeding range it overlaps with the small-billed elaenia *E. parvirostris* without hybridization; in southern Bolivia, where he argued it is not migratory, the two overlap with intergradation. He tentatively retained *E. parvirostris* and *E. a. chilensis* as separate species and suggested that they are "good sympatric species that, for some reason, hybridize in a limited part of their range." An alternative hypothesis to explain the unusually wide breeding and nonbreeding distributions, the dual migration routes which, as we have seen, may be tracking different temperatures, and Traylor's (1982) various observations, would be that *E. a. chilensis* is a composite of two or more, at present, unrecognized cryptic taxa. The migrants involved could have breeding ranges either overlapping in northern Argentina or altitudinally separated, but with distinct nonbreeding ranges, one along the Andes and the other mostly in warm, humid lowland zones of Brazil. Further studies of the details of its migrations are clearly needed.

Overall, these findings suggest quantitative refinements to the long-held view (see Introduction) that there are two major climatically defined migration patterns among austral migrants. These two patterns are probably distinguishable according to whether July DMTs in the nonbreeding distributions reach an upper or lower limit of 20 °C. The findings say nothing, however, of the relative frequencies of these migration patterns, or those of intermediate patterns, among all austral migrants. It is worth stressing that those austral migrants studied here, with nonbreeding distributions in warm, humid lowlands, including *E. a. chilensis* along its Brazilian migration route, are tracking relatively constant temperatures during their migrations, at least between breeding and core, nonbreeding distributions. The degree to which they do this can be impressive. *M. s. swainsoni*, for example, migrates some 2500–3000 km, but shows a nonsignificant shift in mean January and July DMTs of only 0.4 °C. Mirroring these findings, I have elsewhere (Joseph 1996) tentatively suggested that a separation in winter between two broad groups of austral migrant passerine birds either side of ca. 20 °C is indicated by the distribution maps in Ridgely and Tudor (1989, 1994).

In contrast to the purely latitudinal approach of Hayes (1995), the simple climatic patterns noted among the nine migrants studied here, and in Joseph (1996) indicate that seasonal changes in distributions of austral migrants can be *described* in climatic terms on a case-by-case basis. They further hint at the utility of a more sophisticated multivariate climatic approach in *predicting* and modeling seasonal distribution patterns (e.g., Joseph and Stockwell 2000). For some species, prediction of the nonbreeding distribution, for example, may be as simple as defining the climatic envelope of the relatively well known breeding distribution, and asking when and where else on the continent that climatic envelope occurs. For other species, an alternative procedure might be to define core breeding and nonbreeding distributions and then direct field searches to areas with similar climatic envelopes at appropriate times of the year. This could be a powerful tool for predicting distributions of migrants with poorly known nonbreeding ranges and also nonmigrants with, at present, patchily known distributions, especially in regions where habitats have been heavily fragmented. It would help channel scarce resources into well-planned field searches for such species. Marantz and Remsen (1991) drew attention to this problem, noting that for many austral migrants with nonbreeding ranges in northern South America there is a need to distinguish wintering destination(s) and areas where the birds are passage migrants. Discussions of how data on suites of climatic parameters, not just DMTs, from all localities at which a species has been recorded, rather than from climatic stations within the range, can be assembled to assess the likelihood of that species occurring at any other given locality are in Busby (1991), Walker (1990), Walker and Cocks (1991) and McMahon et al. (1996). Nix (1976) also discussed the relevance of environmental parameters to understanding migration patterns in birds. I am currently exploring a refinement of the approach used in this paper using an algorithm with which to climatically model distribution patterns (Joseph and Stockwell 2000).

In conclusion, this study suggests that distributions of austral migrants can be *described* in climatic terms, that this in turn shows the potential for more sophisticated multivariate climatic approaches in *predicting* total distributions, and that this approach to describing and predicting distributions may be of use in the study of migrant birds in some anthropogenically altered environments such as those that have become fragmented.

Acknowledgement. I thank P. Marquet and G. Bradshaw for inviting me to attend the IAI Habitat Fragmentation Workshop at which this work was presented, and its participants for their comments. J. Bates, S. Degnan, J. Rappole and an anonymous reviewer applied strong selection pressure to this study's ancestor and so helped select for the traits, which after some random drift finally coalesced into the focus presented here. E. Lessa provided conditions in which to work, assisted with statistical analyses and commented on the work as it evolved. For help with literature, comments and critical reading of the manuscript I thank D. Agro, J. Bates, L. Belasco, J.P. Cuello, T. Chesser, S. Degnan, F. Hayes, J. Lyons, J. García-Moreno, J.F. Pacheco, J. Reid, R. Slade, M.F. Smith and F.G. Stiles. In particular, T. Chesser provided a most thoughtful critique of the submitted manuscript that improved it considerably. Financial support was provided by the Third World Academy of Sciences (TWAS), the Programa de Desarollo de las Ciencias Basicas (PEDECIBA), the Consejo Nacional de Investigacion Científica y Técnologica (CONICYT), and CSIC, Montevideo.

References

Araya MB, Millie HG (1991) Guia de Campo de las Aves de Chile. Editorial Universitaria, Santiago

Belton W (1994) Aves do Rio Grande do Sul. Editora Unisinos, Sao Leopoldo

Berthold P (1994) Bird migration. A general survey. Oxford University Press, Oxford

Busby JR (1991) BIOCLIM – a bioclimatic analysis and prediction system. In: Margules CR, Austin MP (eds) Nature conservation: cost effective biological surveys and data analysis. CSIRO, Canberra, Australia, pp 64–68

Chesser RT (1994) Migration in South America: an overview of the austral system. Bird Conserv Int 4:91–107

Chesser RT (1998) Further perspectives on the breeding distribution of migratory birds: South American austral migrant flycatchers. J Anim Ecol 67:69–77

Chesser RT, Marín AM (1994) Seasonal distribution and natural history of the Patagonian tyrant (*Colorhamphus parvirostris*). Wilson Bull 106:649–667

da Silva JMC (1995) Seasonal distribution of the Lined Seedeater *Sporophila lineola*. Bull Br Ornithol Club 115:14–21

Escalante R (1968) Notes on the Royal Tern in Uruguay. Condor 70:243–247

De Graaf RM, Rappole JH (1995) Neotropical migratory birds. Natural history, distribution and population change. Comstock, Ithaca, New York

Escalante R (1985) Taxonomy and conservation of austral-breeding Royal Terns. Ornithol Monogr 36:935–942

Gore MEJ, Gepp ARM (1978) Las aves del Uruguay. Mosca Hermanos, Montevideo

Griffiths JF (1976) Applied climatology. An introduction, 2nd edn. Oxford University Press, Oxford

Hagan JM, Johnston DW (eds) (1992) Ecology and conservation of Neotropical migrant landbirds. Smithsonian Institution Press, Washington, DC

Hayes FE (1995) Definitions for migrant birds: what is a Neotropical migrant? Auk 112:521–523

Hayes FE, Scharf PA, Ridgely RS (1994) Austral bird migrants in Paraguay. Condor 96:83–97

Hoffmann JAJ (1975) Climatic atlas of South America I. Maps of mean temperature and precipitation. WMO, Budapest, Hungary

Joseph L (1996) Preliminary climatic overview of migration patterns in South American austral migrant passerines. Ecotropica 2:185–193

Joseph L, Stockwell D (2000) Temperature-based models of the migration of Swainson's flycatcher *Myiarchus swainsoni* across South America: a new use for museum specimens of migratory birds. Proc Natl Acad Nat Sci Philadelphia 150:293–300

Keast A, Morton E (eds) (1980) Migrant birds in the Neotropics. Smithsonian Institution Press, Washington, DC

Lanyon WE (1978) Revision of the *Myiarchus* flycatchers of South America. Bull Am Mus Nat Hist 161:431–627

Marantz CA, Remsen JV (1991) Seasonal distribution of the Slaty Elaenia, a little-known austral migrant of South America. J Field Ornithol 62:162–172

Marini MA, Cavalcanti RB (1990) Migrações de *Elaenia albiceps chilensis* e *Elaenia chiriquensis albivertex* (Aves: Tyrannidae). Bol Mus Para Emilio Goeldi 6:59–67

Martyn D (1992) Climates of the world. Developments in atmospheric science 18. Elsevier, Amsterdam

McMahon JP, Hutchinson MF, Nix HA, Ord KD (1996) ANUCLIM version 1 user's guide. Centre for Research in Environmental Studies, Australian National University, Canberra. Published electronically at http://cres.anu.edu.au/software/anuclim.html

Narosky T, di Giacomo A (1993) Las aves de la Provincia de Buenos Aires: distribución y estatus. Asociacion Ornitologica del Plata, Buenos Aires

Narosky T, Yzurieta D (1993) Birds of Argentina and Uruguay. A field guide. Second English edition. Asociacion Ornitologica del Plata and Vazquez Mazzini, Buenos Aires

Newton I, Dale L (1996a) Relationship between migration and latitude among west European birds. J Anim Ecol 65:137–146

Newton I, Dale L (1996b) Bird migration at different latitudes in eastern North America. Auk 113:626–635

Nix HA (1976) Environmental control of breeding, post-breeding dispersal and migration of birds in the Australian region. In: Frith HJ, Calaby JH (eds) Proceedings of the XVI International Ornithological Congress 1974, Australian Academy of Science, Canberra, Australia, pp 272–305

Paynter RA (1982) Ornithological gazetteer of Venezuela. Museum of Comparative Zoology, Harvard University, Cambridge, MA

Prohaska F (1976) The climate of Argentina, Paraguay and Uruguay. In: Schwerdtfeger W (ed) World survey of climatology, volume 12. Climates of Central and South America. Elsevier, Amsterdam, pp 13–112

Ridgely RS, Tudor G (1989) The birds of South America. Volume I. The oscine passerines. Oxford University Press, Oxford

Ridgely RS, Tudor G (1994) The birds of South America. Volume II. The sub-oscine passerines. Oxford University Press, Oxford

Romanella MMN (1993) Aves de la Provincia de San Luis. Lista y distribución. Museo Privade de Ciencias Naturales, San Luis

Root T (1988) Environmental factors associated with avian distributional patterns. J Biogeogr 15:489–505

Scherer-Neto P, Straube FC (1995) Aves do Paraná: historia, lista anotada e bibliografía. The Authors, Curitiba

Schwerdtfeger W (ed) (1976) Climates of Central and South America. World survey of climatology, volume 12. Elsevier, Amsterdam

StatSoft (1994) Statistica for Windows, Version 5.0. StatSoft, Tulsa, Oklahoma

Stotz D, Fitzpatrick J, Parker T, Moskovits D (1996) Neotropical birds–ecology and conservation. University of Chicago Press, Chicago

Straneck R (1993) Aportes para la unificación de *Serpophaga subcristata* y *Serpophaga munda*, y la revalidación de *Serpophaga griseiceps*. (Aves: Tyrannnidae). Rev Mus Argent Cienc Nat Bernardino Rivadivia Zool 16:51–63

Traylor MA (1982) Notes on tyrant flycatchers (Aves: Tyrannidae). Fieldiana Zool N S 13(1338):1–22

Tuomisto H, Ruokolainen K, Kalliola R et al. (1995) Dissecting Amazonian biodiversity. Nature 269:63–66

Walker PA (1990) Modeling wildlife distributions using a geographic information system: kangaroos in relation to climate. J Biogeogr 17:279–289

Walker PA, Cocks KD (1991) HABITAT: a procedure for modeling a disjoint environ-
    mental envelope for a plant or animal species. Global Ecol Biogeogr Lett 1:108–118
Wetmore A (1926) Observations on the birds of Argentina, Paraguay, Uruguay and Chile.
    Bull US Natl Mus 133:1–448
Zimmer JT (1938) Notes on migrations of South American birds. Auk 55:405–410
Zimmer JT (1941) Studies of Peruvian birds. XXXVI. The genera *Elaenia* and *Myiopagis*.
    American Museum Novitates 1108, pp 1–23

16 Habitat Heterogeneity on a Forest–Savanna Ecotone in Noel Kempff Mercado National Park (Santa Cruz, Bolivia): Implications for the Long-Term Conservation of Biodiversity in a Changing Climate

T.J. Killeen, T.M. Siles, T. Grimwood, L.L. Tieszen, M.K. Steininger, C.J. Tucker, S. Panfil

16.1 Introduction

Noel Kempff National Park is a biological reserve in northeast Santa Cruz, Bolivia; the park is part of a regional setting that includes extractive forest reserves, cattle ranches, indigenous communities, and colonization zones. Future development and migration in the surrounding areas will have an effect on the conservation of the flora and fauna of this important protected area. The region incorporates five distinct ecosystems: humid upland forest, inundated and riparian forest, semi-deciduous and deciduous forest, upland savanna, and savanna wetland. The biodiversity of the park has been extensively studied with respect to species richness of individual habitats and the importance of habitat heterogeneity has been identified as the key factor in contributing to the overall levels of biodiversity (Killeen and Schulenberg 1998). There is no single ecosystem in Noel Kempff National Park that is "super diverse"; nonetheless, the total regional diversity (γ·diversity) is very high because all five ecosystems are distinct and are well represented on the landscape (Table 16.1).

Biodiversity typically is partitioned into components that relate species richness to spatial scale (Scott and Jennings 1998). Species richness is usually reported for specific localities (α diversity) or described by comparing species composition among habitats (β diversity). Most studies of biodiversity in the tropics have focused on measurements of α diversity by measuring species richness at a specific locality as a function of some standard unit of measurement (plot size, trap hours, etc.). There is an extensive literature documenting α diversity and relating it to a variety of environmental variables (i.e., Gentry 1988; Dallmeier and Comiskey 1997). In contrast, very little attention has been focused on habitat diversity in tropical ecosystems and the trop-

Ecological Studies, Vol. 162
G.A. Bradshaw and P.A. Marquet (Eds.)
How Landscapes Change
© Springer-Verlag Berlin Heidelberg 2003

Table 16.1. The total species richness of the major groups of organisms known for Noel Kempff National Park. (Killeen and Schulenberg 1998)

Group	Documented	Estimated	Endangered
Vascular plants	2705	4000	?
Mammals	124	150	33
Birds	597	650	13
Reptiles	75	?	14
Amphibians	62	?	?
Fish	125	250	?
Scarab beetles	97	?	?

ical forest is often portrayed to be species rich, but homogenous in community structure over large areas. Recent research has shown that the distribution of individual species is often related to edaphic factors and that habitat heterogeneity is an important component of tropical forest formations (Clark 1998; Tuomisto 1998). Likewise, there is increasing evidence of the importance of disturbance events in maintaining habitat heterogeneity in the Amazon (Lamotte 1990; Nelson 1994; Nelson and Irmão 1998).

The importance of β diversity in Noel Kempff National Park and the contribution it makes to γ diversity, is demonstrated by the similarity matrix calculated for 25 different forest plots established in four major types of forest formations (Table 16.2). A total of 838 species were found in all of the plots, ranging from a low of 31 species to a high of 123 species (≥10 cm dbh). Only those plots that were adjacent to each other showed similarity indices larger than 50 %, with values ranging from 10–35 % similarity for plots with similar edaphic characteristics. The floristic similarity in contrasting forest types was even more pronounced with typically less than 1–5 % similarity; only the humid forest category showing some 10–20 % similarity to some sites in both the dry and flooded forest categories. The importance of β diversity is consistent among groups of organisms, and species turnover for mammals, birds, and scarab beetles has been shown to be similar to that of vegetation (Killeen and Schulenberg 1998).

The geology of northeastern Santa Cruz is well documented (Litherland and Power 1991) and is remarkable for having a wide variety of contrasting landscapes with easily identifiable rock formations. This has allowed us to relate geomorphology to biological communities, and to facilitate comparison among habitats (Killeen 1998). Geomorphological heterogeneity has been shown to be a good predictor of species diversity, largely because differences in edaphic characteristics translate into differences in plant community structure and composition (Burnett et al. 1998; Nichols et al. 1998). The different vegetation types present in Noel Kempff National Park are strikingly

Table 16.2. The floristic similarity for 25 1-ha forest plots established in Noel Kempff Mercado National Park. The values below the diagonal are Sorensen's Index calculated on presence absence data, while above the diagonal are Sorensen's Index calculated on abundance values for individual species; values greater than 0.25 are boldface, while those greater than 0.5 are highlighted by a box

Plot	Cerrado transition				Dry forest					Humid forest									Inundated forest						
	EN-1	EN-2	LT-1	LT-2	AC-1	AC-2	CP-1	CP-2	SR-1	H1-1	H1-2	H2-1	H2-2	LF-1	LF-2	LG-b	CH-1	CH-2	CR-1	CR-2	LT-3	MV-1	LO-1	LO-2	FO-b
EN-1	–	**0.79**	0.08	0.07	0.02	0.02	–	0.02	0.02	0.02	0.03	0.01	0.01	0.01	0.01	0.04	0.02	0.04	0.01	0.02	0.02	0.02	0.02	0.02	0.02
EN-2	**0.64**	–	0.08	0.07	–	–	–	–	0.04	0.04	0.04	0.01	0.01	0.03	0.03	0.03	0.04	0.06	0.01	0.03	0.03	0.02	0.03	0.01	0.02
LT-1	0.15	0.08	–	0.55	0.07	0.06	–	–	–	0.04	0.04	–	0.01	–	–	0.01	0.01	0.01	0.01	–	0.03	0.03	0.02	0.02	0.02
LT-2	0.09	0.07	**0.60**	–	0.06	0.02	–	–	–	–	–	–	0.01	–	–	0.01	0.01	0.01	–	–	0.01	0.02	0.03	0.02	–
AC-1	0.02	–	0.07	0.06	–	**0.61**	0.34	0.26	0.12	0.03	0.02	0.07	0.09	0.02	0.01	0.02	0.05	0.08	0.01	0.01	0.08	0.09	0.02	0.02	0.02
AC-2	0.02	–	0.06	0.02	**0.64**	–	0.33	0.33	0.12	0.03	0.02	0.08	0.08	0.01	0.01	0.01	0.04	0.09	0.01	0.01	0.10	0.09	0.03	0.01	0.01
CP-1	–	–	–	–	**0.45**	0.44	–	0.35	0.05	0.01	–	0.05	0.05	0.01	0.01	0.02	0.05	0.06	0.01	0.01	–	0.05	0.02	0.01	–
CP-2	0.02	–	–	–	0.34	0.33	**0.50**	–	0.06	0.03	0.03	0.11	0.11	0.01	0.01	0.04	0.04	0.10	0.01	0.01	0.02	0.07	0.03	0.01	–
SR-1	0.07	0.07	–	–	0.27	0.28	0.18	0.20	–	0.04	0.04	0.03	0.02	0.02	0.02	0.01	0.02	0.05	0.01	0.01	0.01	0.05	0.02	0.02	–
H1-1	0.09	0.09	0.08	0.04	0.07	0.08	0.05	0.07	0.09	–	**0.51**	0.08	0.13	**0.26**	0.23	0.17	0.16	0.15	0.01	0.03	0.06	0.04	0.09	0.13	0.01
H1-2	0.07	0.05	0.04	0.01	0.09	0.10	0.04	0.06	0.09	**0.61**	–	0.15	0.23	**0.30**	**0.31**	0.23	0.12	0.12	0.01	0.03	0.12	0.04	0.13	0.19	0.01
H2-1	0.02	0.03	–	–	0.15	0.15	0.11	0.20	0.13	0.18	0.19	–	**0.61**	0.09	0.08	0.15	0.08	0.12	0.02	0.01	0.01	0.06	0.07	0.06	–
H2-2	0.01	0.01	–	–	0.15	0.13	0.13	0.17	0.10	0.19	0.20	**0.54**	–	0.15	0.15	0.21	0.11	0.14	0.01	0.02	0.08	0.05	0.08	0.08	0.01
LF-1	0.06	0.08	0.04	–	0.06	0.04	0.04	0.07	0.11	**0.35**	**0.33**	0.14	0.17	–	**0.70**	**0.34**	**0.29**	0.11	0.01	0.02	0.10	0.04	0.07	0.15	–
LF-2	0.06	0.07	0.02	–	0.05	0.03	0.03	0.05	0.10	**0.28**	**0.27**	0.12	0.16	**0.64**	–	**0.29**	0.24	0.10	0.01	0.03	0.10	0.03	0.07	0.15	–
LG-b	0.13	0.10	0.01	0.02	0.08	0.05	0.04	0.05	0.05	**0.25**	0.22	0.26	**0.26**	**0.35**	**0.28**	–	0.11	0.10	0.01	0.02	0.10	0.02	0.06	0.10	0.01
CH-1	0.05	0.09	0.06	0.03	0.14	0.16	0.11	0.16	0.12	**0.28**	**0.26**	0.22	0.23	**0.28**	**0.28**	0.20	–	0.36	0.02	0.01	0.01	0.05	0.04	0.06	0.01
CH-2	0.10	0.10	0.06	0.04	0.17	0.19	0.14	0.18	0.15	**0.25**	0.23	0.25	**0.26**	0.20	0.23	0.19	**0.51**	–	0.02	0.02	0.02	0.09	0.04	0.05	0.02
CR-1	0.05	0.05	0.03	0.03	0.04	0.06	0.04	0.04	0.04	0.04	0.06	0.05	0.05	0.07	0.08	0.04	0.01	0.06	–	**0.54**	0.04	0.04	0.02	0.03	0.01
CR-2	0.11	0.09	0.01	–	0.03	0.03	0.03	0.03	0.05	0.09	0.10	0.09	0.07	0.11	0.13	0.07	0.04	0.10	**0.55**	–	0.05	0.11	0.10	0.05	0.04
LT-3	0.14	0.09	0.08	0.09	0.05	0.05	0.19	0.19	0.19	0.14	0.13	0.14	0.13	0.15	0.14	0.15	0.06	0.08	0.07	0.10	–	0.04	0.04	0.07	0.02
MV-1	0.07	0.07	0.08	0.08	0.01	0.01	0.21	0.13	0.07	0.13	0.14	0.12	0.16	0.12	0.05	0.14	0.19	0.19	0.07	0.18	0.17	–	0.10	0.05	0.02
LO-1	0.06	0.06	0.03	0.03	0.23	0.23	0.13	0.21	0.17	0.17	0.21	0.14	0.14	0.18	0.14	0.15	0.15	0.15	0.07	0.15	0.14	0.23	–	**0.59**	0.05
LO-2	0.08	0.08	0.02	0.02	0.11	0.11	0.06	0.08	0.11	0.19	0.19	0.11	0.11	0.18	0.16	0.14	0.14	0.14	0.09	0.12	0.16	0.15	**0.52**	–	0.01
FO-b	0.18	0.17	0.09	0.09	0.06	0.07	0.03	0.02	0.03	0.05	0.05	0.02	0.02	0.03	0.02	0.01	0.04	0.04	0.04	0.06	0.11	0.05	0.04	0.04	–

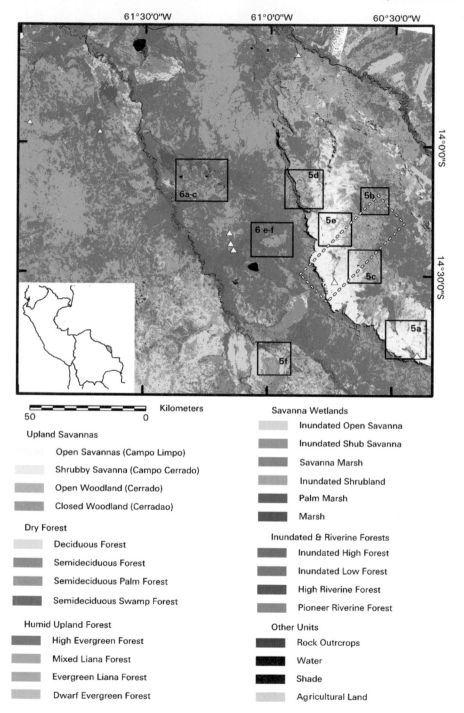

61°30'0"W   61°0'0"W   60°30'0"W

14°0'0"S

14°30'0"S

Kilometers
50   0

Upland Savannas

    Open Savannas (Campo Limpo)

    Shrubby Savanna (Campo Cerrado)

    Open Woodland (Cerrado)

    Closed Woodland (Cerradao)

Dry Forest

    Deciduous Forest

    Semideciduous Forest

    Semideciduous Palm Forest

    Semideciduous Swamp Forest

Humid Upland Forest

    High Evergreen Forest

    Mixed Liana Forest

    Evergreen Liana Forest

    Dwarf Evergreen Forest

Savanna Wetlands

    Inundated Open Savanna

    Inundated Shub Savanna

    Savanna Marsh

    Inundated Shrubland

    Palm Marsh

    Marsh

Inundated & Riverine Forests

    Inundated High Forest

    Inundated Low Forest

    High Riverine Forest

    Pioneer Riverine Forest

Other Units

    Rock Outrcrops

    Water

    Shade

    Agricultural Land

Fig. 16.1. Vegetation map of Noel Kempff Park and surrounding region; the *rectangles* correspond to the areas shown in greater detail in Figs. 16.5 and 16.6; the *rectangle* surrounded by *dashed lines* is the ecotone transect shown in Fig. 16.7 and Table 16.6; *white triangles* are soil pits shown in Tables 16.4 and 16.5. The vegetation map is modified from Killeen et al. (1998) and was produced by a supervised classification using 1996 Landsat TM images for path 230 row 70

←——

different in their structure and physiognomy, facilitating their study using remote sensing technology (Killeen et al. 1998). A vegetation map was compiled from Landsat TM imagery using a supervised classification based on extensive ground and aerial verification (Fig. 16.1). An important attribute of habitat heterogeneity is its spatial arrangement on the landscape and the vegetation map has been analyzed for spatial complexity. The results show that those landscapes dominated by either forest or savanna had the smallest number of patches, the smallest amount of edge, and simpler patch geometry when compared to landscapes dominated by ecotonal associations intermediate to open savanna or high forest. Transitional landscapes and landscapes that incorporate wetland habitats not only have greater levels of habitat diversity, but also are more spatially complex landscapes (Fig. 16.2). However, these landscapes have very little core area and offer a highly fragmented landscape that may represent a less than optimum environment for many forest dwelling organisms.

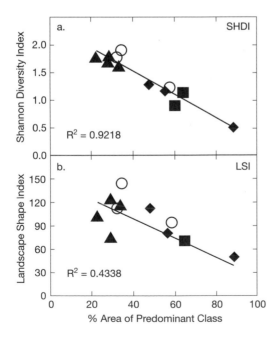

Fig. 16.2. The comparison of 12 landscapes subsets using the spatial statistical program Fragstats. **a** Shannon Diversity Index (SHDI), **b** Landscape Shape Index (LSI). The *x*-axis represents the percent cover of the most abundant class: *squares* are forest-dominated landscapes, *diamonds* are grassland-dominated landscapes, *triangles* are transitional landscapes where neither forest nor savanna predominate, and *circles* are lowland landscapes which contain appreciable areas of wetland vegetation

It is our working hypothesis that current high habitat diversity is the result of past climate change, which caused repeated expansions and contractions of the humid forest and dry forest-savanna biomes. As a result, today's landscape consists of patches of forest and savanna that are often situated in a matrix of a contrasting vegetation type. The resulting habitat mosaic is both species-rich and spatially complex. In this paper, we describe how the extremely high level of habitat heterogeneity in the region is related to the interaction of climatic stress, geomorphology, and disturbance as caused by both fire and flooding. In selected localities, we make a hypothesis with regard to recent movements of the forest-savanna boundary; these hypotheses are based upon the spatial characteristics of the forest savanna mosaic or on direct change as observed in a temporal study of satellite imagery. Finally, we use stable carbon isotope ratios of soil samples to document past vegetation types at some of these localities.

Climate change may have a greater than normal impact on biodiversity within the park due to its position across a climatic tension zone. Climate change in the Amazon basin and the subsequent migration of forest and savanna biomes has been the subject of debate over the past three decades. Haffer (1969) and Van der Hammen (1972) hypothesized that drier conditions during the Pleistocene caused savannas to expand at the expense of rain forest vegetation. This position has been challenged by Collinvaux (1996), who argues that lower temperatures were the primary factor governing plant distributions. Recent evidence for regional drying during the Pleistocene has been produced by paleoecological studies in Rondônia, Brazil, where pollen cores from Katira show clear evidence of an increase in the abundance of grass pollen between 22,000 and 11,000 years B.P. (Absey et al. 1991); a trend that is supported by soil carbon isotope studies conducted in the same region (Pessenda et al. 1998). Ongoing studies within Noel Kempff Mercado National Park likewise point to a recent expansion of Amazonian vegetation, with a sudden increase in Amazonian taxa occurring only between 3000 and 500 years B.P. (Mayle et al. 2000).

Conservation of the biodiversity in the region will depend upon the ability to organize and manage latitudinal corridors, allowing species to shift their distribution in response to climate change. The importance of habitat heterogeneity in the region dictates the need to develop a long-term management philosophy that allows the disturbance phenomena that are responsible for the formation of the broad range of habitat types and the intermediate successional stages that are characteristic of the region.

16.2 Climatic Stress

The abiotic factor underlying the development of the ecosystem mosaic in Noel Kempff National Park is climate, and the key attribute of the regional climate is the pronounced seasonal fluctuations in precipitation, which are the product of two climatic phenomena that influence the weather at these latitudes on the continent. The first phenomenon is the Intercontinental Convergence Zone (ITCZ), a belt of warm moist air over the equator that shifts north during the austral winter (dry season) and south during the austral summer (wet season). The ITCZ brings warm moist winds that originate over the Atlantic Ocean and the Central Amazon. In contrast, the dry season is characterized by cool dry winds that originate due to the interaction of two anticyclonic systems in the South Atlantic and South Pacific (Ronchail 1989). The incursion of these cool, and generally dry, southern air masses into the Bolivian lowlands may occur throughout the year; they are most frequent and penetrate farther north during the austral winter (dry season).

Although detailed climate data are not available for the park, extrapolation based on nearby weather stations does show that the region lies astride a sharp gradient between the tropics and subtropics (Killeen and Schulenberg 1998). The mean annual temperature is between 25 and 26 °C; cool fronts during June may lower temperatures to 10 °C for short periods, but the mean daily temperature is still a relatively warm 18–20 °C during the coolest month of the year (July). The mean annual precipitation is between 1400 and 1600 mm; the dry season lasts from June to September, when mean monthly precipitation is less than 30 mm, while the rainy season lasts from December to April, when mean monthly precipitation is in excess of 200 mm.

Perhaps just as important are the very large differences that can be observed in yearly totals when the available climatic data are evaluated over several decades. The maximum and minimum annual totals of precipitation in the park probably vary from >2200 mm for extreme wet years to <700 mm for drought years (Fig. 16.3). The distribution of this precipitation throughout the year is likewise extremely unpredictable; this is important for its potential impact on disturbance events. An extreme rainy season would lead to maximum levels of inundation on landscapes, which would cause mortality of upland species intolerant of flooding. In contrast, abnormally high rainfall in the dry season could lead to fire suppression and the build up of combustible fuel in savanna habitats that would result in the eventual occurrence of fires of greater than average intensity. These climatic extremes set the stage for the large-scale disturbance events that contribute additional habitat variability.

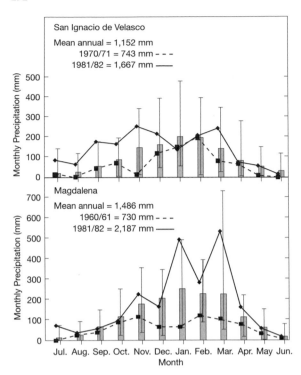

Fig. 16.3. Climatic patterns at the two closest weather stations in Bolivia; the *histogram* is the mean monthly precipitation, with *error bars* showing minimum and maximum values; the *solid* and *dashed lines* show two atypical years

16.3 Geomorphology

The relationship between vegetation and geomorphology is one of the most obvious aspects of the natural history of Noel Kempff National Park. Seasonal climatic stress, or the lack of it, is the result of soil conditions associated with specific landforms and these interactions lead to the development of different plant communities. Figure 16.4 shows two hypothetical transects across the park: one across the northern sector of the plateau and the other across the southern sector. Each shows the position of the major geological formations, their relationship to the topography and the vegetation types that predominate on each landform. Differences in geology and topography result in differences in edaphic characteristics such as soil texture, fertility, and moisture regime (Table 16.3).

The most conspicuous physical feature in the park is a large sandstone table mountain known as the Huanchaca Plateau (Figs. 16.1 and 16.4). The plateau is a broadly concave geological "unconformity" that is composed of Precambrian sandstone and quartzite rocks that were deposited between 900 and 1000 million years ago. The center of the plateau is a shallow basin composed of Cretaceous sandstone, with quartzite rocks situated along the outer escarpments that reach a maximum of 900 m in elevation. Scattered across the

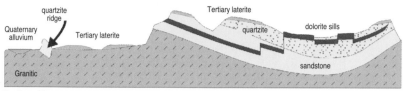

Northern Transect

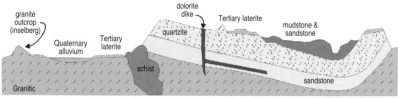

Southern Transect

Lowland Landscape

Quaternary Alluvium
  inundated savanna, palm swamp
  inundated & riverine forests
Tertiary Laterite
  high forest, liana forest
Precambrian Schist
  semideciduous forest, bald outcrops
Precambrian Quartzite
  liana forest
Precambrian Granitoid
  liana forest, semideciduous forest, bald outcrops

Meseta Landscape

Tertiary Laterite
  dwarf evergreen forest, cerradão, cerrado
Cretaceous Mudstone & Sandstone
  evergreen high forest
Precambrian Dolorite Dikes & Sills
  evergreen high forest, liana forest
Precambrian Sandstone & Conglomerate
  cerrado, cerradão, gallery forest
Precambrian Quartzite
  campo rupestre, campo cerrado,
  cerradão, liana forest

Fig. 16.4. Two geological profiles of Noel Kempff National Park; the text lists the predominant vegetation type for each of the geological formations

plateau are numerous dikes and sills where volcanic rocks (dolorite) have been intruded into the sedimentary rocks. To the west of the Huanchaca Plateau is a gently rolling plain with an elevation between 220 and 300 m that is composed of granitic rocks of the Brazilian Shield that have been dated to 1.2–1.4 billion years old. Overlaying large sections of the Huanchaca Plateau and the adjacent lowlands are Tertiary sediments that are estimated at between 6 and 20 million years old (Litherland and Power 1989).

The relationship between geomorphology and vegetation is most pronounced on the plateau. Open upland savanna is restricted mainly to the quartzite strata exposed on the southern and central sectors of the plateau; soils are shallow, sandy, and infertile. Scrub and tree savanna predominate on soils derived from the more easily weathered rocks of sandstone and sandstone conglomerates, as well as on the relatively deep lateritic soils derived from Tertiary sediments that overlie much of the plateau (Fig. 16.4, Table 16.3). Gallery forest and forest islands occur along stream valleys where erosion from surrounding interfluves has formed deeper, more fertile soils,

Table 16.3. Edaphic properties of selected habitats in Noel Kempff National Park

Vegetation type	pH	CEC	N	P	AL	Geology	Water regime Soil class	Dry season	Wet season
Campo seeps	L–H	L	L–H	L–M	M	Sandstone/quartzite	Inceptisol/ultisol	Dry	Saturated
Campo rupestre	L	L	L	L	H	Quartzite	Entisol	Dry	Dry
Campo sujo	L	L	L	L	H	Sandstone/quartzite	Oxisol	Dry	Humid
Campo cerrado	L	L	L	L	H	Sandstone/tertiary	Alfisol	Dry	Humid
Cerrado	L	L	L	L	M–H	Sandstone/tertiary	Ultisol	Dry	Humid
Cerradão	L	L	L	L	M–H	Sandstone/tertiary	Ultisol	Dry	Humid
Cerradão/gallery forest	L	M	M	L	M–H	Sandstone/tertiary	Alfisol	Dry	Humid
Dwarf evergreen forest	L	L	L	L–VH	H–VH	Tertiary		Dry	Humid
Evergreen high forest	L–M	L–M	L	L	H	Tertiary	Oxisols/alfisol	Humid	Humid
Evergreen liana forest	L–M	L	L–M	L	L–H	Tertiary	Oxisols/alfisol	Dry	Humid
Gallery forest (sandstone)	L–M	L–M	L–H	L–M	H–VH	Sandstone	Ultisol	Humid	Humid
Gallery forest (dolorite)	M–VH	M–VH	H–VH	L	L	Dolorite	Entisol/alfisol	Humid	Humid
Semideciduous forest	M–H	H	L–H	L–M	L	Granite/schist	Alfisol/inceptisol	Dry	Humid
Deciduous forest	L	M	H	M–H	H	Sandstone	Entisol	Dry	Humid
Evergreen inundated forest	L	L–H	M–H	L–H	L–H	Quaternary	Entisol	Humid	Flooded
Semideciduous inundated forest	L	L–M	L	L	L	Quaternary	Entisol	Humid	Flooded
Swamp forest	L	L(H)	M	L–M	H	Quaternary	Entisol	Humid	Saturated
Termite savanna	L	L–M	L–H	L	H–VH	Quaternary	Entisol	Humid/dry	Flooded (<1 m)
Open savanna	M	M	M–H	L	L	Quaternary	Alfisol	Humid/dry	Flooded (<1 m)
Savanna marsh	L	(L)H	L(H)	L	L	Quaternary	Ultisol	Wet	Flooded (>1 m)
L	4.0–5.0	0.6–2.5	0.01–0.1	0–2.0	<30% of CEC				
M	5.1–6.0	2.6–5.0	0.11–0.2	2.1–5.0	30–50% of CEC				
H	6.1–7.0	5.1–10	0.21–1.0	5.1–10.0	50–70% of CEC				
VH	>7.0	>10	>1.0	>10	>70% CEC				

and where soil moisture during the dry season is sufficient to support high forest. In many instances, gallery forest is associated with dolorite, a metamorphic volcanoclastic rock that weathers to produce a relatively rich soil type with better structural characteristics (Table 16.3). Dolorite intrusions support forest vegetation in landscapes otherwise dominated by savanna vegetation; these forest patches can be seen as straight lines that are discordant with the regional drainage pattern (Fig. 16.5A).

Geological substrate is also an important factor influencing the development of the vegetation on the lowland plains. Semideciduous forest is restricted to landscapes associated with the exposed basement rocks and metamorphic schist of the Brazilian Shield, with mesotrophic soils (Askew et al. 1970; Killeen et al. 1990). This can be contrasted to evergreen liana forests associated with dystrophic upland peneplains with deep, highly weathered, lateritic soils derived from Tertiary sediments or on low ridges derived from quartzite rocks (Table 16.3). Both semideciduous forest and liana forest have extremely dry soil profiles at the end of the dry season. A curious, almost counter-intuitive, example of the interaction of soil humidity and fertility was observed in the south central portion of the study area. Here, semideciduous forest was found down-slope in less drought-stressed valley bottoms, which also had younger, more fertile soils derived directly from the Precambrian basement rocks. In contrast, the evergreen liana forest or mixed liana forest occupies the more drought-stressed interfluves characterized by weathered lateritic soils derived from Tertiary sediments. This has been interpreted as the result of a high nutrient requirement for deciduous species, when compared to evergreen liana species.

Floodplains are composed of Quaternary sediments and have some of the most fertile soils found within the region (Table 16.3). In moderately well drained soils, high forest predominates, but most Quaternary landscapes are covered by either inundated forest or savanna wetland. These are the only landscapes that have appreciable amounts of loam, and most show evidence of recent sedimentation within soil profiles. Soil fertility is usually greater in forest communities (Table 16.3), but the differences may be the result of the vegetation acting on the soil profile rather than a precondition that led to the development of either forest or savanna. Clay hardpans are evident in savanna habitats, many of which have a perched water table that accentuates seasonal inundation. Microtopographic relief has a notable effect on vegetation, contributing to habitat heterogeneity both within and among habitats. Near rivers, levees support a forest community distinct from flooded forest found in adjacent backwaters. Large areas of the floodplains are characterized by a uniform dispersion of small mounds from 1 to 4 m in diameter that have been formed by the excavation of subsoil particles by termites.

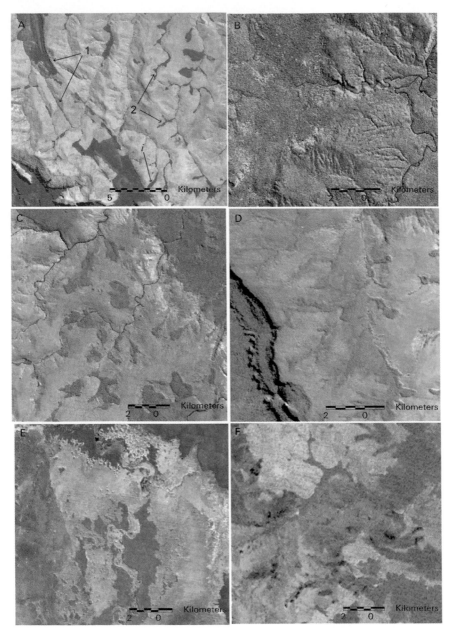

Fig. 16.5. Landsat images of selected landscapes within Noel Kempff Mercado National Park. **A** Open savanna with scattered forest islands associated with linear dolorite intrusions (*arrow 1*) or on sandstone derived valley bottoms (*arrow 2*), *i* shows the site for soils transect data displayed in Table 16.3; **B** forest maximum stage with scattered savanna patches; **C** transitional landscape with mosaic of savanna forest and transitional scrub (11/7/92 TM, WRS 229 × 70); **D** transitional landscape with forest encroaching on savanna; **E** alluvial plain with savanna wetlands and fragmented gallery forest (10/7/92 TM, WRS 230 × 70); **F** fire scar in savanna with gallery forest (27/8/75 MSS, WRS 229 × 70)

16.4 Fire

Fire occurs regularly in both the dry and humid forest ecosystems and is most frequent from August through October at the end of the dry season. Natural fire is known to be caused by lightning, and occurs frequently along plateau escarpments. The distribution of human settlements in the area is unknown prior to the 20th century, but it is likely that the area has long been inhabited by humans and that anthropogenic fire has probably also been an important factor. The Guarasug'we, a local indigenous group, have been extinct for more than 30 years, but there is no historical record or archaeological evidence of large population groups living in the region (Fawcett 1958). Rubber tappers were common throughout the region from the 1880s to the late 1960s, with small outposts situated along the two major rivers of the region. There are currently four small communities near the park; these are inhabited by Chiquitano migrants, who settled in the region following the end of the rubber boom. In recent years, the area has been thinly populated for several months each year by timber crews harvesting mahogany and other valuable tropical timber species. It is probable that all these groups have been a major source of fire.

In dry forest, many trees have fire scars both on the bark and within the growth pattern observed in trunk cross sections (J. Huffman, pers. comm.). In humid forest, the frequency of forest fires is unknown, but is probably increasing due to logging activity; large-scale forest fires were observed in 1987, 1993, and 1995. Satellite imagery was used to identify a burn scar that occurred between July 1992 and May 1994 in a logging concession within the park expansion area (Fig. 16.6A–D). Low intensity ground fires leave a large portion of the forest canopy intact with only a small reduction in canopy cover. At one such burn in high evergreen forest near the Los Fierros experiment station, hemispherical photos taken approximately 3 weeks after a fire showed that the forest canopy was only slightly reduced when compared to an adjacent unburned forest patch (77.6 vs. 89.2 %; $t=-4.53$, $n=17$; $P=0.01$). Isolated forest fires do not lead to a conversion of forest to savanna, as sometimes suggested in the popular and scientific literature. Burned areas will regenerate to forest, as evidenced by the rapid refoliation of the burn scar observed between June 1994 and July 1995 (Fig. 16.6C,D). Even abandoned agricultural fields will develop into a second growth forest. Conversion of forest to savanna is the consequence of repeated frequent fires over many years, particularly along edges or ecotones where grass and other savanna species can invade and replace the forest vegetation.

Savanna fires occur with much greater temporal and spatial frequency, about once every 3 years, on the open grasslands on top of the Huanchaca Plateau. They are an annual occurrence in the savanna wetland complexes situated to the south of the park, where cattle ranchers burn the grasslands to increase the availability of palatable forage (Killeen 1991). Both natural and

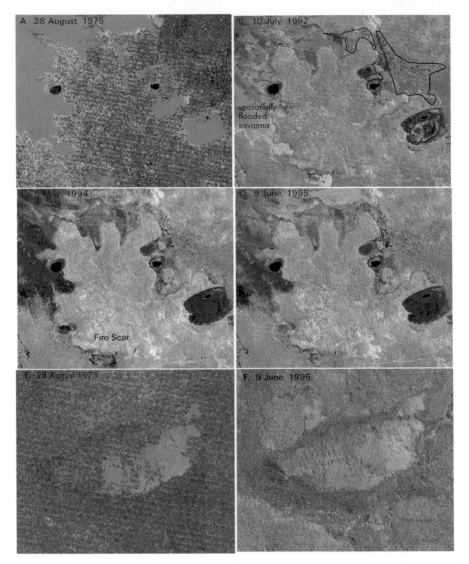

Fig. 16.6. Landsat images (WRS 230 × 70) of selected landscapes within Noel Kempff Mercado National Park. **A** Forest and savanna vegetation prior to disturbance events (28/8/75 MSS); **B** *polygons* identify areas where a change in inundation has led to forest die back (10/7/92 TM); **C** fire scar in high forest habitat (5/5/1994 TM); **D** regrowth of forest in area of fire scar (6/9/98 TM); **E** and **F** forest expansion in inundated termite savanna (28/8/75 MSS and 9/6/95 TM)

anthropogenic fires are most frequent from August through October at the end of the dry season when fire conditions are at their peak. Nonetheless, fires can occur in virtually any month of the year, as the savanna grasses are highly combustible and quickly dry out. Park guards using fire to clear a remote airstrip in March 1993 lost control of the fire and burned several hundred hectares of wooded savanna vegetation at a time of the year with almost daily afternoon rainstorms.

Aspect is also important, and this can be viewed as an interaction between geomorphology and climatic events. Lightning is most common on the highest promontories of the plateau, which are also the landscapes with the most resistant rocks and the poorest soils. These areas support the most open savanna vegetation found in the region (Figs. 16.1 and 16.4). In contrast, nearby forest patches associated with dolorite intrusions escape fire, because the dolorite has weathered more quickly than the surrounding sandstone or quartzite leading to the development of deep, steep-sided gullies that are less susceptible to invasion by wildfire (Fig. 16.5A). The prevailing winds are from the north during most of the year, but southern fronts during the dry season tend to foster the movement of fires from south to north. The topography of the plateau promotes temporal and spatial variability; gallery forest and marshes associated with streams can impede the spread of fire (Fig. 16.5F). Likewise, small forest islands on the flooded plain to the south often have a teardrop shape, with the tapered end facing north in the direction of the prevailing dry season winds.

Aerial observation and satellite images of the plateau reveal a mosaic of savanna patches with different fire histories. Even during very dry years when the majority of savanna landscapes burn, patches of unburned vegetation have been observed on the plateau and in the lowland savanna wetlands. The different patches have similar floristic composition, but the amount of standing litter, shade, and woody cover varies between patches depending upon the length of time since the last burn. This heterogeneity is further compounded by the intensity of fire on the different faces of the fire (upwind, downwind, lateral) contributing to the spatial complexity and structural diversity present within a specific habitat.

16.5 Flooding

Floodplains have very little relief, but even slight topographic differences produce noticeable changes in vegetation structure and composition. Maximum water level fluctuations are approximately 3 m and many different plant communities occur along a gradient from permanently flooded to seasonally saturated soils. Different types of forest communities occur near rivers ranging from almost well drained on natural levees to deeply flooded sites in depressions situated adjacent to levees. Frequently, areas with prolonged flooding are dominated by palm species, both in savanna and forest ecosystems. There are also differences in wetland communities relating to water flows across the surface of the landscape: open savannas without termite mounds experience diffuse laminar flows across the surface of the landscape; in contrast, shallowly inundated savannas with a well-developed shrub stratum are characterized by stagnant water.

Inundated shrub savannas have a unique physiognomy due to the presence of termite mounds that provide sufficient relief from inundation to allow for woody species to become established. The microtopography of termite mounds is responsible for the development of the clumped arrangement of woody species that contributes to the spatial complexity of savanna wetlands (Diniz de Araujo Neto et al. 1986). Termite mound platforms also occur in inundated forests and produce a microhabitat suitable for species intolerant of flooding. This seasonal fluctuation, and the spatial variability within this ecosystem can be observed by comparing satellite images from different months (Fig. 16.6B, C). Seasonal inundation can be viewed as a constant edaphic characteristic of wetland ecosystems; however, it can also function as a disturbance event when an extremely wet year causes greater than average flooding leading to plant mortality around the edges of wetland communities. Figure 16.6A–C also shows a landscape where a changing drainage pattern has led to a reduction in forest cover. Over-flights of this locality revealed thousands of dead standing trees and a low, dense canopy dominated by vines and emergent aquatic vegetation.

16.6 Succession on the Savanna-Forest Interface

Despite the influence of edaphic conditions on the distribution of vegetation, it has long been recognized that the relative position of the border between savanna and forest habitats is dynamic. For many parts of the world, there has been speculation that savanna is encroaching on forest vegetation due to the interference of man and the increased frequency and periodicity of fire. However, field botanists with extensive experience in central Brazil have postulated that forests have been encroaching upon savanna in the Holocene (Ratter 1992); these researchers base their hypothesis on the recurrent observation of mature (senescent) savanna species well within the boundaries of forest communities. The dynamic between savanna and forest in upland landscapes is relatively straightforward and depends largely on fire regime and the slow accumulation of nutrients that occurs simultaneously with forest succession (Braithwaite 1996).

There are three possible scenarios of vegetation change at any given locality: forest encroachment on savanna, savanna expansion into forest, and stasis with forest and savanna boundaries in equilibrium. All three situations are occurring in Noel Kempff National Park under present conditions and, in some cases, evaluating the spatial patterns of forest and savanna patches can identify the direction of change. Selected landscape windows on the Huanchaca Plateau and the adjacent floodplain represent different stages on the continuum of plant communities between open grassland and closed forest (Fig. 16.1). The following criteria were used to arrive at a preliminary evaluation regarding the direction of change for each landscape window.

1. Savanna expansion was recognized as a "frontal movement" with the source of fire originating in savanna and invading forest along existing savanna-forest interfaces. Infrequent fires have been observed in the middle of forest formations, but these do not lead to the formation of savanna.

2. Forest encroachment was recognized as an "enveloping movement" with a gradual expansion of gallery forest along river courses with an eventual filling in of isolated savanna patches along interfluves separating gallery forests.

3. Sharp boundaries were judged to be typical of savanna expansion. Since forest species are not fire-resistant and mortality following fire is high, fire along a forest-savanna edge leads to an invasion by grasses, leading in turn to hotter future fires that result in a rapid change in species composition.

4. Broad ecotones were judged to be typical of forest expansion. Since savannas occur over degraded soils, fertility is frequently a limiting factor in the growth of woody vegetation (Medina 1996). A gradual closing of the woody canopy will lead to a reduction in the grass sward and fires that are less hot; as soil characteristics improve, forest species can become established.

Figure 16.5A represents a landscape where savanna predominates near the escarpment on quartzite substrate in an area with a high incidence of lightning. This landscape represents a savanna expansion maximum, where forest patches have been reduced to only those areas where edaphic and topographic conditions are most favorable. The edges between savanna and forest are sharp and well defined and correspond to the limits between geological units, or are associated with a well-defined firebreak, such as a stream or the crest of a gully or gorge. In contrast, Fig. 16.5B shows a forest-dominated landscape, where the geomorphology favors forest development; nonetheless, the presence of numerous small scrub patches indicates that at some time in the past, savanna vegetation may have been more widespread. Transitional landscapes (Fig. 16.5C) are more complex mixtures of forest, scrubland and grassland; these landscapes show intermediate situations that may represent either savanna encroachment on forest, forest succession into savanna, or a stasis where forest and savanna are in equilibrium. Figure 16.5D shows an example of forest encroachment on savanna; savanna patches are surrounded by early successional forest on the interfluves with high forest expanding outward from the valleys.

The movement of the savanna and forest frontier on inundated landscapes is a poorly understood phenomenon that is mediated by both flooding regime and fire. Sarmiento (1996) has shown that savannas develop on landscapes that experience seasonal extremes of flooding and drought. This seems to be the case in termite savannas where woody vegetation is restricted to the small, elevated platforms that escape flooding. However, many savanna wetlands do not experience true seasonal drought; even when large cracks appear on the soil surface, the profile remains relatively moist. Fire, which is frequent in these ecosystems, also plays a role in the formation and maintenance of these grassland ecosystems. The occurrence of termite mound platforms in both forest and savanna with similar inundation regimes is suggestive of a successional process where one habitat replaces the other over time. There may be some inundation regimes, when combined with certain soil characteristics (e.g., clay pan) that favor savanna formation. Once grasses become established, the necessary fuel is available for recurrent fire, thus establishing ecosystem structure; likewise, the absence of fire will eventually lead to loss of the savanna habitat. Figure 16.6E and F show the outlines of a savanna patch with successional sequences ranging from inundated shrub savanna to mixed liana forest; over-flights have verified the existence of termite savanna patches and adjacent wetland habitat with a continuous shrub stratum.

The formation of forest islands in seasonally inundated savanna landscapes is also dependent upon the interaction of river movements and fire. Figure 16.5E shows a large expanse of open inundated savanna with forest islands that are arranged along an abandoned channel of the Paraguá River. This river is currently without a clearly definable channel in this area, and water movement is occurring via laminar flow from south to north. The relictual gallery forest shows evidence of once being continuous and is, apparently, becoming fragmented with time due to the action of fire. The floodplain is heavily impacted by fire almost every year as ranchers burn grassland to provide forage for cattle during times of scarcity. This anthropogenic fire invades the gallery forest along the abandoned river course leading to the fragmentation of the gallery forest forming smaller detached forest islands.

16.7 Direct Evidence for Past Climate Change

Evidence to support the hypothesis that the forest savanna borders fluctuate in both directions was provided by a soil survey conducted across a vegetation gradient in the Las Gamas study area (Fig. 16.5A) and at other selected areas around the park (Fig. 16.1). An analysis of stable carbon isotope ratios in soil organic matter was made to ascertain the history of the forest-savanna border at this locality. A high $^{13}C/^{12}C$ ratio is indicative of a past vegetation type dominated by species that use the C_4 photosynthetic pathway, while a low

Table 16.4. The $\delta^{13}C$ values of soil profiles sampled along a vegetation transect between cerrado and gallery forest at the Las Gamas study site; the analysis was replicated on three samples with a variance of only 0.71 % ($P=0.84$)

Position on transect Vegetation type	T1 Cerrado	T3 Cerrado/ forest	T5 Forest/ forest	T7 Low forest	T9 High forest
Soil organic matter					
0–10	−23.03	−24.65	−27.22	−26.77	−26.61
20–30	−18.92	−21.89		−26.15	−24.98
30–50			−24.82	−25.09	−25.98
80–100					−23.09
Laterite nodules					
0–10	−17.30				−16.83
20–30					
30–50					−16.80
80–100	−14.53		−16.55		−17.86

ratio is produced by trees and shrubs that fix atmospheric carbon by the C_3 photosynthetic pathway. Expressed as $\delta^{13}C$ values, numbers between −28 and −30 are typical for C_3 plants and soil organic matter in habitats dominated by C_3 plants; $\delta^{13}C$ values between −11 and −14 are characteristic of C_4 grasses and grassland soils (Boutton 1996). Table 16.4 shows the $\delta^{13}C$ values from soil profiles taken along the vegetation gradient. Surprisingly, all five soil profiles have a signature that lacks a strong signal typical of C_4 grasses, indicating that forest covered the entire transect in the very recent past.

The strongest C_4 signals were associated with laterite nodules that were present throughout the soil profile and as a solid massive reef at the bottom of each soil profile. Laterite (sometimes referred to as plynthite) is an iron-rich clay that hardens to form a recalcitrant, rock-like substance. Laterite is known to form in soils that experience pronounced seasonal fluctuations of soil moisture and temperature (Young 1976). Laterite nodules from the B horizon of the pits sampled at the two extremes of the vegetation transect were selected for dating using radioisotope ^{14}C analysis. The results indicate that the laterite was formed, and that savanna conditions existed, approximately 9000 (gallery forest) and 6650 (cerrado savanna) years before the present time. Evidence from laterite deposits at two other forested sites on top of the plateau (Enano-2 and Huanc-2) support the hypothesis that savanna-forest vegetation has pulsed back and forth across existing ecotones (Table 16.5).

Carbon isotope values for laterite nodules were surprisingly variable and several sites had laterite nodules with carbon isotope values characteristic of forest vegetation. This demonstrates that laterite can also develop under forest vegetation, and the presence of laterite cannot be used as a diagnosis for past presence of savanna vegetation. The movement of savanna and forest

Table 16.5. The $\delta^{13}C$ values for soil profiles sampled in selected forest localities in the study area; a smaller negative $\delta^{13}C$ value is characteristic of savanna soils; the analysis was replicated on eight samples with a variance of only 0.49 % (NS). Note: Eleven samples from both Tables 16.4 and 16.5 were randomly selected for a duplicate carbon isotope analysis; none varied more than 1.5 % from the original value and a t-test of the 11 paired samples was insignificant

Locality Vegetation	Chore-1 Liana forest	Chore-2 Liana forest	Paraguá Liana forest	Porvenir Inundated forest	Enano-2 Dwarf forest	Huan-2 Gallery forest
Soil organic matter						
0–10	−27.68	−27.44	−26.83	−25.51	−27.32	−26.79
20–30	−26.93			−26.07	−24.43	−25.77
30–50		−25.44	−26.00	−21.85	−25.91	−22.23
80–100	−25.14	−24.31			−23.24	
Laterite nodules						
0–10		−25.28			−18.82	−22.43
20–30					−16.89	−15.04
30–50				−18.47	−23.33	−14.66
80–100				−16.61	−20.36	

ecotones in wetland habitats was confirmed from a soil pit in an inundated forest, where laterite nodules found at the bottom of the soil pit had a carbon isotope ratio typical of savanna vegetation (Table 16.5, Porvenir). The development of liana forest is a poorly understood phenomenon. The distribution of this habitat on the landscape suggests that edaphic factors, particularly drought stress and soil infertility, are responsible for its development. The carbon isotope ratios show that forest has been the predominant vegetation type for a considerable period of time at the three sites where soil pits were analyzed for carbon isotope values (Chore-1, Chore-2, Paraguá). The locality that had a laterite reef showed no evidence of past savanna vegetation and the isotope values of that laterite were typical of an ecosystem dominated by C_3 plants (Table 16.5).

To test the hypothesis that forest and savanna vegetation pulse back and forth at the landscape level, 15 soil pits were sampled along a 60-km transect established in the central portion of the plateau (see Figs. 16.1 and 16.7). The vegetation along this transect ranged from grassland landscapes to transitional scrubland and high closed forest. Figure 16.8 shows the $\delta^{13}C$ values for those 15 soil pits and provides evidence that sites currently under both forest and savanna vegetation previously supported the contrasting vegetation type. Sites #1, #4 and #5 are relatively open grassland associations and their upper soil profiles strongly reflect the recent carbon inputs from the dominant C_4 grasses, with a slight shift that corresponds to an increase in the cover of C_3 forbs and woody shrubs (See Table 16.6). All three of these sites show similar

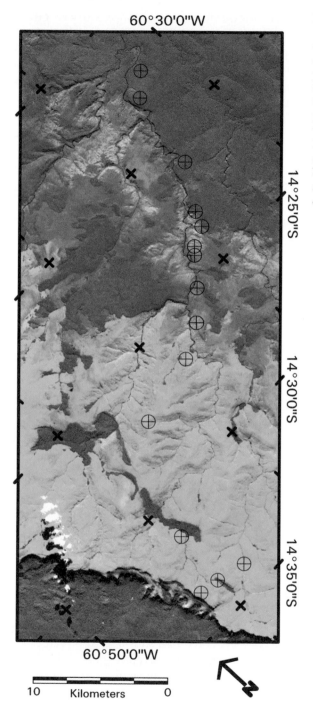

Fig. 16.7. Landsat image (WRS 229 × 70) showing 15 soil pits characterized for stable carbon isotopes sampled along an ecotonal transect between open grassland to high forest in the central part of the plateau (see box with dashed white lines in Fig. 16.1); the soil pits are sequentially numbered from bottom to top and correspond to the identification codes in Fig. 16.8

$\delta^{13}C$ values throughout their soil profile indicating a long history of grass-dominated vegetation. Site #2 is a small forest island in an otherwise grassland-dominated landscape, while soil pit #3 was open grassland in a humid valley bottom with very fertile soils situated downstream from site #2. Both of these sites show evidence of past vegetation change. The subsoil of the forest island site has elevated levels of ^{13}C in the 80–100-cm portion of the soil profile indicating inputs from C_4 grasses. The reverse was true of the valley bottom site, which was also characterized by humid soil, good nutrient status (0.25 % N) and elevated levels of organic matter (4.2 % C) when compared to other savanna soil pits (0.11 % N and 0.5–1.2 % C). The juxtaposition of these two sites in an otherwise stable grassland-dominated landscape indicates that the local distribution of forest islands along streams has changed over time.

Sites 6–11 were deemed to represent transitional vegetation, including moderately wooded C_4-dominated savanna to dense bamboo groves and cerradão, a dense scrubland dominated by the same woody species found in more open cerrado and campo cerrado habitats. In general, these sites show intermediate levels of $\delta^{13}C$ values, although sites dominated by the C_3 grasses *Rhipidocladum verticillata* and *Ichnanthus inconstans* (Killeen and Hinz 1992) had $\delta^{13}C$ values similar to forest sites. Soil pits #9 and #11 showed evidence of past vegetation change with an increase in the abundance of ^{13}C in

Table 16.6. Vegetation structure for 15 sample sites analyzed for $\delta^{13}C$ analysis (see Figs. 16.7 and 16.8); the analysis was replicated on 8 samples with a variance of only 0.25 % (NS)

Position on transect	Vegetation type	C_4 grass cover (%)	C_3 grass cover (%)	Woody cover (%)	Mean canopy height (m)	Maximum tree height (m)
1	Campo rupestre	68		5	0.5	1
2	Forest island			>100	10	35
3	Valley campo	58		6	1	1
4	Campo sujo	68		33	1	2
5	Campo cerrado	63		29	1	4
6	Campo cerrado	55		48	1	4
7	Cerrado/bamboo	9	65	28	2	4
8	Cerrado		31	>100	4	8
9	Cerradão	11	30	51	3	8
10	Cerradão/bamboo		83	38	2	8
11	Cerradão		83	85	5	10
12	Open forest			>100	13	20
13	High forest			>100	15	30
14	High forest			>100	10	30
15	High forest			>100	20	40

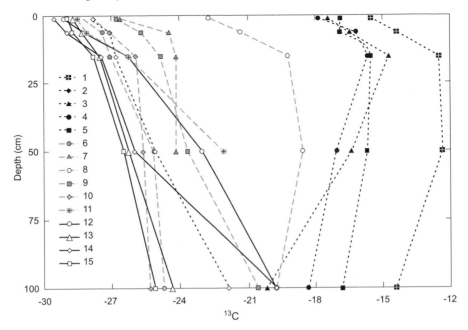

Fig. 16.8. The $\delta^{13}C$ values of soil profiles of 15 soil pits characterized for stable carbon isotopes sampled along an ecotonal transect between open grassland to high forest in the central part of the plateau; soil pits from grassland landscapes have *dotted lines* (*1–5*), transitional landscapes are *gray solid lines* (*6–11*), and forested landscapes have *solid black lines* (*12–15*); soil pit #2 was a forest island situated on a grassland landscape

both the 35–50 cm and the 80–100 cm-soil profiles; site #9 also had C_4 grasses in the herbaceous stratum, #11 did not.

The forest sites are near the edge of a large forest patch situated to the south and east of the access trail used to collect soil samples; all were protected from fires originating in a nearby savanna by the presence of a wide (30–100 m) permanent river. Neither site #13 nor #15 showed any evidence of past C_4 grasses and the slight enrichment in ^{13}C can be attributable to the action of bacterial and invertebrate decomposers (Boutton 1996). In contrast, both sites #12 and #14 had higher levels of ^{13}C in the lower soil profiles. The forest community at site #12 had numerous attributes of an early successional forest community, including a lack of large diameter trees and an abundance of immature palm trees; it was also situated close to the current scrubland–forest edge (Fig. 16.7). However, the vegetation at site #14 was evidently a mature forest with a complex, multi-layer canopy and numerous large diameter trees.

In conclusion, the carbon isotope ratio of soil organic matter indicates that there has been change and stability along different portions of the transect. Savanna landscapes near the escarpment have been relatively stable, but forest islands have probably changed their local distribution within the narrow

confines of the valley bottoms. The transitional habitats in the very broad ecotone show both stability and change, as do the forest sites sampled on the edge
of the very large patch of humid tropical forest situated in the middle of the
Huanchaca Plateau.

16.8 Conservation Issues

Conservation of the biodiversity in Noel Kempff National Park and adjacent
areas will depend upon maintaining a heterogeneous habitat mosaic. Particularly important are the grassland and woodland formations, which are habitat
for some of the most threatened species known to occur in the park (Redford
and Fonseca 1986; Killeen and Schulenberg 1998). The pampas deer (*Ozoto-
ceros bezoarticus*) has only been observed in the open upland savannas of the
Huanchaca Plateau, while the marsh deer (*Blastocerus dichotomous*) is
restricted to the savanna marshes and palm swamps associated with the
Paraguá River. The maned wolf (*Chrysocyon brachyurus*), the greater rhea
(*Rhea americana*) and the giant anteater (*Myrmecophaga tridacyla*) are
somewhat less selective in their habitat preferences, and have been observed
in both upland and wetland savannas.

Surveys of the avifauna also demonstrate the importance of savanna habitats, with more that 100 species of birds that are considered to be characteristic of or restricted to cerrado savanna; of these, 20 are considered to be threatened or endangered. In addition, there are several species that have been
observed only in gallery forest on top of the plateau, while an additional 15
species have been registered only in the wetland savannas in the adjacent lowland (Bates et al. 1998). Perhaps the most endangered species in the park is the
black and tawny seedeater (*Sporophila nigrarufa*), a species known to breed
only in shallowly inundated termite savannas and which is apparently intolerant of cattle grazing (S. Davis, pers. comm.). Given the size of Noel Kempff
National Park and the widespread degradation of cerrado habitats in Brazil,
this may well be the single most important reserve for many of these species.

Future climate change may have a greater than normal impact on the
region due to its position across a climatic tension zone where conditions
change over relatively short distances. The possible scenarios resulting from
climate change range from a potential increase in precipitation due to a
warmer earth, to a decrease in precipitation due to regional effects resulting
from deforestation in adjacent regions of Brazil. The park is at the extreme of
the ranges of many plant and animal species and climate change could lead to
the local extinction of many of these species. Latitudinal displacement of the
distribution of species in response to these potential changes is one possible
solution for the long-term conservation of these species. Movement to the
north is already limited due to the extensive deforestation in Rondônia (Skole

and Tucker 1993). However, there is a potential biodiversity corridor situated to the northeast of the park along the Río Iténez; this area currently contains a mosaic of savanna and humid forest habitats similar to those found within the park.

On both sides of the Bolivia-Brazil border there exists a series of forest reserves, provincial parks, indigenous territories and other conservation units (Fig. 16.9) and preliminary initiatives have been undertaken to work towards the integrated management of these reserves. To the south, there remains a great deal of relatively unaltered landscape in Bolivia and there are several protected areas which could form important links that connect Noel

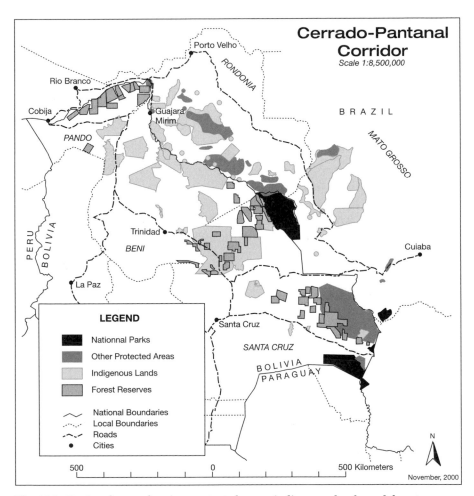

Fig. 16.9. Regional map showing protected areas, indigenous lands, and forest concessions in Bolivia and Brazil with the potential to function as a latitudinal corridor; prepared by Lisete Correa at the GIS facility at the Noel Kempff Mercado Natural History Museum

Kempff with the dry forest and wetland formations of the Gran Pantanal. Unfortunately, there is a large gap in the reserve system between Noel Kempff Mercado National Park and the San Matias Natural Integrated Management Area, which borders the Bolivian Pantanal. Although largely undeveloped, most of this land is privately held and a development corridor will inevitably emerge along the road between the city of Santa Cruz and Cuiabá, Brazil. Development in Santa Cruz has accelerated considerably in recent years and currently deforestation rates there are among the highest in the world (Steininger et al. 2001), while the landscape in adjacent regions of Mato Grosso, Brazil is already deforested and highly fragmented. Conservation organizations should work towards acquiring key areas in the region, while working with government and landowners to ensure development scenarios that foster biodiversity conservation.

Regardless of the development of latitudinal corridors, the long-term conservation of the biodiversity within the park will depend upon maintaining the high levels of habitat heterogeneity for which it is famous. Central to this philosophy must be a willingness to allow, or even promote, disturbance events that promote habitat heterogeneity. Fortunately, current park managers are aware of the importance of fire as a component of natural ecosystems, but until public perceptions change, future governmental managers may tend to adopt contrary positions with respect to fire. The experience in North America regarding the importance of fire as a management tool in maintaining certain types of habitats is illustrative. The suppression of fire in many areas led to a loss of habitat or led to high intensity fires that altered habitat structure more significantly than frequent low intensity fires. Environmental education programs should distinguish between the different types of fires and their impact, including fires set by cattle ranchers, fires caused by the negligence of timber crews, fires prescribed by ecologists, and natural wildfires. Future researchers should focus on questions relating to the frequency and intensity of fire in different habitats, and the impact of fire on species that are considered to be of special conservation importance.

References

Absey ML, Cleef AF, Fournier M et al. (1991) Mise en évidence quatre phases d'ouverture de la forêt dense de sud-est de l'Amazonie au cours des 60.000 dernières années. C R Acad Sci Paris Ser II 312:673–678

Askew GP, Moffat DJ, Montgomery RF, Searl PL (1970) Interrelationships of soils and vegetation in the savanna-forest boundary zone of northeastern Mato Grosso. Geogr J 1436:370–376

Bates JM, Stotz DF, Schulenberg TS (1998) Avifauna. In: Killeen TJ, Schulenberg T (eds) A biological assessment of Noel Kempff Marcado National Park. RAP Working Papers, vol 10

Braithwaite R (1996) Biodiversity and fire in the savanna landscape. In: Solbrig OT, Medina E, Silva JF (eds) Biodiversity and savanna ecosystem processes: a global perspective. Ecological studies, vol 121. Springer, Berlin Heidelberg New York, pp 121–143

Boutton TW (1996) Stable carbon isotope ratios of soil organic matter and their use as indicators of vegetation and climate change. In: Boutton TW, Yamasaki S (eds) Mass spectrometry of soils. Marcel Dekker, New York, pp 47–82

Burnett MR, August PV, Brown JH Jr, Killingbeck KT (1998) The influence of geomorphological heterogeneity on biodiversity. I. A patch scale perspective. Conserv Biol 12:363–370

Clark DA (1998) Deciphering landscape mosaics of neotropical trees: GIS and systematic sampling provides new views of tropical rainforest diversity. Ann Mo Bot Gard 85:18–33

Collinvaux PA (1996) Quaternary environmental history and forest diversity in the Neotropics. Evolution and environment in tropical America. In: Jackson JBC, Budd AF, Coates AG (eds) Evolution and environment in tropical America. University of Chicago Press, Chicago, pp 359–405

Dallmeier F, Comiskey JA (eds) (1997) Forest biodiversity research, monitoring and modeling. Man and biosphere series, vol. 20. Parthenon Publishing, New York

Diniz de Araujo Neto M, Haridisan MPA, Johnson CE (1986) The murundus of central Brazil. J Trop Ecol 2:17–35

Fawcett PH (1958) Exploration Fawcett. Arranged from his manuscripts and letters, and records (edited by B. Fawcett). Hutchinson, London

Gentry A (1988) Changes in plant community diversity and floristic composition on environmental end geographical gradients. Ann Mo Bot Gard 75:1–34

Hafer OR, Haffer J (1969) Speciation in Amazonian forest birds. Science 165:131–137

Killeen TJ (1991) The effect of grazing on native Gramineae in Concepción, Santa Cruz, Bolivia. Trop Grasslands 25:12–19

Killeen TJ (1998) Vegetation and flora. In: Killeen TJ, Schulenberg T (eds) A biological assessment of Noel Kempff Mercado National Park. RAP working papers, vol 10. Conservation International, Washington, DC

Killeen TJ, Hinz PN (1992) Grasses of the Precambrian Shield region in eastern lowland Bolivia. II. Life-form and C_3-C_4 photosynthetic types. J Trop Ecol 8:409–433

Killeen TJ, Schulenberg T (eds) (1998) A biological assessment of Noel Kempff Mercado National Park, Santa Cruz, Bolivia. RAP Working Papers, vol 10. Conservation International, Washington, DC

Killeen TJ, Louman BT, Grimwood T (1990) La ecología paisajística de la región de Concepción y Lomerio, Santa Cruz, Bolivia. Ecol Bolivia 16:1–45

Killeen TJ, Steininger M, Tucker CJ (1998) Formaciones Nacionales del Parque Nacional Noel Kempff Mercado (vegetation map). In: Killeen TJ, Schulenberg T (eds) A biological assessment of Noel Kempff Mercado National Park. RAP Working Papers, vol 10. Conservation International, Washington, DC

Lamotte S (1990) Fluvial dynamics and succession in the lower Ucayali River basin, Peruvian Amazonia. For Ecol Manage 33/34:141–156

Litherland M, Power G (1989) The geological and geomorphological evolution of the Serranía Huanchaca, eastern Bolivia: the legendary "Lost World". J S Am Earth Sci 2:1–17

Mayle FE, Burbridge R, Killeen TJ (2000) Millennial-scale dynamics of Southern Amazonian Rainforests. Science 290:2291–2294

Medina E (1996) Biodiversity and nutrient relations in savanna ecosystems: interactions between primary producers, soil microorganisms, and soils. In: Solbrig OT, Medina E, Silva JF (eds) Biodiversity and savanna ecosystem processes: a global perspective. Ecological studies, vol 121. Springer, Berlin Heidelberg New York, pp 45–60

Nelson BW (1994) Natural forest disturbance and change in the Brazilian Amazon. Remote Sensing Rev 10:105–125

Nelson BW, Irmão MN (1998) Fire penetration in standing Amazon forests. Proc Ninth Brazilian Remote Sensing Symposium, Santos, SP, 13–18 September, 1998

Nichols WF, Nichols KT, August PV (1998) The Influence of geomorphological heterogeneity on biodiversity. II. A landscape perspective. Conserv Biol 12:371–380

Pessenda LCR, Gomes BM, Aravena R et al. (1998) The carbon isotope record in spoils along a forest-cerrado ecosystem transect: implications for vegetation changed in the Rôndonian State, southwestern Amazon region. Holocene 8:599–603

Ratter JA (1992) Transitions between cerrado and forest vegetation in Brazil. In: Furley PA, Proctor J, Ratter JA (eds) Nature and dynamics of forest-savanna boundaries. Chapman and Hall, London, pp 417–430

Redford KH, da Fonseca GAB (1986) The role of gallery forest in the zoogeography of the Cerrado's non-volant mammalian fauna. Biotropica 18:115–125

Ronchail J (1989) Variaciones climaticas en Bolivia: Caracterización de los efectos climáticos. Bill Inst Fr Et And 18:65–73Sarmiento (1996) The ecology of neotropical savannas. Harvard University Press, Cambridge, MA

Scott JM, Jennings MD (1998) Large-area mapping of biodiversity. Ann Mo Bot Gard 85:34–47

Skole D, Tucker CJ (1993) Tropical deforestation and habitat fragmentation in the Amazon: satellite data from 1978 to 1988. Science 260:1905–1910

Steininger MK, Tucker CJ, Ersts P, Killeen TJ, Villegas Z, Hecht SB (2001) Clearance and fragmentation of tropical deciduous forest in the Tierras Bajas, Santa Cruz, Bolivia. Conserv Biol 15(4):127–134

Tuomisto H (1998) What satellite imagery and large-scale field studies can tell about biodiversity patterns in Amazonian forests. Ann Mo Bot Gard 85:48–62

Van der Hammen T (1972) The Pleistocene changes of vegetation and climate in tropical South America. J Biogeogr 1:3–26

17 Bandages for Wounded Landscapes: Faunal Corridors and Their Role in Wildlife Conservation in the Americas

S.G.W. Laurance and W.F. Laurance

17.1 Introduction

The loss and fragmentation of natural habitats is probably the single greatest threat to the world's biological diversity. Fragmentation has a variety of effects including the isolation of habitat remnants, a sharp increase in the amount of habitat edge and, often, a disproportionate loss of certain habitat types – such as accessible areas on fertile, well-drained soils that are most productive for agriculture (Laurance et al. 1999).

Wildlife corridors have been advocated as a strategy to lower extinction rates in fragmented landscapes since at least the 1970s (e.g., Willis 1974; Diamond 1975; Wilson and Willis 1975; Wegner and Merriam 1979). By definition, a wildlife corridor is a linear remnant that differs from the surrounding vegetation and connects patches of similar habitat that were more extensively connected in the recent past (Saunders and Hobbs 1991). It is important to emphasize that corridors are not an artificial feature of the landscape, but are intended to help maintain historical habitat connectivity (Noss 1991; Bennett 1999). By facilitating movements of individuals among habitat remnants, corridors can increase population persistence in two ways. First, the demographic and genetic contributions of immigrants can bolster small, dwindling populations in fragments, providing a buffer against local extinction (Brown and Kodric-Brown 1977). Second, if a fragment population should go extinct, immigrants may eventually recolonize the fragment and reestablish the population.

In addition to facilitating animal movements, corridors can provide important habitats for some species (e.g., Laurance 1996; Downes et al. 1997; Laurance and Laurance 1999) and thereby help to maintain population continuity across the landscape. They can also fulfill other ecosystem functions, such as protecting riparian habitats and aquatic ecosystems, stabilizing erodible soils, and maintaining water quality (Naiman et al. 1993). Furthermore, corri-

Ecological Studies, Vol. 162
G.A. Bradshaw and P.A. Marquet (Eds.)
How Landscapes Change
© Springer-Verlag Berlin Heidelberg 2003

dors can have economic benefits, such as providing windbreaks for crops or shelterbelts for livestock (Forman and Godron 1986; Noss 1987).

Despite their potential benefits, corridors – especially poorly designed corridors – could have unacceptable costs or impacts in some contexts, and these need to be considered before the "corridor paradigm" is accepted uncritically. For example, if animals in corridors suffer high rates of mortality, then corridors could act as population sinks, "draining" individuals from nearby fragments and possibly reducing population persistence. Corridors might also facilitate the spread of pathogens among reserves (Simberloff and Cox 1987). Finally, corridors may be expensive to establish or maintain, consuming scarce conservation dollars that would be better spent elsewhere (Simberloff et al. 1992).

Below we consider some implications of studies from the Americas and elsewhere for the design of faunal corridors, and then put forward generic recommendations for corridor design in open- and closed-forest ecosystems. We conclude by advocating a proactive strategy for landscape management in regions experiencing rapid deforestation.

17.2 Considerations in Corridor Design

17.2.1 Corridor Width

Perhaps the most fundamental question when designing a faunal corridor is "How wide should it be?" Because they are linear habitats surrounded by a matrix of crops, pastures, or other modified habitats, faunal corridors will often be prone to edge effects – diverse physical and biotic changes associated with artificial ecotones between structurally different habitats. In tropical rainforests, hotter, drier microclimatic conditions more typical of surrounding cattle pastures penetrate at least 15–60 m into the interior regions of forest remnants (Kapos 1989; Williams-Linera 1990; Kapos et al. 1993). Disturbances from windstorms can occur over considerably larger spatial scales. Studies of fragmented forests in Amazonia and Queensland, for example, indicate that rates of treefelling and forest structural damage increase within 100–300 m of forest edges (Laurance 1991a, 1997; Bierregaard et al. 1992; Ferreira and Laurance 1997; Laurance et al. 1997, 1998a). These fundamental changes in forest dynamics can markedly alter the structure, floristic composition, and microclimatic conditions of forest remnants.

Some animal species respond positively to forest edges, whereas others tend to be edge-avoiders (Yahner 1988). In Amazonia, for example, many faunal groups – leaf-litter insects (Didham 1997a), beetles (Didham et al. 1998), butterflies (Brown and Hutchings 1997), birds (Lovejoy et al. 1986), and small mammals (Malcolm 1997) – are influenced by forest edges, with some species declining or increasing within 50–250 m of edges (Fig. 17.1). The strong responses of many fauna to edges suggests that corridors that are too narrow will tend to act as selective filters, facilitating movements mostly for edge-tolerant species (Hill 1995). Nevertheless, some forest animals will use suboptimal habitats for dispersal (e.g., Weddell 1991; Downes et al. 1997; Newell 1999), and thus, even narrow corridors are likely to facilitate movements for a subset of forest species (Crome et al. 1995; Isaacs 1995; Laurance and Laurance 1999; Lima and Gascon 1999).

Certain forest-dependent animals may avoid corridors because of competitive interactions with generalist species that become abundant in modified habitats. For example, Catterall et al. (1991) found that the noisy miner (*Manorina melanocephala*), a highly territorial honeyeater that favors forest edges in eastern Australia, excludes many smaller birds from the vicinity of forest edges. For this reason, they concluded that corridors in eastern Australia should be at least 500 m wide to ensure that they can be used by the majority of forest bird species. Predation intensity may also be elevated near edges. A number of studies have demonstrated that omnivorous mammals and avian brood parasites increase their activity near edges, and can sharply elevate rates of egg or nestling loss for birds in fragmented habitats (e.g., Gates and Gysel 1978; Andren and Angelstam 1988; Martin 1992; Paton 1994).

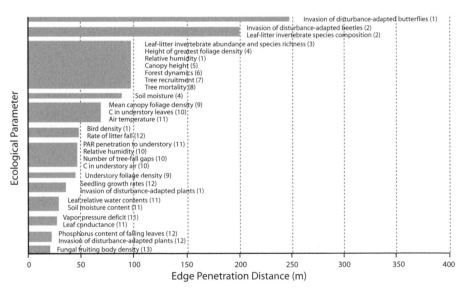

Fig. 17.1. The distances (*d*) to which various edge effects penetrate into the interiors of rainforest fragments in central Amazonia (after Laurance et al. 2002). *Numbers in parentheses* indicate data sources: *1* Lovejoy et al. (1986); *2* Didham (1997b), Carvalho and Vasconcelos (1999); *3* Didham (1997a); *4* Camargo and Kapos (1995); *5* Camargo (1993); *6* Laurance et al. (1998a); *7* Laurance et al. (1998b); *8* Ferreira and Laurance (1997); *9* Malcolm (1994); *10* Kapos et al. (1993); *11* Kapos (1989); *12* Bierregaard et al. (1992); *13* R.K. Didham (pers. observ.)

The importance of edge effects is partly determined by the degree of structural differences between adjoining habitats (Laurance and Yensen 1991). Closed-forest communities adjacent to cleared areas are likely to be more prone to edge effects than are open habitats such as grasslands or woodlands, because the physical and microclimatic gradients between closed forests and the adjoining matrix are very steep. For this reason, we assert that wildlife corridors for closed forests should be wider than those in open communities (see below). Topography may also affect the vulnerability of corridors to edge-related wind damage and microclimatic changes. Riparian corridors, for example, often occur in sheltered gullies where they are partly protected from wind shear and lateral light penetration (Forman 1995). Thus, corridors in exposed areas, such as slopes or ridge tops, should generally be wider than those in sheltered locations.

Corridor width can also affect the availability of food resources, shelter, microhabitat diversity, and space for animal home ranges or territories. In forest communities – especially tropical forests – many resources such as fruits, flowers, specific host plants, and den cavities are scarce or patchily distributed (Gilbert 1980; Gibbs et al. 1993). In these habitats, narrow corridors may provide too few resources to support resident individuals of some species, thereby reducing their effectiveness.

17.2.2 Corridor Length

As landscapes become more developed, the distances among habitat remnants increase. When distances are short, many animals can disperse between fragments in a single movement through corridors of natural or even disturbed habitat (e.g., Machtans et al. 1996; Downes et al. 1997; St. Clair et al. 1998). However, as corridors become longer, individuals are unable to traverse the corridor in a single movement or dispersal event, in which case it must provide shelter, food, and other resources for longer-term survival. For species such as amphibians, reptiles, and small mammals, which often have limited vagility, corridors exceeding a kilometer in length could require days or weeks for traversal. Hence, longer corridors should also be increasingly wide, because wide corridors will generally provide more resources and higher-quality habitat than narrow corridors. The most effective corridors are likely to be those that can support breeding populations of forest-dependent species, which can in turn introduce dispersing individuals into the system (Bennett 1990a,b; Laurance 1996).

As a general rule of thumb, we propose that corridors should adhere to the "10% rule": the width of a corridor at its narrowest point should not fall below 10% of its length (Fig. 17.2). While this ratio is based merely on informed guesswork, it does provide a simple way of adjusting the widths of corridors to account for variation in their length. Empirical data on the inter-

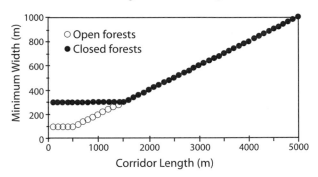

Fig. 17.2. Recommended minimum widths for faunal corridors of various lengths in open and closed forests. The minimum width for open forests is 100 m, while that for closed forests is 300 m. For long corridors, corridor width is increased to ensure the minimum width does not fall below 10 % of corridor length

action of corridor width and length for forest-dependent species would obviously be valuable.

17.2.3 Faunal Stepping Stones

Another design feature that may improve the performance of corridors is to incorporate patches of similar habitat – faunal "stepping stones" – at regular intervals along the corridor route (cf. Hanski et al. 1994; Keitt et al. 1997). For example, if a 5-km-long corridor were needed to link two nature reserves, then several moderately large (e.g., 50–200 ha) fragments along the route could provide needed food and shelter for animals traversing the corridor, as well as breeding habitat for some species (e.g., Schultz 1998). In addition to improving corridor effectiveness, the diversity and conservation values of small fragments will be enhanced if they are linked to larger forest tracts (e.g., Laurance 1990), and overall connectivity of the landscape will be increased.

17.2.4 Primary Versus Secondary habitat

Another important consideration is habitat type. Wherever possible, corridors should be comprised of primary rather than degraded or secondary habitats, because many species that are vulnerable to fragmentation are confined to primary habitat (e.g., Laurance 1990, 1991b, 1994; Saunders and de Rebeira 1991; Bierregaard et al. 1992; Goerck 1997). In landscapes where primary vegetation between reserves has disappeared, reforestation can be used to establish faunal corridors (Lamb et al. 1997), but this will be prohibitively expensive or impractical in many cases, and may be of limited effectiveness for the most vulnerable species. Nevertheless, a corridor of regenerating forest is likely to be better than no corridor at all. In tropical Queensland, for example, Herbert River ringtail possums (*Pseudocheirus herbertensis*) and Lumholtz's tree-kangaroos (*Dendrolagus lumholtzi*) are moderately vulnerable to fragmentation and will use even narrow (30–50 m) corridors of regenerating rainforest, but the most vulnerable arboreal mammal, the lemuroid ringtail possum (*Hemibelideus lemuroides*), occurs only in wide (>200 m) corridors of primary rainforest (Laurance and Laurance 1999).

17.2.5 Topographic Position

Topographic position is another key consideration. In largely deforested landscapes, remnant vegetation often persists along streams and rivers, and such riparian strips can function as wildlife corridors (e.g., Forman and Godron 1986; Naiman et al. 1993; Crome et al. 1995; Mares and Ernest 1995; Machtans et al. 1996; Lima and Gascon 1999). Riparian corridors have several advantages: water is present for wildlife; they tend to be productive and buffered from wind disturbance; they help stabilize stream banks and maintain water quality; and they can provide habitat for aquatic and semi-aquatic species. However, some terrestrial wildlife species avoid or rarely use riparian areas (Crome and Moore 1989; Lindenmayer and Nix 1993; McComb et al. 1993), and thus riparian corridors should generally be wide enough to include continuous strips of adjoining, nonriparian vegetation.

17.2.6 Nonterrestrial Corridors

The preceding discussion has focused on corridors of terrestrial vegetation. It should be noted, however, that many temperate and tropical fish, shrimps, and other species rely on migratory movements during their life cycles, either between the freshwater and marine biomes or among different streams and rivers (e.g., Holmquist et al. 1998; Policansky and Magnuson 1998). Freshwater rivers are often fragmented by hydroelectric dams, and thus in a broad sense, fish ladders, seasonal water flows, and other mitigation measures designed to permit some animal movements constitute faunal corridors. Marine systems can also become fragmented. Movements of crabs among intertidal oyster reefs, for example, were reduced when such reefs lacked vegetated estuarine habitats, salt marshes, and sea grass beds, all of which functioned as landscape corridors (Micheli and Peterson 1999). The study of corridors in freshwater and marine systems is still in its infancy.

17.3 Design and Management of Faunal Corridors

17.3.1 Conclusions About Corridor Effectiveness

Although no studies have conclusively demonstrated that corridors increase persistence of fragmented populations (cf. Nicholls and Margules 1991; Hobbs 1992; Inglis and Underwood 1992; Beier and Noss 1998), evidence is rapidly accumulating that many species will use corridors for habitat or movement. Evidence that corridors could have negative effects on populations is less compelling. For example, mathematical models suggest corridors are likely to facilitate the transmission of disease between nature reserves under only a very limited set of conditions (Hess 1994). Other potential problems, such as increased predation pressure in corridors due to an influx of generalist predators or brood parasites along forest edges (Wilcove 1985; Simberloff and Cox 1987), can be minimized if corridors are adequately designed and managed. We believe, therefore, that the preponderance of available evidence suggests corridors are likely to have positive benefits for wildlife in fragmented landscapes in the majority of contexts (Wegner and Merriam 1979; Harris 1984; Fahrig and Merriam 1985; Lovejoy et al. 1986; Noss 1987; Bennett 1990a,b, 1999; Laurance 1990, 1991b; Saunders and Hobbs 1991; Beier 1993; Haas 1995; Isaacs 1995; Machtans et al. 1996; Downes et al. 1997; Beier and Noss 1998; Haddad 1999; Haddad and Baum 1999; Laurance and Laurance 1999; Lima and Gascon 1999). We conclude that corridors should be regarded as beneficial in degraded landscapes unless specific local evidence suggests otherwise.

17.3.2 Guidelines and Principles for Corridor Design

Because land managers often must make decisions about wildlife corridors in the absence of detailed, local ecological information, there is a real need for general guidelines and principles for corridor design. In this spirit, we present the following eight recommendations:
1. Corridors will have a positive effect on wildlife in most contexts.
2. Wherever possible, corridors should be designed to accommodate the needs of target species, which will commonly be those that are strongly dependent on primary habitat and sensitive to fragmentation. This will entail at least two design features: that the corridor habitat is of appropriate quality (e.g., primary forest, with few or no disjunctions or disturbances), and that it is of sufficient width to provide forest-interior habitat that minimizes important edge effects.
3. In habitats where the edge contrast is steep, we recommend wider corridors. In closed forests (e.g., rainforests and dense temperate forests), edge effects are often important within ca. 100 m of forest edges (Fig. 17.1).

Hence, corridors in closed forests should be at least 300 m wide to preserve a core of forest-interior habitat (Laurance and Gascon 1997).

4. In habitats where the physical and climatic gradients between the corridor and adjoining matrix are less steep, such as open forests, shrublands, and grasslands, corridors of at least 100 m width may prove effective.

5. For longer corridors (i.e., >1 km), corridor width should be adjusted upward using the 10 % rule (see Fig. 17.2).

6. Streams and rivers are often excellent locations for corridors, but in general the corridor should be wide enough to include continuous strips of nonriparian habitat.

7. Faunal stepping stones – habitat patches – at regular intervals along corridors are likely to improve corridor effectiveness and enhance overall landscape connectivity.

8. Corridors that traverse severely modified habitats (e.g., pastures, crops) should generally be wider than those that traverse habitats that are structurally similar to the corridor (e.g., plantations, regrowth forest).

17.3.3 Proactive Landscape Management

We conclude with a plea for proactive land management in regions experiencing rapid deforestation and habitat modification. In the Americas, this includes extensive areas in Amazonia and Central America, as well as parts of western North America and southern South America. The urgency of this task is highlighted not only by the alarming contemporary rates of deforestation, but also by the realization that few nations have designated more than a small fraction (3–10 %) of their total land area as nature reserves. It is becoming apparent that the vast areas of semi-natural and managed lands that dominate most regions will play a vital role in the conservation of threatened populations, species, and ecological processes (Laurance and Bierregaard 1997).

As discussed above, the most vulnerable species in fragmented landscapes are very often those that require primary habitat for movement and survival (Laurance 1990, 1991b, 1994; Saunders and Hobbs 1991; Gascon et al. 1999; Laurance and Laurance 1999). For these species, there is an urgent need to retain corridors of primary habitat between planned and existing nature reserves, and other habitat remnants, to reduce the impacts of ongoing fragmentation. We assert that such efforts will be vastly more effective before and during – rather than after – the deforestation process. Once severed by development, effective linkages among reserves and habitat remnants will be exceedingly difficult to reestablish, both because of the slow rate of regeneration of forest in degraded landscapes and the considerable economic cost of removing lands from production for the purpose of nature conservation (Laurance and Gascon 1997).

Deceptively simple land-use guidelines can have a profound effect on deforestation patterns. In tropical regions such as Amazonia, for example, the protection of 300-m-wide strips of primary forest along streams and rivers would dramatically enhance ecosystem connectivity in areas undergoing rapid deforestation (Laurance and Gascon 1997). There is a dire need for legislation, educational initiatives, and other measures to protect forest linkages during the crucial development process. Such strategies will clearly enhance the survival prospects of species that are at greatest risk from habitat loss and fragmentation.

Acknowledgement. We thank John P. Hayes, Claude Gascon, and the editors for commenting on drafts of the manuscript. This is publication number 259 in the BDFFP technical series.

References

Andren H, Angelstam P (1988) Elevated predation rates as an edge effect in habitat islands: experimental evidence. Ecology 69:544–547

Beier P (1993) Determining minimum habitat areas and habitat corridors for cougars. Conserv Biol 7:94–108

Beier P, Noss RF (1998) Do habitat corridors provide connectivity? Conserv Biol 12:1241–1252

Bennett AF (1990a) Habitat corridors: their role in wildlife management and conservation. Department of Conservation and Environment, Melbourne, Australia

Bennett AF (1990b) Habitat corridors and the conservation of small mammals in a fragmented forest environment. Landscape Ecol 4:109–122

Bennett AF (1999) Linkages in the landscape: the role of corridors and connectivity in wildlife conservation. IUCN, Gland, Switzerland

Bierregaard RO Jr, Lovejoy TE, Kapos V, Santos AA, Hutchings RW (1992) The biological dynamics of tropical rainforest fragments. Bioscience 42:859–866

Brown JH, Kodric-Brown A (1977) Turnover rates in insular biogeography: effect of immigration on extinction. Ecology 58:445–449

Brown KR, Hutchings RW (1997) Disturbance, fragmentation, and the dynamics of Amazonian forest butterflies. In: Laurance WF, Bierregaard RO Jr (eds) Tropical forest remnants: ecology, management, and conservation of fragmented communities. University of Chicago Press, Chicago, pp 91–110

Camargo JL (1993) Variation in soil moisture and air vapour pressure deficit relative to tropical rain forest edges near Manaus, Brazil. MPhil Thesis, Cambridge University, England

Camargo JL, Kapos V (1995) Complex edge effects on soil moisture and microclimate in central Amazonian forest. J Trop Ecol 11:205–211

Carvalho K, Vasconcelos HL (1999) Forest fragmentation in central Amazonia and its effects on litter-dwelling ants. Biol Conserv 91:151–157

Catterall CP, Green RJ, Jones DN (1991) Habitat use by birds across a forest-suburb interface in Brisbane: implications for corridors. In: Saunders DA, Hobbs RJ (eds) Nature conservation 2: the role of corridors. Surrey Beatty, Sydney, Australia, pp 247–258

Crome FHJ, Moore LA (1989) Display site constancy of bowerbirds and the effects of logging on Mt Windsor Tableland, north Queensland. Emu 89:47–52

Crome FHJ, Isaacs J, Moore LA (1995) The utility to wildlife of remnant riparian vegetation and associated windbreaks in the tropical Queensland uplands. Pac Conserv Biol 2:328–343

Diamond JM (1975) The island dilemma: lessons of modern biogeographic studies for the design of nature reserves. Biol Conserv 7:129–146

Didham RK (1997a) The influence of edge effects and forest fragmentation on leaf-litter invertebrates in central Amazonia. In: Laurance WF, Bierregaard RO Jr (eds) Tropical forest remnants: ecology, management, and conservation of fragmented communities. University of Chicago Press, Chicago, pp 55–70

Didham RK (1997b) An overview of invertebrate responses to habitat fragmentation. In: Watt A, Stork NE, Hunter M (eds) Forests and insects. Chapman and Hall, London, pp 201–218

Didham RK, Hammond PM, Lawton JH, Eggleton P, Stork N (1998) Beetle species responses to tropical forest fragmentation. Ecol Monogr 68:295–323

Downes SK, Handasyde KA, Elgar MA (1997) The use of corridors by mammals in fragmented Australian eucalypt forests. Conserv Biol 11:718–726

Fahrig L, Merriam G (1985) Habitat patch connectivity and population survival. Ecology 66:1762–1768

Ferreira LV, Laurance WF (1997) Effects of forest fragmentation on mortality and damage of selected trees species in central Amazon. Conserv Biol 11:797–801

Forman RTT (1995) Land mosaics. Cambridge University Press, Cambridge, UK

Forman RTT, Godron M (1986) Landscape ecology. Wiley, New York

Gascon C, Lovejoy TE, Bierregaard RO et al. (1999) Matrix habitat and species persistence in tropical forest remnants. Biol Conserv 91:236–241

Gates JE, Gysel LW (1978) Avian nest predation and fledgling success in field-forest ecotones. Ecology 59: 871–883

Gibbs JP, Hunter ML, Melvin SM (1993) Snag availability and communities of cavity nesting birds in tropical versus temperate forests. Biotropica 25:236–241

Gilbert LE (1980) Food web organization and the conservation of neotropical diversity. In: Soulé ME, Wilcox BA (eds) Conservation biology: an ecological-evolutionary perspective. Sinauer, Sunderland, MA, pp 11–33

Goerck JM (1997) Patterns of rarity in the birds of the Atlantic forest of Brazil. Conserv Biol 11:112–118

Haas CA (1995) Dispersal and use of corridors by birds in wooded patches on an agricultural landscape. Conserv Biol 9:845–854

Haddad NM (1999) Corridor and distance effects on interpatch movements: a landscape experiment with butterflies. Ecol Appl 9:612–622

Haddad NM, Baum KA (1999) An experimental test of corridor effects on butterfly densities. Ecol Appl 9:623–633

Hanski I, Kuussaari M, Nieminen M (1994) Metapopulation structure and migration in the butterfly *Melitaea cinxia*. Ecology 75:747–762

Harris LD (1984) The fragmented forest: island biogeographic theory and the preservation of biotic diversity. University of Chicago Press, Chicago

Hess GR (1994) Conservation corridors and contagious disease: a cautionary note. Conserv Biol 8:109–112

Hill CJ (1995) Linear strips of rain forest vegetation as potential dispersal corridors for rainforest insects. Conserv Biol 9:1559–1566

Hobbs RJ (1992) The role of corridors in conservation: solution or bandwagon? Trends Ecol Evol 11:389–392

Holmquist JG, Schmidt-Genenbach JM, Yoshioka BB (1998) High dams and marine-freshwater linkages: effects on native and introduced fauna in the Caribbean. Conserv Biol 12:621–630

Inglis G, Underwood AJ (1992) Comments on some designs proposed for experiments on the biological importance of corridors. Conserv Biol 6:581–586

Isaacs J (1995) Species composition and movement of birds in riparian vegetation in a tropical agricultural landscape. MSc Thesis, James Cook University, Townsville, Australia

Kapos V (1989) Effects of isolation on the water status of forest patches in the Brazilian Amazon. J Trop Ecol 5:173–185

Kapos V, Ganade G, Matsui E, Victoria RL (1993) ^{13}C as an indicator of edge effects in tropical rain forest fragments. J Trop Ecol 8:425–432

Keitt T, Urban BDL, Milne BT (1997) Detecting critical scales in fragmented landscapes. Conserv Ecol 1 (online, URL: http://www.consecol.org/vol1/iss1/art4)

Lamb D, Parotta J, Keenan R, Tucker N (1997) Rejoining habitat remnants: restoring degraded rainforest lands. In: Laurance WF, Bierregaard RO Jr (eds) Tropical forest remnants: ecology, management, and conservation of fragmented communities. University of Chicago Press, Chicago, pp 366–385

Laurance SGW (1996) The utilisation of linear rainforest remnants by arboreal marsupials in north Queensland, Australia. MNatRes Thesis, University of New England, Armidale, NSW, Australia

Laurance SGW, Laurance WF (1999) Tropical wildlife corridors: use of linear rainforest remnants by arboreal mammals. Biol Conserv 91:231–239

Laurance WF (1990) Comparative responses of five arboreal marsupials to tropical forest fragmentation. J Mammal 71:641–653

Laurance WF (1991a) Edge effects in tropical forest fragments: application of a model for the design of nature reserves. Biol Conserv 57:205–219

Laurance WF (1991b) Ecological correlates of extinction proneness in Australian tropical rain forest mammals. Conserv Biol 5:79–89

Laurance WF (1994) Rainforest fragmentation and the structure of small mammal communities in tropical Queensland. Biol Conserv 69:23–32

Laurance WF (1997) Hyper-disturbed parks: edge effects and the ecology of isolated rainforest reserves in tropical Australia. In: Laurance WF, Bierregaard RO Jr (eds) Tropical forest remnants: ecology, management, and conservation of fragmented communities. University of Chicago Press, Chicago, pp 71–83

Laurance WF, Bierregaard RO Jr (eds) (1997) Tropical forest remnants: ecology, management and conservation of fragmented communities. University of Chicago Press, Chicago

Laurance WF, Gascon C (1997) How to creatively fragment a landscape. Conserv Biol 11:577–579

Laurance WF, Yensen E (1991) Predicting the impacts of edge effects in fragmented habitats. Biol Conserv 55:77–92

Laurance WF, Laurance SG, Ferreira LV et al. (1997) Biomass collapse in Amazonian forest fragments. Science 278:1117–1118

Laurance WF, Ferreira LV, Rankin-de Merona JM, Laurance SG (1998a) Rain forest fragmentation and the dynamics of Amazonian tree communities. Ecology 79:2032–2040

Laurance WF, Ferreira LV, Rankin-de Merona JM et al. (1998b) Effects of forest fragmentation on recruitment patterns in Amazonian tree communities. Conserv Biol 12:460–464

Laurance WF, Gascon C, Rankin-de Merona JM (1999) Predicting effects of habitat destruction on plant communities: a test of a model using Amazonian trees. Ecol Appl 9:548–554

Laurance WF, Vasconcelos H, Lovejoy TE (2000) Forest loss and fragmentation in the Amazon: implications for wildlife conservation. Oryx 34:39–45

Lima MG, Gascon C (1999) The conservation value of linear forest remnants in central Amazonia. Biol Conserv 91:241–247

Lindenmayer DB, Nix HA (1993) Ecological principles for the design of wildlife corridors. Conserv Biol 7:627–630

Lovejoy TE, Bierregaard RO, Rylands AB et al. (1986) Edge and other effects of isolation on Amazon forest remnants. In: Soulé ME (ed) Conservation biology: the science of scarcity and diversity. Sinauer, Sunderland, MA, pp 257–285

Machtans CS, Villard MA, Hannon SJ (1996) Use of riparian buffer strips as movement corridors by forest birds. Conserv Biol 10:1366–1379

Malcolm JR (1994) Edge effects in central Amazonian forest fragments. Ecology 75:2438–2445

Malcolm JR (1997) Biomass and diversity of small mammals in Amazonian forest fragments. In: Laurance WF, Bierregaard RO Jr (eds) Tropical forest remnants: ecology, management, and conservation of fragmented communities. University of Chicago Press, Chicago, pp 207–221

Mares MA, Ernest KA (1995) Population and community ecology of small mammals in a gallery forest of central Brazil. J Mammal 76:750–768

Martin TE (1992) Breeding productivity considerations: what are the appropriate habitat features for management? In: Hagan JM, Johnson DW (eds) Ecology and conservation of neotropical migrant landbirds. Smithsonian Institution Press, Washington, DC, pp 455–473

McComb W, McGarigal K, Anthony RG (1993) Small mammal and amphibian abundance in streamside and upslope habitats of mature Douglas-fir stands, western Oregon. Northwest Sci 67:7–15

Micheli F, Peterson CH (1999) Estuarine vegetated habitats as corridors for predator movements. Conserv Biol 13:869–881

Naiman RJ, Decamps H, Pollock M (1993) The role of riparian corridors in maintaining regional biodiversity. Ecol Appl 3:209–212

Newell GR (1999) Responses of Lumholtz's tree-kangaroo *Dendrolagus lumholtzi* to loss of habitat within a tropical rainforest fragment. Biol Conserv 91:181–189

Nicholls AO, Margules CR (1991) The design of studies to demonstrate the biological importance of corridors. In: Saunders DA, Hobbs RJ (eds) Nature conservation 2: The role of corridors. Surrey Beatty, Sydney, Australia, pp 46–64

Noss RF (1987) Corridors in real landscapes: a reply to Simberloff and Cox. Conserv Biol 1:159–164

Noss RF (1991) Landscape connectivity: different functions at different scales. In: Hudson W (ed) Landscape linkages and biodiversity. Island Press, Washington, DC, pp 27–39

Paton PWC (1994) The effect of edge on avian nest success: how strong is the evidence? Conserv Biol 8:17–26

Policansky D, Magnuson JJ (1998) Genetics, metapopulations, and ecosystem management of fisheries. Ecol Appl 8 (suppl):S119–S123

Saunders DA, Hobbs RJ (eds) (1991) Nature conservation 2: the role of corridors. Surrey Beatty, Sydney, Australia

Saunders DA, de Rebeira CP (1991) Values of corridors to avian populations in a fragmented landscape. In: Saunders DA, Hobbs RJ (eds) Nature conservation 2: the role of corridors, Surrey Beatty, Sydney, Australia, pp 221–240

Schultz CB (1998) Dispersal behavior and its implications for reserve design in a rare Oregon butterfly. Conserv Biol 12:284–292

Simberloff D, Cox JA (1987) Consequences and costs of conservation corridors. Conserv Biol 1:63–71

Simberloff D, Farr JA, Cox J, Mehlman DW (1992) Movement corridors: conservation bargains or poor investments? Conserv Biol 6:493–503

St Clair CC, Bélisle M, Desrochers A, Hannon S (1998) Winter responses of forest birds to habitat corridors and gaps. Conserv Ecol 2 (13 pages, online)

Weddell BJ (1991) Distribution and movements of Columbian ground squirrels (*Spermophilus columbianus* (Ord)): are habitat patches like islands? J Biogeogr 18:385–394

Wegner J, Merriam G (1979) Movement by birds and small mammals between a wood and adjoining farmland habitats. J Appl Ecol 16:349–357

Wilcove DS (1985) Nest predation in forest tracts and the decline of migratory songbirds. Ecology 66:1211–1214

Williams-Linera G (1990) Vegetation structure and environmental conditions of forest edges in Panama. J Ecol 78:356–373

Willis EO (1974) Populations and local extinctions of birds on Barro Colorado Island, Panama. Ecol Monogr 44:153–169

Wilson EO, Willis EO (1975) Applied biogeography. In: Cody ML, Diamond JM (eds) Ecology and evolution of communities. Belknap Press, Cambridge, MA, pp 522–534

Yahner RH (1988) Changes in wildlife communities near edges. Conserv Biol 2:333–339

18 Management of the Semi-Natural Matrix

J.H. Brown, C.G. Curtin, R.W. Braithwaite

18.1 Introduction

Conservation efforts have traditionally focused on two main themes: preservation of endangered species and establishment of biological reserves (Fox 1981). Yet, there are too many species to be saved one-at-a-time, and any reserve system is inadequate to preserve the biotas of entire continents, especially in the face of global changes in climate and land use.

In this paper, we call attention to the importance of what we call the "semi-natural matrix", the forest, woodland, savanna, grassland, desert, and tundra in which both the intensively used urban and agricultural land and the relatively pristine reserves are embedded. With increasing human population, these lands are of central importance for three reasons: (1) they support the majority of the earth's biological diversity; (2) they are subject to conflicting pressures to be set aside as natural reserves and to be used more intensively; and (3) they represent the real challenge of sustainable development.

18.2 Definition

We suggest that it is useful to divide the lands of the world into three functional categories: (1) agro-urban lands, where the sustained provision of living space and carbohydrates for the human population is the primary goal; (2) semi-natural matrix, where sustained human harvest of natural resources is dependent upon, and must be balanced with maintenance of ecological function and biological diversity; and (3) natural reserves, where the primary goal is to preserve ecological systems minimally affected by human activities. This three-way classification represents several gradients. Most importantly it is a gradient of human impacts including: destruction of habitat, extirpation of native biota, importance of exotic and domestic species, pollution, and human-domination of ecosystems (Peet et al. 1983; Clark and Munn 1986; Turner et al. 1990). It is also a gradient of productivity, reflecting the limiting

Ecological Studies, Vol. 162
G.A. Bradshaw and P.A. Marquet (Eds.)
How Landscapes Change

effects of climate, topography, and soils on the capacity of land to produce biomass and to support human populations (Tansley 1939; Auclair 1976; Clark and Munn 1986; Turner et al. 1990; Zonneveld and Forman 1990). Finally, it is a gradient of size of patches of remaining natural ecosystems and of management goals for these lands (Curtis 1956; Burgess and Sharpe 1981; Forman and Godron 1986; Hobbs et al. 1993; Forman 1995).

Central to this classification is the concept of semi-natural matrix. We use this term to describe ecosystems, such as forest, woodland, savanna, grassland, tundra, and desert, which retain their essential natural features despite being lightly to moderately used by humans to harvest natural resources: timber, firewood, grazing for domestic livestock, wild crops, game, water, minerals, and fossil fuels. We consider freshwater streams and lakes to be part of the semi-natural matrix, and while we do not consider the oceans, they satisfy most of the criteria for inclusion in this category. These lands and waters comprise a landscape matrix in which both the more intensively used urban and agricultural ecosystems and the relatively pristine natural reserves are embedded.

Just a few thousand years ago the entire world was a semi-natural matrix (Tansley 1939; Curtis 1956; Crosby 1986; Whitney 1994). The success of modern humans in converting highly productive lands to agro-urban ecosystems has fueled the explosive growth of human populations (Thomas 1956; Clark and Munn 1986; Turner et al. 1990). Although the living space and carbohydrate production of agro-urban lands support the vast majority of the contemporary human populations, other essential resources have always been harvested from the semi-natural matrix (Crosby 1986). Pre-literate societies had restricted areas that effectively operated as natural reserves (e.g., Polunin 1991). Colonial powers depended on the matrix for expansion. For example, in the eighteenth century tall and straight white pine (*Pinus strobus*) or "mast trees", made the New England colonies essential for the maintenance of the British Navy and woodlots were protected and preserved to conserve this vital resource (Cronon 1983). Today the fossil fuels that largely support the human population and its technological economy are extracted almost exclusively from the semi-natural matrix. Thus, the world's major super-powers have been maintained, in part, by the harvesting of resources from the semi-natural matrix.

As the human population has grown, increased demand for natural resources and degradation of the semi-natural matrix has resulted in pressures from conservationists to preserve native species and the integrity of natural ecosystem processes (e.g., Soule and Wilcox 1980; Peet et al. 1983; Soule 1986; Wright 1987; Wilson 1988; Vitousek 1994; Ludwig et al. 1997). This has been accomplished primarily by setting aside relatively pristine lands as nature reserves in which the only activities permitted are "compatible", as in the case of the National Refuge System (Curtin 1993), or complimentary (Organic Legislation of 1899, 1912 and 1964 Wilderness Act) with the reserves enabling legislation (Bean and Rowland 1998). We draw the line between nature reserves and the semi-natural matrix at the point where preservation of the ecosystem in its natural state ceases to be the primary goal. According to these criteria, both Categories I-III of the IUCN's (1990) classification of "National Parks and Protected Areas" and the "Core" of MAB Biosphere Reserves (Maldague 1984) would be considered reserves. Most protected areas in IUCN's Categories IV-VIII and Buffer Zones of Biosphere Reserves would be classified as semi-natural matrix. We recognize that the majority of parks in the IUCN Categories I-III also serve other functions, but they meet our criteria for natural reserves so long as conservation remains their dominant goal.

18.3 Land Area and Use

We estimate that the semi-natural matrix comprises more than 85 % of land area worldwide, with approximately 11 % of the world's land devoted to agriculture (CIA Factbook 2000). Although there is some heterogeneity in the distribution of land use categories around the world, it is less than one might expect. Table 18.1 shows data from the US Central Intelligence Agency (2000) on land use for some developed and developing countries. These countries have been selected to show a range of values as well as to provide figures for representative countries and geographic regions. The extremes of agricultural development range from more than 30 % for India, Italy, France, and Nigeria, to less than 10 % for Australia, Brazil, Canada, and the Republic of the Congo. The substantial variation in land use that does exist among the different countries appears to be related more to geographic location (and associated climate and habitat) than to degree of economic development. Furthermore, many countries with large fractions of agro-urban lands have converted extensive areas of semi-natural matrix to more intensive uses and are relying on imports from other countries for supplies of essential natural products (e.g., timber, meat) produced by the semi-natural matrix (Clark and Munn 1986; Turner et al. 1990).

The percentages in Table 18.1 also suggest that nonagricultural land is approximately evenly divided between meadows and pastures, forests and woodlands, and deserts and tundra. All of this land except for the few percent in urban areas or natural reserves is the semi-natural matrix, but the intensity and nature of human exploitation vary greatly. Some pastures and forests are intensively managed and regularly harvested, whereas other forests and many deserts and tundra are lightly used (Turner et al. 1990). Except for fossil fuels and minerals, the resources harvested from these lands are at least potentially renewable, and except for water they are biotic. The semi-natural matrix is also used with varying intensities for other human activities, ranging from recreation to transportation to disposal of wastes (Turner et al. 1990).

Table 18.1. Percentages of the land area of selected developed and developing countries in different categories of land use (Central Intelligence Agency 2000)

	Developed countries					
	Australia	Canada	France	Russia	USA	UK
Arable land	6	5	33	8	19	25
Permanent crops	0	0	2	0	0	0
Meadow and pasture	54	3	20	4	25	46
Forest and woodland	19	54	27	46	30	9
Other	21	38	18	42	26	19

	Developing countries					
	Brazil	China	Congo	India	Mexico	Nigeria
Arable land	5	10	3	56	12	33
Permanent crops	1	0	0	1	1	3
Meadow and pasture	22	43	7	4	39	44
Forest and woodland	58	14	7	23	26	12
Other	14	33	13	16	28	8

18.4 Role in Conservation

The semi-natural matrix plays three crucial roles in the conservation of the earth's biotic diversity and natural resources. First, it contains and maintains most of the existing populations and species of wild plants and animals. Quantitative data are difficult to obtain, but it is clear that most wildlife have been excluded from agricultural and urban areas and that most reserves are too small and fragmented to support many species (Soule and Wilcox 1980; Soule 1986; Turner et al. 1990; Forman 1995).

Second, the semi-natural matrix serves as a buffer between the intensively used agro-urban ecosystems and the relatively undisturbed reserves. On one hand, to the extent that reserves are surrounded by extensive patches of habitat where native species are able to survive or by corridors of habitat through which individuals are able to disperse, persistence of native species can be enormously enhanced (e.g., Hobbs et al. 1993; Forman 1995; but see Simberloff et al. 1992). On the other hand, the matrix surrounding agricultural and urban habitats buffers these ecosystems from deleterious effects of both concentrated human activities and wild species. Thus, the matrix serves as depositories and dilution spaces for pollutants and wastes, it contains roads and railways used for transportation and parks and waterways used for recreation, and it facilitates the large-scale control of pest species which have detrimental effects on agricultural production and human health.

The third important role of the semi-natural matrix is production of natural resources for human consumption. As discussed above, while most of the carbohydrates consumed by the human population come from intensively managed agricultural ecosystems, most of the other essential resources come from the semi-natural matrix. These include wood and other wild plant products, meat, hides, and milk products from grazing livestock, game, fish, water, oil, gas, coal, and minerals. Most of these natural resources are so essential that the very existence of civilization depends on their continued supply.

18.5 Case Study: Temperate Ecosystems – Conflicts Between Traditional Conservation Goals and Management of the Matrix

Within many parts of the midwestern USA, the semi-natural matrix has undergone a profound conversion from grazed open woodlands to closed woodlots bounded by intensive agriculture (Curtis 1956; Dunn and Stearns 1993; Whitney 1994). The closing of woodlots as a result of farm abandonment has led to large changes in species composition and declines in species diversity (Lanyon 1981; Wilcove et al. 1986; Curtin 1997). At the same time, remnant prairies have been almost eliminated, leading to large declines in native grassland species (Graber and Graber 1963; Bowles 1981; Hoffman and Sample 1988; Panzer 1988; Orwig 1992). Orchards, hedgerows, and brushy, tree-filled fencerows, once a habitat of major importance in supporting wildlife (Leopold 1945; Graber and Graber 1963) have been greatly reduced because they interfered with the efficient operation and mechanization of farms as row crops were increasingly planted (Vance 1976; Burger 1978; Carlson 1985; Whitney 1994).

Using farming lands embedded in the semi-natural matrix of southwestern Wisconsin as an example, we illustrate how the interaction of conservation policies and land use reform, agricultural economics, and background climate and disturbance regimes have led to profound, unanticipated changes within this system. Even in regions with access to the extensive research and conservation base associated with large land grant universities and a history of conservation and progressive land use planning, traditional conservation policies and practices have been largely ineffectual, and perhaps detrimental, to preserving the semi-natural matrix.

Prior to recent trends in land use, southwestern Wisconsin already contained a highly modified landscape. Before European settlement in the early 1800s the landscape was covered predominately with oak savanna, characterized as stands of open-growing oaks at densities ranging from one tree per acre up to a maximum of 50 % canopy cover, and predominantly herbaceous ground layer of native forbs and grasses (Curtis 1959; Nuzzo 1985). Savanna is typically considered a fire-dependent ecosystem with the presence of savanna determined by local environmental factors such as fire, climate, topography, and soil. In the absence of fire, or other factors such as dry sandy soils or grazing, savanna typically reverts to forest (Cottam 1949; Bray 1958; Grimm 1983; Leitner et al. 1991).

After European settlement, plowing, grazing, and building of roads and railroads created functional firebreaks, drastically reducing both the frequency and size of wildfires. In the 1830s, with the cessation of the natural fire regime, pine barrens, bush prairies, and oak savanna attained closed canopies, eliminating light-dependent prairie species from the ground layer. The naturalist John Muir (1913) described this phenomenon, which he experienced during his childhood: "As soon as the oak openings in our neighborhood were settled, and the farmers had prevented running grass-fires, the shrub grew up into trees, and formed tall thickets so dense it was difficult to walk through them and every trace of the sunny 'openings' vanished." Many of southern Wisconsin's current forests date from this period (Curtis 1959).

These initial changes in vegetation were relatively temporary as farming radically altered the landscape. While in 1833 oak savanna occupied 74 % of a two-county area in central Wisconsin, by 1934 it had essentially disappeared. In these counties 42 % of savanna (primarily *Quercus velutina* savannas associated with sandy soils) were converted to cropland and 36 % (predominantly *Quercus macrocarpa*) was converted to pasture. The remaining 23 % (primarily *Quercus alba*, associated with unglaciated forest soil), underwent successional transformation into oak-hickory forests (Auclair 1976). In southern Wisconsin, from 1830 to 1880, land in crop production increased from 400,000 to 15,300,000 acres. Crop acreage leveled off in 1910 at about 21,000,000 acres. By the 1950s only 0.2 % of the land in southern Wisconsin remained in a semi-natural state (Curtis 1956).

The impacts of these habitat changes on forest, savanna, and prairie landscapes are well documented. In Cadiz Township, in Green County, Wisconsin, the permanent flowing streams had decreased 26 % by 1902 and 36 % by 1935. This was due largely to the drying up of springs at stream headwaters, which in turn was due to decreased subsoil water storage and reduced infiltration on agricultural fields and pastured woodlots (Shriner and Copeland 1904; Curtis 1956). A decline of forest interior birds also occurred in southern Wisconsin due to the reduction in woodlot size (Ambuel and Temple 1983). Another impact of this habitat fragmentation is on seed dispersal and regeneration of hardwoods. In Cadiz Township, the average nearest-neighbor distance between woodlots is 450 m, but the seed dispersal distances of the major tree species – basswood (*Tilia americana*), sugar maple (*Acer saccharum*), and white ash (*Fraxinus americana*) – range from 40 to 300 m, indicating that this landscape has fragmented beyond the capacity of trees to move between landscape patches (Dunn and Stearns 1993).

While habitat fragmentation, during the first 100 years of settlement, undoubtedly altered the macro composition of Wisconsin's semi-natural matrix, the last 40 years have seen very different, though equally dramatic environmental changes (Fig. 18.1). Work by Wisconsin ecologist John T. Curtis in the early 1950s examined long-term changes in land-use in Cadiz Township in Green County, Wisconsin (Curtis 1956). By comparing Curtis's map with current aerial photos and ground surveys, we found that while the extent of remaining forest area has changed little since the 1950s, the characteristics of the forest cover had been significantly altered (Fig. 18.2). Curtis recorded grazing in 77 % of the woodlots, indicating that the majority of the woodlots had open savanna-like qualities. Observations in 1991 indicated only 11 % of all woodlots were grazed, resulting in nearly a 70 % decrease in the amount of open, savanna-like habitat. This is not an isolated example; in Vermont Township in Dane County, Wisconsin, almost all forests have attained dense understories since the 1960s (Wisconsin Department of Transportation Air Photos; Curtin 1995; E.A. Howell, pers. comm. 1990). Open meadows and savanna-like open oak forests, until 25 years ago the most prevalent community type, are now almost nonexistent. While historical savannas and current forests would both be classified as semi-natural matrices, the former was more effective at preserving native and rare species, while the latter has a different structure, altered ecosystem processes, lower diversity, and more exotic species.

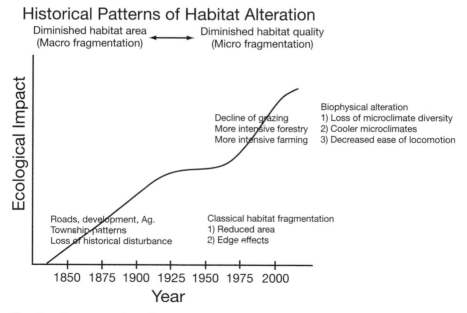

Fig. 18.1. Summary of southwestern Wisconsin's land use history. The first 100 years of settlement result in large structural changes in the landscape. While post-World War II shifts in patch structure have altered the structure and function of the habitat itself

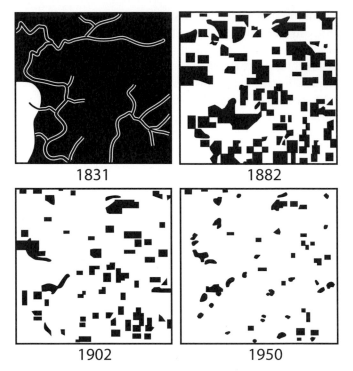

Fig. 18.2: While the first clearings were confined to what was considered the best agricultural land, by 1882 the major influence on clearing patterns was the land survey, which resulted in square land holdings independent of topography. Forest cover decreased from 30 % in 1882 to 4 % in 1954 (Curtis 1956). Since the 1950s, while the amount of forest has not changed, the amount of open savanna-like habitat, important for the conservation of native species, has precipitously declined. This pattern is typical of southwestern Wisconsin landscapes

Thus, while overgrazing early in the twentieth century often extirpated native species and led to soil erosion, light to moderate grazing was important in preserving the open character of the semi-natural matrix (Curtis 1959; Nuzzo 1985). Most prairie grasses and savanna species are tolerant of grazing, but intolerant of low light levels in the forest understory (Bray 1958). Grazing by domestic livestock often resulted in a modified savanna. These savanna-like habitats were composed of trees growing singly, in small open clumps, or dense groves with an understory of exotic pasture grasses or remnants of native flora (Curtis 1959). Through the late 19th and early to mid-twentieth century the open savanna-like forests, maintained by grazing, provided temporary refugia for many prairie and savanna species (Curtis 1956; Pleyte 1975; S.A. Temple, pers. comm. 1991; R. Hay, pers. comm. 1992; Curtin 1997).

The relatively subtle interaction of changes in forest and cropland legislation, land management policy and practices, and farm economics led to profound changes in the composition of the semi-natural matrix. Problems of overgrazing were widespread by early into this century (Cohee 1934; Zeasman and Hembre 1963). In 1937 soil conservation districts started to address problems of lack of vegetation regeneration and soil erosion (Jacobs et al. 1990; Dunn and Stearns 1993). The Forest Crop Law (1927) stipulated that landowners follow "sound Forestry practices," while the Woodland Tax Law (1954), and Managed Forest Law (1985), stated that land owners must follow signed management plans to receive tax credits (Solberg 1961; Zeasman and Hembre 1963; Jacobs et al. 1990). Through these management plans State Foresters had direct control over land use. Yet these direct controls were rarely applied to more that 15 or 20 % of the forest land in southern Wisconsin (Wisconsin State Forester R. Amiel, pers. comm. 1992). More important were the indirect effects of State Foresters encouraging voluntary changes in land-use. Federal cost-sharing programs helped facilitate these changes by funding between 50 and 90 % of the costs for fencing woodlands (Wisconsin State Forester R. Amiel, pers. comm. 1992). Economic factors also played an important role as changes in the dairy industry decreased the economic viability of grazing cows in woodlands (Etgen and Reeves 1978; Castle and Watkins 1979).

Starting in the 1950s, cessation of widespread grazing caused a dramatic closing of the forest with a shift from open prairie and savanna-like habitat to a dense understory dominated by exotics such as buckthorn (*Ramnus* spp.), honeysuckle (*Lonicera tatarica* and *L. morrowi*), barberry (*Berberis vulgaris*), and native blackberries and saplings (Curtis 1959; McCune and Cottam 1985; Nuzzo 1985). Increased plant cover and shading due to shifts in grazing regime resulted in a 3–5 °C decrease in microclimatic temperature at the ground surface, and this in turn caused declines in many grassland and savanna vertebrate populations living in these habitat remnants (Pleyte 1975; Curtin 1997).

Yet the most serious impact of this habitat change is the sheer magnitude of the alteration. Whereas older, less intensive farming practices, combined with local environmental factors (sandy soils, drought, rock outcrops), appear to have preserved a mosaic of different community types (forest, open forest, grazed meadow, etc.), current land uses have further altered and simplified the landscape (Whitney 1994). For example, in a single 5-year period from 1982 to 1987 the amount of pastureland in Wisconsin dropped by over 10 %, as upland areas were rapidly converted to cropland or allowed to succeed into dense forest (USDA Soil Conservation Service 1987). On the one hand, the remnants protected as State Natural or Scientific areas may be preserved, especially if conservation-motivated management practices can prevent closing of the forest understory and maintain the native or native-like communities. On the other hand, these protected areas are relatively small, accounting for less than 5 % of the State (Wisconsin DNR, unpubl.). Ironically, as state conservation and land preservation programs continue to grow, they may be coming too late for many prairie and savanna habitats and their native species. Corridors between prime habitats, buffer areas, and other secondary habitats have been virtually eliminated, isolating remaining populations in increasingly small fragments of increasingly degraded habitat.

The success of midwestern land use policies of the 1930s through 1950s in reducing overgrazing and expanding tree plantations is a hallmark of progressive land use policy (Cohee 1934; Zeasman and Hembre 1963; Jacobs et al. 1990; Dunn and Stearns 1993). Yet, ironically, it was these same conservation policies that have unwittingly led to much of the widespread degradation now seen in southwestern Wisconsin. Despite the apparent success in achieving their immediate goals of land use reform, these policies and practices failed to preserve the unique structure, function, and biotic composition of the oak savanna ecosystem. While a substantial area of semi-natural matrix still remains, the degraded ecosystems are less valuable for both economic and conservation purposes.

18.6 Management Goals and Methods

The obligate dependence of humans on the semi-natural matrix requires active management and partially human-oriented, rather than strictly preservationist strategies. The guiding principles should be the twin cornerstones of traditional natural resource management: sustained yield and multiple use (e.g., Leopold 1933; Christensen et al. 1996). Yet, as we illustrated above, these policies must take into account the role of abiotic conditions and biotic interactions in maintaining the structure and dynamics of the ecosystem. To be successful, wildfire, grazing, and other natural disturbances must be maintained to preserve the key functional attributes of the system. Multiple use must include more than just concern for maximizing short-term yields and minimizing variation in production of harvestable resources for direct human exploitation, because this approach has been shown repeatedly to lead to system collapse (Gunderson et al. 1995). Instead, management policies must emphasize flexible strategies of resource use that take account of both short-term fluctuations and long-term trends. It is necessary to maintain the "resilience" of ecosystems in response to natural fluctuations and human-caused perturbations (e.g., Holling 1973, 1986). A long-horizon management strategy is also necessary for the continued long-term productivity of renewable resources required to sustain viable social and economic systems (Clark and Munn 1986; Gunderson et al. 1995). If the matrix is to remain "semi-natural," the demands of humans and their dependent plants and animals must be balanced with the requirements of wild organisms. To reach the objective of dynamically sustaining ecosystems that provide "ecological goods and services" we suggest that the semi-natural matrix be managed to attain three major goals: (1) preservation of ecosystem function; (2) maintenance of species diversity; and (3) sustained yields of resources for human use (see also Franklin 1990; Gillis 1990).

Both conservation of wild organisms and sustained production of resources for human consumption depend on the maintenance of "ecosystem function" (Clark and Munn 1986; Gunderson et al. 1995). Because of its size, the semi-natural matrix plays a major role in regulating biogeochemical processes throughout the biosphere (e.g., Schlesinger 1990; Vitousek 1994). Essential functions include the fertility of the soil, the quality of the water and air, the recycling of nutrients, the dispersal of organisms sufficient to permit recovery from natural and human-caused disturbance, and the presence of plant, animal, and microbe species that play crucial roles in such processes as nitrogen fixation and biological control of pests and pathogens. Although these attributes may be difficult to define and measure, there are usually signs when serious degradation occurs. These include changes in soil erosion, vegetation physiognomy, air and water chemistry, and populations of certain kinds of organisms. Increased attention should be devoted to the development of "ecological thermometers" that will permit inexpensive monitoring of variables that are likely to change rapidly when important ecosystem processes have been altered (e.g., Cairns et al. 1977; Cairns 1980).

Maintenance of species diversity within the semi-natural matrix is essential if some substantial fraction of global biotic diversity is to be saved. Initially, the priority should be to maintain *local* species diversity. This gives local officials and resource managers an immediate, realistic goal. Management plans can be initiated before national and international inventories of rare and endangered species are completed, and their success can be monitored with minimum effort and expense. The semi-natural matrix surrounding natural reserves can be managed with attention to maintaining critical habitats and corridors for dispersal of endangered species. It is not necessary to invoke a global extinction crisis to point out the local ecological, economic, and esthetic benefits of maintaining diverse assemblages of organisms.

Sustained yields of natural resources for human use provide a powerful economic motivation for preserving and managing the semi-natural matrix (e.g., Pearce et al. 1989; Bruntland 1990; Franklin 1990; Gillis 1990). It is important to make the distinction clear, on the one hand, between natural reserves, where the emphasis rightly should be on minimizing the effects of all kinds of human activities and, on the other hand, semi-natural areas, where human exploitation must be continued, but closely monitored and regulated. The renewal of the plant and animal products and water harvested from the semi-natural matrix is dependent on maintenance of both ecosystem function and biotic diversity.

The techniques for managing the semi-natural matrix will necessarily be local and system-specific. They will need to rely more on adaptive management than general ecological theory. They may also include practices that make conservation biologists and environmentalists uncomfortable. For example, although it is important to maintain ecosystem function and local species diversity, preservation of all native species may not be possible or compatible with other multiple use goals. Also, native species that once filled certain key ecological roles may be extinct or unable to perform these functions in human-modified ecosystems. It may be desirable to use exotic species to fill some important roles. In the example of southwestern Wisconsin, in the absence of fire, grazing by dairy cattle played a crucial role in preserving the open savanna-like qualities of the landscape, maintaining the structural integrity of the system, retaining the functional components of the ecosystem, and retarding the loss of biological diversity. Similarly, in much of the southwestern US and northern Mexico, livestock grazing may offer the best opportunity for preserving wild rangelands while providing a sustainable rural economy. These two examples illustrate the importance of identifying which native or exotic organisms play crucial roles in ecosystems, as both ecological "keystones" and economic resources (e.g., Jones et al. 1994).

18.7 Priorities

Unfortunately, increased attention to the rather unglamorous semi-natural matrix must, to some extent, come at the expense of spectacular wild areas and emotionally appealing endangered species. We do not suggest that other conservation strategies be abandoned. We do suggest that traditional efforts to save endangered species and establish biological reserves will be largely ineffectual unless there is a complementary effort to monitor and manage the semi-natural matrix. Because the manpower and the governmental and private moneys that are available for conservation are severely limited, these resources must be allocated wisely. This is facilitated through a combination of top-down and bottom-up approaches to conservation. A good example of a progressive, top-down approach to conservation is the Ecoregional Planning being undertaken by The Nature Conservancy and World Wildlife Fund. Under this approach critical elements, both common and rare, are identified in order to preserve species and habitats of special interest and to maintain large-scale landscape processes within discrete geographic areas.

The bottom-up approach is typified by community-based conservation in which local groups or individuals take responsibility, often with the help of federal agencies or nongovernment organizations (NGO), in managing and protecting their landscape (e.g., Western et al. 1994). The declines in East African wildlife (Western 1997) and conflicts over North American rangelands (Donahue 1999), despite substantial government investment and land ownership, illustrate the problems associated with centralized, government-directed management of natural resources. Developing countries are increasingly adopting decentralized management strategies based in local communities. However, a review of recent annual meeting proceedings and journal contents of the Society for Conservation Biology indicates that developed countries are lagging behind developing countries in adopting this powerful approach to conservation. In the US, for example, most government agencies and NGOs still practice almost exclusively a classical preservationist approach to conservation in which conservation agendas are influenced largely by political activists, government employees, and scientific "experts," who often have little experience with local ecosystems and little interaction with local people. An exception is The Nature Conservancy, which has recently made community-based conservation a priority.

The Malpai Borderlands Group (MBG) is a good example of the benefits that can come from applying local scientific expertise to inform community-based management of the semi-natural matrix. The MBG is a diverse association of local ranchers, agency personnel, conservationists, and scientists who are dedicated to preserving the open space, ranching economies, and biodiversity in an area of nearly 1 million acres along the US–Mexican Border (McDonald 1995). The common goal is to manage arid rangelands to maintain ecological structure and function in a highly variable environment. Major efforts are devoted to developing sustainable grazing practices and reintroducing fire after nearly 100 years of suppression. Initial estimates are that this local, private-sector approach to conservation will cost the US taxpayer only about 5 % of the total cost for the US Government to preserve these ecosystems by purchasing the land and managing it as a system of parks and reserves (B. McDonald and D. Western, pers. comm. 2000).

18.8 Concluding Remarks

The goal of this paper is to emphasize the critical role of the semi-natural matrix in any effort to sustain the vitality of the earth's ecological systems and to staunch the loss of biological diversity. It is not sufficient to save individual endangered species and to set aside relatively pristine reserves. Improved management of the semi-natural matrix is one priority for more interventionist approaches to conservation. Others include restoring damaged habitats so that they again become productive for both wildlife and natural resources (Cairns et al. 1977; Cairns 1980, 1988; Janzen 1986; Jordan et al. 1987), developing models for long-term land use and natural resource management that are both ecologically sound and economically realistic (Clark and Munn 1986; Lubchenco et al. 1991; Gunderson et al. 1995), and most important of all, reversing the growth and resource consumption of the human population.

References

Ambuel B, Temple SA (1983) Area-dependent changes in the bird communities and vegetation of southern Wisconsin forests. Ecology 64:1057–1068

Auclair AN (1976) Ecological factors in the development of intensive-management ecosystems in the midwestern United States. Ecology 57:431–444

Bean MJ, Rowland MJ (1998) The evolution of national wildlife law. Environmental Defense Fund, Washington, DC

Bowles JB (1981) Iowa's mammal fauna: an ear of decline. Proc Iowa Acad Sci 88:38–42

Bray JR (1958) The distribution of savanna species in relation to light intensity. Can J Bot 36:671–681

Brundtland GH (1990) Our common future. Oxford University Press, Melbourne, Australia

Burger GV (1978) Agriculture and wildlife. In: Brokaw HP (ed) Wildlife in America: contributions to an understanding of American wildlife and its conservation. US Council on Environmental Quality, Washington, DC, pp 89–107

Burgess RL, Sharpe DM (1981) Forest island dynamics in man-dominated landscapes. Springer, Berlin Heidelberg New York

Cairns J Jr (1980) The recovery process in damaged ecosystem. Ann Arbor Science, Ann Arbor, MI

Cairns J Jr (1988) Rehabilitating damaged ecosystems. CRC Press, Boca Raton

Cairns J Jr, Dickson KL, Herricks EE (1977) Recovery and restoration of damaged ecosystems. University Press of Virginia, Charlottesville, VA

Carlson CA (1985) Wildlife and agriculture: can they coexist? J Soil Water Conserv 40:263–266

Central Intelligence Agency (2000) The world factbook. United States Government Printing Office, Washington, DC

Castle ME, Watkins P (1979) Modern milk production. Faber and Faber, Boston

Christensen NL, Bartuska AM, Brown JH et al. (1996) The report of the Ecological Society of America Committee on the Scientific Basis for Ecosystem Management. Ecol Appl 6:665–691

Clark WC, Munn RE (1986) Sustainable development of the biosphere. Cambridge University Press, Cambridge

Cohee ME (1934) Erosion and land utilization in the driftless area of Wisconsin. J Land Public Utility Econ 9:272–282

Cottam G (1949) The phytosociologyy of an oak woods in southwestern Wisconsin. Ecology 30:271–287

Cronon W (1983) Changes in the land: Indians, colonists, and the ecology of New England. Hill and Wang, New York

Crosby AW (1986) Ecological imperialism: the biological expansion of Europe, 900–1900. Cambridge University Press, Cambridge

Curtin CG (1993) The evolution of the National Wildlife Refuge system and the doctrine of compatibility. Conserv Biol 7:29–38

Curtin CG (1995) Latitudinal compensation in ornate box turtles: implications for species response to shifting land use and climate. PhD Thesis. University of Wisconsin, Madison

Curtin CG (1997) Biophysical analysis of the impact of shifting land use on ornate box turtles, Wisconsin, USA. In: Van Abbema J (ed) Conservation, restoration, and management of tortoises and turtles. The New York Turtle and Tortoise Society, New York, pp 31–36

Curtis JT (1956) The modification of mid-latitude grasslands and forests by man. In: Thomas WL Jr (ed) Man's role in changing the face of the earth. University of Chicago Press, Chicago, pp 721–736

Curtis JT (1959) The vegetation of Wisconsin. University of Wisconsin Press, Madison

Donahue DL (1999) The western range revisited: removing livestock from public lands to conserve native biodiversity. University of Oklahoma Press, Norman

Dunn CP, Stearns F (1993) Landscape ecology in Wisconsin: 1830–1990. In: Fralish JS, McIntosh RP, Loucks OL (eds) J.T. Curtis: fifty years of Wisconsin landscape ecology. University of Wisconsin Press, Madison

Etgen WM, Reaves PM (1978) Dairy cattle feeding and management. Wiley, New York

Forman RTT (1995) Land mosaics: the ecology of landscapes and regions. Cambridge University Press, Cambridge

Forman RTT, Godron M (1986) Landscape ecology. Wiley, New York

Fox S (1981) The American conservation movement: John Muir and his legacy. University of Wisconsin Press, Madison

Franklin JF (1990) Thoughts on application of sylviculture systems under new forestry. For Watch 10:8–11

Gillis AM (1990) The new forestry. Bioscience 40:558–562

Graber RR, Graber JW (1963) A comparative study of bird populations in Illinois, 1906–1909 and 1956–1958. Ill Nat Hist Surv Bull 28:383–528

Grimm EC (1983) Chronology and dynamics of vegetation change in the prairie woodland region of southern Minnesota. New Phytol 93:311–350

Gunderson LH, Holling CS, Light SS (1995) Barriers and bridges: to the renewal of ecosystems and institutions. Columbia University Press, New York

Hobbs RJ, Saunders DA, Arnold GW (1993) Integrated landscape ecology: a western Australian perspective. Biol Conserv 64:231–238

Hoffman RM, Sample D (1988) Birds of wet-mesic and wet prairies in Wisconsin. Passenger Pigeon 50:143–152

Holling CS (1973) Resilience and stability in ecological systems. Annu Rev Ecol Syst 4:1–23

Holling CS (1986) The resilience of terrestrial ecosystems: local surprise and global change. In: Clark WC, Munn RE (eds) Sustainable development of the biosphere. University of Cambridge Press, Cambridge

Jacobs HM, Jordahl HC Jr, Roberts JC (1990) Land resource policy in Wisconsin: an interpretive history. In: Born SM, Harkin DA, Jacobs HM, Roberts JC (eds) Future issues facing Wisconsin's land resources. Inst for Environmental Studies Report 138, University of Wisconsin, Madison

Janzen DH (1986) Guanacaste National Park: tropical ecological and cultural restoration. EUND-FPN-PEA, San Jose, Costa Rica

Jones CG, Lawton JH, Shachak M (1994) Organisms as ecosystem engineers. Oikos 69:373–386

Jordan WR III, Gilpin ME, Aber JD (1987) Restoration ecology. Cambridge University Press, Cambridge

Lanyon WE (1981) Breeding birds and old field succession on fallow Long Island farmland. Bull Am Mus Nat Hist 168:1–60

Leitner LA, Dunn CP, Guntenspergen GR, Stearns F, Sharpe DM (1991) Effects of site, landscape features, and fire regime on vegetation patterns in presettlement southern Wisconsin. Landscape Ecol 5:203–217

Leopold A (1933) Game management. University of Wisconsin Press, Madison

Leopold A (1945) The outlook for farm wildlife. Trans North Am Wildl Conf 10:165–168

Lubchenco J, Olson AM, Brubaker LB et al. (1991) The sustainable biosphere initiative: an ecological research agenda. Ecology 72:371–412

Ludwig D, Walker B, Holling CS (1997) Sustainability, stability, and resilience. Conserv Ecol 1:7(On Line)2

Malldague M (1984) The biosphere reserve concept: its implementation and its potential as a tool for integrated development. In: DiCastri F, Baker FWG, Hadley M (eds) Ecology in practice. Part 1. Ecosystem management. Tycooy International Publishing, Dublin and UNESCO, Paris, pp 376–401

McCune B, Cottam G (1985) The successional status of southern Wisconsin oak woods. Ecology 66:1270–1278

McDonald B (1995) The formation and history of the Malpai Borderlands Group. In: DeBano L, Ffolliott PF, Rubio-Ortera A et al. (eds) Biodiversity and management of the madrean archipelago: the sky islands of southwestern United States and northwestern Mexico. 19–23 Sept 1994, Tucson, AZ. General Technical Report RM-GTR-264. USDA, Forest Service, Rocky Mountain Research Station, Fort Collins, CO, pp 483–486

Muir J (1913) The story of my boyhood and youth. Houghton Mifflin, Boston

Nuzzo VA (1985) Extent and status of Midwest oak savanna: presettlement and 1985. Nat Areas J 6:6–36

Orwig T (1992) Loess Hills prairies as butterfly survival: opportunities and challenges. In: Smith DD, Jacobs CA (eds) Proceedings of the 12th North American Prairie Conference: recapturing the vanishing heritage. University of Northern Iowa, Cedar Falls, IA

Panzer R (1988) Managing prairie remnants for insect conservation. Nat Areas J 8:83–90

Pearce D, Markandya A, Barbier EB (1989) Blueprint for a green economy. Earthscape Publications, London

Peet RK, Glenn-Lewin DC, Wolfe JW (1983) Prediction of man's impact on vegetation. In: Holzer W, Werger MJA, Ikusima I (eds) Man's impact on plant species diversity. W Junk Publishers, The Hague

Pleyte TA (1975) The slender glass lizard (*Ophisaurus attenuatus*) in Waushara Co., Wisconsin. MS Thesis, University of Wisconsin, Milwaukee

Polunin NVC (1991) Delimiting nature: regulated area development in the coastal zone of Malaysia. In: West PC, Brechin SR (eds) Resident peoples and national parks: social dilemmas and strategies in international conservation. University of Arizona Press, Tucson, pp 107–121

Schlesinger WH, Reynolds JF, Cunningham GL et al. (1990) Biological feedbacks in global desertification. Science 247:1043–1048

Shriner FA, Copeland EB (1904) Deforestation and creek flow about Monroe, Wisconsin. Bot Gaz 37:139–143

Simberloff D, Far JA, Cox J, Mehlman DW (1992) Movement corridors: conservation bargains or poor investments? Conserv Biol 6:493–504

Solberg ED (1961) New laws for new forests. University of Wisconsin Press, Madison

Soulé ME (1986) Conservation biology. Sinauer Associates, Sunderland, MA

Soulé ME, Wilcox BA (1980) Conservation biology: an evolutionary-ecological perspective. Sinauer Associates, Sunderland, MA

Tansley AG (1939) The British islands and their vegetation. Cambridge University Press, London

Thomas WL Jr (1956) Man's role in changing the face of the Earth. University of Chicago Press, Chicago

Turner BL, Clark WC, Kates RW, Richards JF, Meyer WB (1990) The Earth as transformed by human action: global and regional changes in the biosphere over the past 300 years. Cambridge University Press, Cambridge

USDA Soil Conservation Service (1987) National Resources Inventory: land use in Wisconsin. USDA Soil Conservation Service, Washington, DC

Vance DR (1976) Changes in land use and wildlife populations in southeastern Illinois. Wildl Soc Bull 4:11–15

Vitousek PM (1994) Beyond global warming: ecology and global change. Ecology 75:1861–1876

Western D (1997) In the dust of Kilimanjaro. Island Press, Washington, DC

Western D, Wright RM, Strum SC (1994) Natural connections: perspectives in community-based conservation. Island Press, Washington, DC

Whitney GG (1994) From coastal wilderness to fruited plain: a history of environmental change in temperate North America 1500 to the present. Cambridge University Press, Cambridge

Wilcove DS, McLellan CH, Dobson AP (1986) Habitat fragmentation in the temperate zone. In: Soulé ME (ed) Conservation biology. Sinauer Associates, Sunderland, MA

Wilson EO (1988) Biodiversity. Harvard University Press, Cambridge, MA

Wright DH (1987) Estimating human effects on global extinction. Int J Biometeorol 31:293–299

Zeasman OR, Hembre IO (1963) A brief history of soil erosion in Wisconsin. University of Wisconsin Extension, Madison

Zonneveld IS, Forman RTT (1990) Changing landscapes: an ecological perspective. Springer, Berlin Heidelberg New York

Human Disturbance and Ecosystem Fragmentation in the Americas

Synthesis and Final Reflections

P.A. MARQUET and G.A. BRADSHAW

As we stated in our introduction, this collection of chapters was intended to characterize human disturbance in relation to the fragmentation of American landscapes and to help integrate concepts from several disciplines including ecology, anthropology, theoretical ecology, and conservation biology, all of which are necessary in order to create a common and comprehensive framework for future research and conservation policy formulation. Our basic tenet is that human-driven changes in the environment, habitat fragmentation in particular, are multivariate in terms of their causal processes as well in their resulting effects upon biodiversity. Within this context, the chapters in this volume, and the conference that promoted its existence, gave rise to several emerging issues and conclusions on which we want to reflect in this closing chapter.

How Landscapes Change:
The Need of a Framework for Understanding

Theory and experience remind us again and again that our world is a tightly interconnected system where changes driven by humans can have far-reaching consequences across time and space. Habitat loss and fragmentation represent the most direct effects of humans on landscapes around the world and have been shown to have a cascade of impacts on ecosystems locally and distally in time and in space. As shown throughout this volume, the extent, rate, and consequences of these changes depend on the local and regional ecological, economic, and social contexts within which they occur (Fig. 1). The ecological context is represented by the web of interacting species that sustain the fluxes of matter and energy that give rise to ecosystem functions within a particular environment. The ecological context defines the conditions under which a particular ecosystem performs and reacts to changes. The economic context, on

Ecological Studies, Vol. 162
G.A. Bradshaw and P.A. Marquet (Eds.)
How Landscapes Change
© Springer-Verlag Berlin Heidelberg 2003

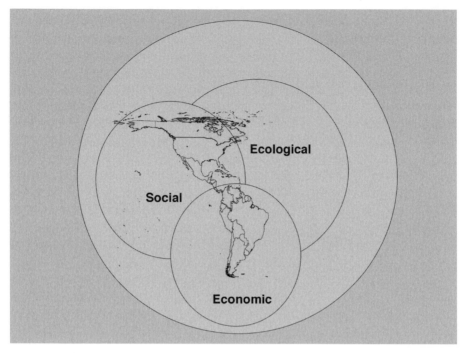

Fig. 1. Graphic representation of the framework that this book endorses to understand the causes and consequences of landscape alteration in the Americas (see text for further discussion)

the other hand, is represented by fluxes of capital and labor that manifest themselves in the exchange, distribution and use of goods and services provided by ecosystems. Finally, the social context, represented by human behavior, organizational structures, institutional arrangements, and value systems, mediate the relationships between people and nature, by fostering, constraining, or setting limits to the interaction between economic forces and ecological systems. Thus, in order to understand, predict and remediate the consequences of habitat loss and fragmentation all these three contexts should be considered (Fig. 2). However, this might not be an easy task, as pointed out by Siebert, Alaback and Nuñez and Grosjean in this volume, since all these three contexts are dynamic and nonlinear, interacting with each other through positive and negative feedbacks, thresholds, and synergisms (e.g., Holling 2001; Laurance and Cochrane 2001; Scheffer et al. 2001) such that changes in one context are mapped, at different rates, in changes in the others. This adds complexity and uncertainty to the end results of the human transformation of ecosystems (Levin 1999; Dybas 2001), and represents a challenge to the decision making and policy formulation (Lempert 2002).

As much as the consideration of ecological, social, and economic contexts provide a framework for understanding, they also represent a framework for

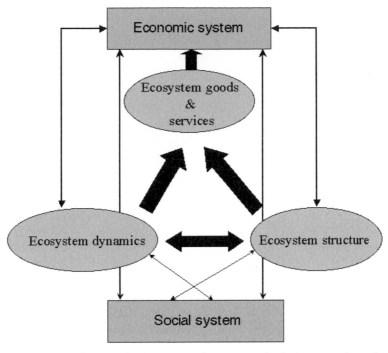

Fig. 2. Diagram showing the interaction between ecological, economic, and social systems

action to revert our current environmental crisis. The solution to the problems posed by global changes in the environment not only require us to deepen our scientific understanding of the problems at issue, but foremost it demands economic and social changes, changes in the way we dwell and interact with the environment. Solutions to the severe and continuing environmental degradation require integrated approaches which re-connect humans and nature as co-adaptive and interacting systems. As such, ecologists must also learn to work in new epistemological frameworks, such as post-normal science (Funtowicz and Ravetz 1999), designed to bring attention toward "aspects of problem solving that have been neglected in traditional accounts of science practice: system complexity, uncertainty, value loading, and a plurality of legitimate perspectives" (Funtowicz and Ravetz 1999). As a theory of science, post-normal science seeks to link epistemology and policy in the analysis and management of complex social and ecological systems (Gallopin et al 2001).

Humans and Landscape Changes in the Americas:
A Plea for Integration

Throughout this volume we have aimed at providing an interdisciplinary overview of the major disruptions sustained by ecosystems in the Americas and argued by example that understanding these relationships and phenomena is best served by integrating ecological knowledge with economic and social-anthropological evidence on the role of humans as drivers of change.

Most ecosystems on earth are in different states of degradation as a consequence of the direct and indirect effect of humans (Vitousek 1994; Hannah et al. 1995; Leemans and Zuidema 1995). This degradation is manifested in many ways, including substantial modifications in the composition, structure, and functioning of ecosystems linked to local and global extinction of species (e.g., the biodiversity crisis; Wilson 1992), alteration of biogeochemical cycles, land transformation, and overexploitation of resources. The impact of humans is so strong and pervasive that current estimates suggest that between one-third to one half of the earth's land surface has been transformed by human action (Vitousek et al. 1997). In fact, most of the earth's land surface is composed of what Brown, Curtin and Braithwaite called "semi-natural matrix", a "land neither intensively used for cities or agriculture nor set aside as natural reserves", which is where most conflicts between human uses and conservation issues will likely arise in the future.

These and other findings confirm the pervasive nature of human impacts on ecosystems and have contributed to a significant re-framing of the study of ecology. In contrast to previous decades, human behavior and activities have become an important factor in ecological studies. In fact, the inclusion of the human dimension represents a significant shift for most biophysical sciences where, until recently, humans and their actions were considered external to ecosystem dynamics (Bradshaw and Bekoff 2001; McDonnell and Pickett 1993). From this perspective, the history of the use of land and water by humans provides important clues for understanding current ecological patterns, and make it possible for us to foresee future courses of landscape change and ecosystem responses. In order to address the problems posed by the action of humans on ecosystems two things are necessary: first, we must consider humans as components of ecosystems and secondly, we must realize that understanding the role of humans in affecting ecological changes requires developing and making explicit how the ecological, economic, and social contexts have interacted through time; that is the historical framework of how such changes have occurred (McDonnell and Pickett 1993; Cronon 1983; Russell 1997). History brings insights, but also controversy.

While the importance of the relationship between humans and their landscapes is now considered integral to deriving any inference concerning the state and dynamics of a given ecosystem, there is much debate among ecologists and social scientists alike as noted by Meggers in this volume. The historical density and impacts of indigenous groups upon Amazonian rainforest ecosystems are highly debatable and, at present, they have yet to be quantified. A similar situation is described by Alaback and colleagues with regard to the effect of Native Americans upon fire regimes in North America. The comparison among the Americas has given scientists further appreciation for the extent and degree to which humans have etched their signature on the landscape (Deloria 1997).

Reaching further back in time indicates that pre-Columbian cultures likely interacted with their environment in a more sustainable way than current populations, as a consequence of their small sizes and different ways of dwelling in the natural world. This view is supported in the chapter by Nuñez and Grosjean who reconstructed the history of human-environment interaction in the desert ecosystems of northern Chile showing that social processes and changes can be mapped onto environmental changes. Nuñez and Grosjean distinguish several discrete stages of this interaction; not surprisingly, major changes occurred as a consequence of the European invasion that brought diseases, exotic species, and a different land tenure system that contrasted sharply with indigenous patterns. European land-use patterns and methods disrupted the sustainability of the human environment interaction in the area by altering the patterns and flows of ecosystems. The disruption of natural ecosystem processes was further exacerbated with the emergence of cities associated with large-scale mining activities.

A similar historical approach is taken by Alaback et al. in this volume, where they reconstruct the history of the fire regime in temperate ecosystems in the Americas, pointing out dramatic changes from pre-Columbian to modern times. It is certain that the impacts of the human enterprise on the Americas have deep historical roots that need to be evaluated because of their social and ecological impact. We can learn from our past in order to manage our future. Some would argue we must own the truths of the past in order to learn. Our ability to "own" our past and relate it to our present and future will critically depend on our understanding of the long-term consequences of human-driven changes. In a recent article, Haila (2002) has emphasized that habitat fragmentation should be viewed as a specific form of habitat degradation contextualized within the natural disturbance regime and heterogeneity of the region under study. This point is also emphasized by Killeen et al. in this volume. While we agree with this conceptualization, we suggest that a great part of the context for understanding and predicting the effects of habitat fragmentation, in addition to the natural disturbance regime, is the social and economic history of humans within the region under study. Furthermore, the explicit consideration of the interaction between nature and society in developing sustainable initiatives represents one of the big challenges ahead (Holling 2000; Kates et al. 2001; Bradshaw and Bekoff 2001; Western 2001).

Fragmentation in the Americas:
On the Road to Ecosystem Disruption?

In general, the impact of humans on terrestrial and marine ecosystem function has had negative effects upon biodiversity by creating widespread species extinction and disrupting the delicate and intricate linkages which connect all life. Major ecosystem types have already been reduced to fragments, and numerous species and genetically distinct populations have been lost in the process (Ehrlich and Wilson 1991; Wilson 1992; Hughes et al. 1997; Ceballos and Ehrlich 2002). For most terrestrial and marine biomes, destruction and degradation of natural habitats, the spread of exotic species and emergent diseases, and the alteration of ecosystem functions and services are well documented (Western 2001). However, the long-term consequences of these changes have yet to be seen (Science 1997; Chapin et al. 1998; Ayensu et al. 1999; Sala et al. 2000). Further, as pointed out repeatedly by the authors in this volume, the rate at which natural habitats, species, and ecosystems services are being lost, as well as the consequences of these losses, put us on the road to ecosystem disruption.

Much of this volume deals with one of the major drivers (habitat loss and fragmentation) that can potentially affect such ecosystem disruptive changes in the Americas (Fig. 3) by precipitating a chain reaction that might involve: (1) an increase in the distribution and establishment of invasive species, which in turn will likely alter ecosystem functioning. For example, habitat fragmentation promotes the penetration of the Argentine ant in coastal southern California (Suarez and Case, this Vol.), and exotic bumblebees in Andean temperate forests (Aizen and Feinsinger, this Vol.), (2) the alteration of landscape pattern and connectivity that might result in an increased rate of local extinction for some species (Keitt; Marquet, Velasco-Hernández and Keymer; Laurance and Laurance; Gascon, Laurance and Lovejoy, this Vol.), (3) the alteration of fire regimes (Alaback et al., this Vol.), (4) the alteration of migration patterns and the genetic structure of local populations, which might be difficult to predict beforehand (Joseph; Joseph, Cunningham and Sarre, this Vol.), and (5) the loss of functional redundancy in ecosystems and the consequent loss of ecosystem resilience and capacity for responding to environmental changes (Jaksic, this Vol.). Similarly, economic incentives and trends towards globalization can lead to widespread land transformation (Siebert, this Vol.) and to the deliberate introduction of exotics for a sustained exploitation of ecosystems (Bustamante, Serey and Picket, this Vol.). The widespread and sustained impact of these processes associated to the human-driven disruption of ecosystems in the Americas threaten the structures and function of natural ecosystems globally, jeopardizing the services they provide to humankind.

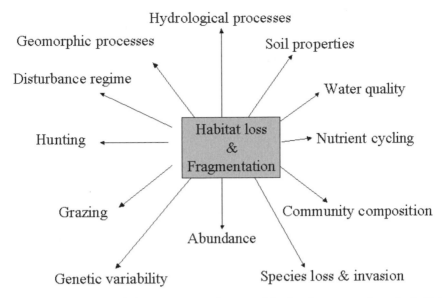

Fig. 3. Habitat loss and fragmentation affect a multitude of processes from the individual to the ecosystem level. The interactions among these processes are not shown to simplify representation

What to do next?

During this new millennium, the tragedy of the commons is likely to become a common tragedy to all, but in particular to developing countries. In the context of a social and ecological landscape in crisis, environmental scientists – ecologists perhaps most of all – face the challenge of providing the necessary scientific foundation and rationale upon which viable and timely social development needs to be based, such that ecosystem disruption is minimized. Different authors in this volume have emphasized different aspects of the solution to this complex problem, which can be summarized in the following:

1. The need to develop a transdisciplinary context for understanding the transformation of landscapes. This entails ecologists and other environmental scientists to interact more closely with anthropologists, sociologists, and economists in order to characterize the drivers responsible for present landscape patterns. These interactions will improve our understanding of the causal processes underlying changes in human-dominated landscapes, as well as contributing to the identification of the important factors to be monitored and modified in order to assure future landscape sustainability (Nuñez and Grosjean; Meggers; Siebert; Alaback et al., this Vol.)

2. The need for more experimental analyses of the effects of habitat fragmentation (Holt and Debinski, this Vol.) and for the development of a predictive theoretical framework that takes explicit account of the history of the fragmentation process itself and its effect upon processes and phenomena operating at different spatial and temporal scales (Kattan and Murcia, this Vol.).
3. The importance of considering the rate at which the landscape is being modified by human action (i.e., landscape dynamics) in addition to focusing on habitat loss and the resulting landscape pattern. As emphasized by Marquet, Velasco-Hernández and Keymer in this volume, species persistence is an emergent phenomenon resulting from the interaction between landscape pattern, landscape dynamics, and species life-history attributes.
4. Increase in the research focused on disentangling the complex web of interactions between different drivers of disturbance, and landscape modification and the consequent effects upon ecosystem functioning. Of paramount importance among these are the linkages among habitat loss and fragmentation and (a) the invasion of exotic species, (b) the alteration of fire regimes, and (c) the social and economic context wherein this interaction takes place (Aizen and Feinsinger; Alaback et al.; Suarez and Case, this Vol.).

Fortunately, several of these key issues are currently the focus of major joint research initiatives such as the Millennium Ecosystem Assessment (Ayensu et al. 1999) and are continuously being brought to the attention of the academic community (e.g., Lawton 2001, Western 2001, Bradshaw and Bekoff 2001).

References

Ayensu E, van Claasen DR, Collins M et al. (1999) International ecosystem assessment. Science 286:685–686
Bradshaw GA, Bekoff M (2001) Ecology and social responsibility: a re-embodiment of science. Trends Ecol Evol 16:460–465
Ceballos G, Ehrlich PR (2002) Mammal population losses and the extinction crisis. Science 296:904–907
Chapin FS, Sala OE, Burke IC, Grime JP, Hooper DU, Lauenroth WK, Lombard A, Mooney HA, Mosier AR, Naeem S, Pacala SW, Roy J, Steffen WL, Tilman D (1998) Ecosystem consequences of changing biodiversity. BioScience 48:45–52
Cronon W (1983) Changes in the land: Indians, colonists, and the ecology of New England. Hill and Wang, New York
Deloria V Jr (1997) Red earth, white lies: Native Americans and the myth of scientific fact. Fulcrum Press, Golden, Colorado, pp 271
Dybas CL (2001) From biodiversity to biocomplexity: a multidisciplinary step towards understanding our environment. Bioscience 51:426–430

Ehrlich PR, Wilson EO (1991) Biodiversity studies: science and policy. Science 253:758–766

Funtowicz S, Ravetz JR (1999) Post-normal science – an insight now maturing. Futures 31:641–646.

Gallopin GC, Funtowicz S, O'Connor M, Ravetz J (2001) Science for the twenty-first century: from social contract to the scientific core. UNESCO. Blackwell, Oxford

Haila Y (2002) A conceptual genealogy of fragmentation research: from island biogeography to landscape ecology. Ecol Appl 12:321–334

Hannah L, Carr JL, Lankerani A (1995) Human disturbance and natural habitat: a biome level analysis of a global data set. Biodiv Conserv 4:128–155

HughesJB, Daily GC, Ehrlich PR (1997) Population diversity: its extent and extinction. Science 278:689–692

Holling CS (2000) Theories for sustainable futures. Conserv Ecol 4(2):7 [online] URL: http://www.consecol.org/vol4/iss2/art7

Holling CS (2001) Understanding the complexity of economic, ecological, and social systems. Ecosystems 4:390–405

Kates RW, Clark WC, Corell R et al. (2001) Sustainability science. Science 292:641–642

Lawton JH (2001) Earth system science. Science 292:1965

Laurance WF, Cochrane MA (2001) Introduction: synergistic effects in fragmented landscapes. Conserv Biol 15:1488–1489

Leemans R, Zuidema G (1995) Evaluating changes in land cover and their importance for global change. Trends Ecol Evol 10:76–81

Lempert RJ (2002) A new decision sciences for complex systems. Proc Natl Acad Sci USA 99:7309–7313

Levin SA (1999) Fragile dominion. Complexity and the commons. Perseus Books, Reading, Massachusetts

McDonnell MJ, Pickett MJ (eds) (1993) Humans as components of ecosystems. The ecology of subtle human effects and populated areas. Springer, Berlin Heidelberg New York

Russell EWB (1997) People and the land through time. Linking ecology and history. Yale University Press, New Haven

Sala OE, Chapin FS, Armesto JJ, Berlow E, Bloomfield J, Dirzo R, Huber-Sanwald E, Huenneke LF, Jackson RB, Kinzig A, Leemans R, Lodge DM, Mooney HA, Oesterheld M, Poff NL, Sykes MT, Walker BH, Walker M, Wall DH (2000) Global biodiversity scenarios for the year 2100. Science 287:1770–1774

Science (1997) Human dominated ecosystems. Science 277:485–525

Scheffer M, Carpenter S, Foley JA, Folke C, Walker B (2001) Catastrophic shifts in ecosystems. Nature 413:591–596

Vitousek PM (1994) Beyond global warming: ecology and global change. Ecology 75:1861–1876

Vitousek PM, Mooney HA, Lubchenco J, Melillo JM (1997) Human domination of earth's ecosystems. Science 277:494–499

Western D (2001) Human-modified ecosystems and future evolution. Proc Natl Acad Sci USA 98:5458–5465

Wilson EO (1992) The diversity of life. Belknap Press of Harvard University Press, Cambridge, Massachusetts.

Subject Index

Ecological Studies

Volumes published since 1997

Printing: Druckhaus Berlin-Mitte
Binding: Buchbinderei Stein & Lehmann, Berlin